Approximation Theory and Spline Functions

NATO ASI Series

Advanced Science Institutes Series

A series presenting the results of activities sponsored by the NATO Science Committee, which aims at the dissemination of advanced scientific and technological knowledge, with a view to strengthening links between scientific communities.

The series is published by an international board of publishers in conjunction with the NATO Scientific Affairs Division

A	Life Sciences	Plenum Publishing Corporation
B	Physics	London and New York
C	Mathematical and Physical Sciences	D. Reidel Publishing Company Dordrecht, Boston and Lancaster
D	Behavioural and Social Sciences	Martinus Nijhoff Publishers
E	Engineering and Materials Sciences	The Hague, Boston and Lancaster
F	Computer and Systems Sciences	Springer-Verlag
G	Ecological Sciences	Berlin, Heidelberg, New York and Tokyo

Series C: Mathematical and Physical Sciences Vol. 136

Approximation Theory and Spline Functions

edited by

S.P. Singh

J.W.H. Burry

and

B. Watson

Department of Mathematics and Statistics,
Memorial University, St. John's, Newfoundland, Canada

D. Reidel Publishing Company

Dordrecht / Boston / Lancaster

Published in cooperation with NATO Scientific Affairs Division

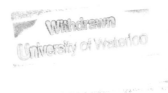

Proceedings of the NATO Advanced Study Institute on
Approximation Theory and Spline Functions
St. John's, Newfoundland, Canada
August 22-September 2, 1983

Library of Congress Cataloging in Publication Data

NATO Advanced Study Institute on Approximation Theory and Spline Functions (1983: St. John's, Nfld.)
Approximation theory and spline functions.

(NATO ASI series. Series C, Mathematical and physical sciences; vol. 136)
"Proceedings of the NATO Advanced Study Institute on Approximation Theory and Spline Functions, St. John's, Newfoundland, Canada, August 22–September 2, 1983"–T.p. verso.
"Published in cooperation with NATO Scientific Affairs Division."
Bibliography: p.
Includes index.
1. Approximation theory–Congresses. 2. Spline theory–Congresses. I. Singh, S. P. (Sankatha Prasad), 1937– . II. Burry, J. H. (John H.), 1938– . III. Watson, B., 1946– . IV. North Atlantic Treaty Organization. Scientific Affairs Division. V. Title. VI. Series: NATO ASI series. Series C, Mathematical and physical sciences; no. 136.
QA221.N36 1983 511'.4 84–15136
ISBN 90–277–1818–0

Published by D. Reidel Publishing Company
P.O. Box 17, 3300 AA Dordrecht, Holland

Sold and distributed in the U.S.A. and Canada
by Kluwer Academic Publishers,
190 Old Derby Street, Hingham, MA 02043, U.S.A.

In all other countries, sold and distributed
by Kluwer Academic Publishers Group,
P.O. Box 322, 3300 AH Dordrecht, Holland

D. Reidel Publishing Company is a member of the Kluwer Academic Publishers Group

---

Printed in The Netherlands.

TABLE OF CONTENTS

PREFACE

A NATO Advanced Study Institute on Approximation Theory and Spline Functions was held at Memorial University of Newfoundland during August 22-September 2, 1983. This volume consists of the Proceedings of that Institute.

These Proceedings include the main invited talks and contributed papers given during the Institute. The aim of these lectures was to bring together Mathematicians, Physicists and Engineers working in the field. The lectures covered a wide range including Multivariate Approximation, Spline Functions, Rational Approximation, Applications of Elliptic Integrals and Functions in the Theory of Approximation, and Padé Approximation.

We express our sincere thanks to Professors E. W. Cheney, J. Meinguet, J. M. Phillips and H. Werner, members of the International Advisory Committee. We also extend our thanks to the main speakers and the invited speakers, whose contributions made these Proceedings complete.

The Advanced Study Institute was financed by the NATO Scientific Affairs Division. We express our thanks for the generous support.

We wish to thank members of the Department of Mathematics and Statistics at Memorial University who willingly helped with the planning and organizing of the Institute.

Special thanks go to Mrs. Mary Pike who helped immensely in the planning and organizing of the Institute, and to Miss Rosalind Genge for her careful and excellent typing of the manuscript of these Proceedings.

St. John's, Newfoundland, Canada
April 1984

S.P. Singh
J.H.W. Burry
B. Watson

PRODUCTS OF POLYNOMIALS

Bernard Beauzamy

In this survey paper, we shall present several results concerning estimates for products of polynomials, in one or in several variables.

The results concerning polynomials in one variable are taken from a joint paper of Per Enflo and the author [1]; the results dealing with polynomials in many variables are due to Per Enflo [2]. Though they were proved earlier, it seemed preferable to put them into the last section, since they are technically more complicated. However, the methods of proofs are of different nature, and there is no interdependence between the case of a single variable and the case of several variables.

In the following pages, only outlines of proofs will be given: we refer the reader to [1] and to [2] for detailed proofs.

What we are looking for is estimates of the following type:

$$\|PQ\| \geq \lambda \|P\| \cdot \|Q\| , \qquad (1)$$

where $\|\cdot\|$ is some norm on the space of polynomials (in one or in many variables), and λ is a constant, depending only on the choice of $\|\cdot\|$, and on the choices of the classes C_1 and C_2, in which we will take P and Q respectively.

Let us first deal with polynomials in one variable. There are many norms which are commonly used. Let us mention some of them.

S. P. Singh et al. (eds.), Approximation Theory and Spline Functions, 1–22.

Put $P(n) = a_0 + a_i x + \ldots + a_N x^N$. Then, we define:

$$\|P\|_1 = \int_0^{2\pi} |P(e^{i\theta})| \frac{d\theta}{2\pi} ,$$

$$\|P\|_2 = (\int_0^{2\pi} |P(e^{i\theta})|^2 \frac{d\theta}{2\pi})^{1/2}$$

$$\|P\|_\infty = \max_{0\leq\theta\leq 2\pi} |P(e^{i\theta})| .$$

For these three norms, P is considered as a (continuous) function on the Torus Π (that is, the interval $[0, 2\pi]$, mod 2π). So these norms are just the norms of the spaces $L_1(\Pi, \frac{d\theta}{2\pi})$, $L_2(\Pi, \frac{d\theta}{2\pi})$, $L_\infty(\Pi, \frac{d\theta}{2\pi})$.

Another type of norms is obtained the following way. Put:

$$|P|_1 = \sum_0^N |a_j|$$

$$|P|_2 = (\sum_0^N |a_j|^2)^{1/2}$$

$$|P|_\infty = \max_{0\leq j\leq N} |a_j| .$$

This time, these norms can be viewed as norms on sequences spaces: the first one is the norm in ℓ_1, the second in ℓ_2, the third in ℓ_∞, when P is identified with the sequence $(a_0, a_1, \ldots, a_N)$.

The norm $|P|_1$ is sometimes written $\|P\|_{A(\Pi)}$, that is, the norm in the algebra $A(\Pi)$, of functions with absolutely summable Fourier series.

The norm $|P|_\infty$ is also written $\|P\|_{PM}$, norm in the space of pseudo-measures (distributions on Π, the Fourier coefficients of which are bounded; see J. P. Kahane [4]).

The relations between these six norms are as follows:

$$|P|_\infty \le \|P\|_1 \le \|P\|_2 = |P|_2 \le \|P\|_\infty \le |P|_1.$$

Let us now come back on the estimate (1). Even if we take the simplest norm, $|P|_1$, we cannot hope to have (1) for all poloynomials, that is, without any restriction on the classes C_1 and C_2. Let us give two examples, of independent interest, which will make this assertion clear:

Example 1. Take $P = 1 - x$, $Q_n = \frac{1}{n+1}(1 + x + \dots + x^n)$. Then $PQ_n = \frac{1}{n+1}(1 - x^{n+1})$. Therefore:

$$|P|_1 = 2, \ |Q_n|_1 = 1, \ |PQ_n|_1 = \frac{2}{n+1} \underset{n\to+\infty}{\longrightarrow} 0 .$$

So no estimate like (1) is possible, involving, uniformly, P and the Q_n's.

Example 2. Let $A(\Pi) = \{f = \sum_{j \in \mathbb{Z}} c_j e^{ij\theta}$, with $\sum |c_j| < +\infty\}$, and put $\|f\|_A = \sum_{j \in \mathbb{Z}} |c_j|$.

Take in $A(\pi)$ two functions f and g with disjoint supports:

$$f = \sum_{j \in \mathbb{Z}} c_j e^{ij\theta}, \quad g = \sum_{\ell \in \mathbb{Z}} d_\ell e^{i\ell\theta},$$

so the product fg vanishes identically. Now put:

$$P_m = \sum_{-m}^{m} c_j \, e^{ij\theta}, \quad P_m' = \sum_{|j|>m}' c_j \, e^{ij\theta}$$

$$Q_n = \sum_{-n}^{n} d_\ell \, e^{i\ell\theta}, \quad Q_n' = \sum_{|\ell|>n} d_\ell \, e^{i\ell\theta},$$

so $(P_m + P_m')(Q_n + Q_n') = 0$, and therefore:

$$P_m Q_n = - P_m' g - P_m Q_n',$$

$$\|P_m Q_n\|_A \leq \|P_m'\|_A \|g\|_A + \|P_m\|_A \|Q_n'\|_A$$

(using the algebra property of $A(\Pi)$)

$$\leq \|P_m'\|_A \|g\|_A + \|f\|_A \cdot \|Q_n'\|_A,$$

and this last quantity tends to 0 when $m, n \to +\infty$, though

$$\|P_m\|_A \underset{m\to+\infty}{\longrightarrow} \|f\|_A, \quad \|Q_n\|_A \underset{n\to+\infty}{\longrightarrow} \|g\|_A.$$

This example can be used to bind from above the constants which will be given later on. Indeed, let $\rho = (\rho_n)_{n\geq 0}$ an increasing sequence of real numbers, satisfying

$$\rho_n \geq 1, \ \rho_{m+n} \leq \rho_m \cdot \rho_n, \quad m, n \in \mathbb{N}.$$

One knows that the Beurling Algebra

$$A_\rho = \{f \in A(\pi), \ f = \sum c_j e^{ij\theta}, \ \sum_{j\in\mathbb{Z}} |c_j| \rho_{|j|} < +\infty\}$$

is non quasi-analytic if and only if (see for example Y. Domar [3]):

$$\sum_{n\geq 0} \frac{\log \rho_n}{1+n^2} < +\infty .$$

If $(\rho_n)_{n\geq 0}$ satisfies this condition, we obtain from the previous computation $|P_n Q_n|_1 \leq G/f_n$, with

$$G = \|f\|_A \cdot \|g\|_{A_\rho} + \|g\|_A \cdot \|f\|_{A_\rho} .$$

This is the case, for example, with $\rho_n = e^{n^\alpha}$, $\alpha < 1$. So we see that the rate of decrease of PQ is much faster than in example 1: we had $|PQ_n| \sim 2/n$, and we have here $|P_n Q_n| \sim G/e^{n^\alpha}$.

There are some simple cases in which estimate (1) is fulfilled. Indeed, fix two integers $m, n \in \mathbb{N}$, and take the classes:

$$C_1 = \{P, d^\circ P \le m\}$$

$$C_2 = \{Q, d^\circ Q \le n\}.$$

Then the estimate (1) holds, λ depending on m and n. This is clear by the following compactness argument:

Let S_m be the unit sphere of C_1, equipped with the $|\cdot|_1$ norm, and, similarly let S_n the unit sphere of C_2. The product $P, Q \to P \cdot Q$ is a continuous mapping on the product $S_m \times S_n$, which is compact. So it reaches its minimum, that is, there are two polynomials $P^o \in C_1$, $Q^o \in C_2$, such that $|PQ|_1 \ge |P^o \cdot Q^o|_1$, for all $P \in C_1$, $Q \in C_2$. Put $\lambda = |P^o Q^o|_1$, then $\lambda \ne 0$ (since P^o, Q^o are polynomials of given degree, satisfying $|P^o|1 = 1$, $|Q^o|_1 = 1$). Since (1) is homogeneous, it is proved.

Of course, this type of argument does not give the exact value of λ. But it applies just as well to the other norms: they are all equivalent on the space of polynomials of given degree.

A precise estimate, in this case, of the constant λ is given by the well-known Gelfond's Theorem (see for example M. Waldschmidt [6]):

Gelfond's Theorem. Let P, Q be polynomials of degree m, n respectively. Then $|PQ|_\infty \ge e^{-(m+n)} |P|_\infty \cdot |Q|_\infty$.

Using the equivalence of the norms, one deduces easily from this theorem estimates using other norms.

We shall look mainly at larger classes than those consisting of polynomials of given degree. Indeed, we shall consider polynomials only having some concentration at low degrees. To define this notion, if $P = a_0 + a_1 x + \ldots + a_N x^N$, and if $m \in \mathbb{N}$, put:

$$P|^m = a_0 + a_1 x + \dots + a_m x^m.$$

We say that P has concentration δ (with $0 < \delta \leq 1$) at degrees $\leq m$ if

$$\|P|^m\|_2 \geq \delta \, \|P\|_2, \tag{2}$$

which means that

$$\sum_0^m |a_j|^2 \geq \delta^2 \sum_0^N |a_j|^2.$$

For given δ, m, the set of polynomials satisfying (2), with $\|P\|_2 = 1$, is not compact, so no compactness argument can apply.

We need first estimates for a single polynomial, considered as a function on the Torus Π.

1. Estimates for a Single Polynomial

We let m denote the normalized Haar measure on Π, that is $\frac{d\theta}{2\pi}$. The lemma which follows says that, in any set of prescribed measure, there is a large subset on which $|P(e^{i\theta})|$ is large, depending on its first coefficient a_0:

Lemma 1. For each measurable subset $E \subset \Pi$, with $m(E) = a$, and each $\alpha \geq 1$ we have, for every polynomial P:

$$m\{E \cap \{|P(e^{i\theta})| \geq \left(\frac{|a_0|}{\sqrt{e}}\right)^{\alpha/a} \|P\|_2^{1-\alpha/a}\}\} \geq a(1 - 1/\alpha).$$

Proof. We may assume $\|P\|_2 = 1$. We observe that for every measurable subset $A \subset \Pi$, we have:

$$\int_{CA} \log |P(e^{i\theta})|^2 \frac{d\theta}{2\pi} \leq \int_{CA} |P(e^{i\theta})|^2 \frac{d\theta}{2\pi}$$

$$\leq \int |P(e^{i\theta})|^2 \frac{d\theta}{2\pi} = 1,$$

and so, by the classical Jensen's Inequality (see for example

W. Rudin [5]):

$$\int_A \text{Log } |P(e^{i\theta})| \frac{d\theta}{2\pi} \geq \text{Log } \frac{|a_0|}{\sqrt{e}} .$$

Taking $A = E \cap \{|P| < (\frac{|a_0|}{\sqrt{e}})^{\alpha/a}\}$, we get

$$M(A) \leq a/\alpha,$$

from which the Lemma follows.

It should be noted that the result concerns only the measure of the set $E \geq \{|P| \geq (\frac{|a_0|}{\sqrt{e}})^{\alpha/a} \|P\|^{1-\alpha/a}\}$, and not the geometry of this set. For example, take $E = \Pi$ (so $a = 1$), take $\alpha = 2$, and $P = 1 + x^N$, so $\|P\|_2 = \sqrt{2}$, and $a_0 = 1$. We get that

$$m\{|P| \geq e^{-1}\sqrt{2}^{-1}\} \geq \frac{1}{2},$$

which is independent of the degree N. But the set $\{|P| \geq (e\sqrt{2})^{-1}\}$ consists in a number of intervals which dependes on N.

The Lemma 1 has an extension, involving the first m coefficients of the polynomial P.

<u>Lemma 2</u>. Let $m \in \mathbb{N}$, $\delta > 0$, $a > 0$, $\alpha \geq 1$. For every polynomial P satisfing

$$\||P|^m\|_2 \geq \delta \|P\|_2 , \tag{2}$$

and every measurable subset $E \subset \Pi$, with $m(E) = a$, we have

$$m\{E \cap \{|P| \geq \delta^{\alpha/a} (2e)^{-(m+2)(a/\alpha)^{-(m+1)}} \|P\|_2\}\} \geq a(1 - \frac{1}{\alpha}).$$

Outline of the Proof

Assume $\||P|^m\|_2 = 1$. If a_0 is substantial, apply lemma 1.

If it is small, look at

$$P_1 = a_1 x + a_2 x^2 + \dots,$$

If a_1 is substantial, apply lemma 1 to P_1: the estimate obtained for P_1 will apply to P, because a_0, being too small, cannot destroy it. If a_1 is too small, pass to

$$P_2 = a_2 x^2 + a_3 x^3 + \dots$$

and so on. By the assumption (2), this process will stop after a number of cases depending only on m. The reader is referred to [1] for details.

Corollary. Under the same assumptions, we have:

$$\|P\|_\infty \geq \delta(2e)^{-(k+2)} \|P\|_2 .$$

We shall now use these results to obtain information about the products of two polynomials.

2. Products of Polynomials

We first consider the case where both polynomials have some concentration at low degrees, thus obtaining an estimate of the type (1):

Theorem 1. Let $m, n \in \mathbf{N}$, $\delta > 0$, $\delta' > 0$. There is a constant $\lambda(\delta, \delta', m, n)$ such that for every polynomial P, Q satisfying:

$$\||P|^m\|_2 \geq \delta \|P\|_2 , \ \||Q|^n\|_2 \geq \delta' \|Q\|_2 ,$$

we have

$$\|PQ\|_2 \geq \lambda \|P\|_2 \cdot \|Q\|_2 .$$

Proof. Take for E the Torus Π and $\alpha = 3$ in Lemma 1.2. Then, on a set of measure $\frac{2}{3}$, one has:

$$|P| \geq \delta^3 (2e)^{-(m+2)(1/3)^{-(m+1)}} \|P\|_2 ,$$

and on another set of measure $\frac{2}{3}$,

$$|Q| \geq \delta^3 (2e)^{-(n+2)(1/3)^{-(n+1)}} \|Q\|_2 ,$$

the result follows.

There is a case when one knows a priori that the polynomial P will be large on a large set. This is given by the following Lemma:

Lemma 2. If $\|P\|_2 \geq \delta \|P\|_\infty$, then

$$m\{|P| \geq \frac{\delta}{2} \|P\|_2\} \geq \frac{\delta^2}{2} .$$

Proof. Assume $\|P\|_2 = 1$. We have

$$\|P\|_2 \leq \|P\|_1^{1/2} \cdot \|P\|_\infty^{1/2}$$

and so $\|P\|_1 \geq \delta$, and

$$\int_{\{|P| \geq \delta/2\}} |P(e^{i\theta})| \frac{d\theta}{2\pi} \geq \frac{\delta}{2} ,$$

from which the result follows.

Using again Lemma 1.2 and Lemma 2 above, we get

Theorem 3. Let $k \in \mathbb{N}$, $\delta, \delta' > 0$. For every polynomial P satisfing

$$\|P\|_2 \geq \delta \|P\|_\infty \tag{3}$$

and every polynomial Q with

$$\||Q|^k\|_2 \geq \delta' \|Q\|_2 , \tag{4}$$

we have:

$$m\{|PQ| \geq \frac{\delta}{2}\,\delta'^{4/\delta^2}(2e)^{-(k+2)(\delta^2/4)^{-(k+1)}}\|P\|_2 \cdot \|Q\|_2\} \geq \frac{\delta^2}{4}.$$

Corollary 4. For every polynomial P satisfying

$$|P|_\infty \geq \delta\,|P|_1 \tag{5}$$

and every Q satisfying (4), we have:

$$\|PQ\|_2 \geq \lambda(\delta, \delta', k)\,\|P\|_2 \cdot \|Q\|_2 .$$

Indeed, the assumption (5) is stronger than (3).

Assume now that P satisfies (5), and put

$$|a_{j_0}| = |P|_\infty .$$

If we put $P' = \sum_{j \in A} a_j x^j$, where $A \not\ni j_0$ (that is, P' is any part of P not containing the index j_0), and if we put $Q_0 = Q|^k$, and Q' is any part of Q not meeting Q_0 (that is, $Q' = \sum_{j \in B} b_j x^j$, where $B \cap \{0, \ldots, k\} = \phi$), then, by Corollary 4, we get:

$$\|(a_{j_0} + P')(Q_0 + Q')\|_2 \geq \lambda\,\|P\|_2 \cdot \|Q\|_2 .$$

Making use of this argument, by an inductive process, we get:

Theorem 5. Let $\delta, \delta' > 0$, $k \in \mathbb{N}$. There exists a constant $\lambda(\delta, \delta'; k) > 0$ such that for every P satisfying (5), every Q satisfying (4), we have:

$$|PQ|_\infty \geq \lambda\,\|P\|_2 \cdot \|Q\|_2 .$$

This theorem means that if P has a large coefficient and Q some concentration at low degrees, the product PQ will have a large coefficient. But there is no information about the place where this large coefficient is: it does not depend only on δ, δ', k, and the rank of the large coefficient of P, it really depends on the particular choice of P and Q. See [1].

A simultaneous version of Theorem 5 holds for a family of P's: if all of them have a large coefficient, at the same place, all the products PQ will have a large coefficient, at the same place. This is the basic ingredient for the proofs of the next paragraph.

3. Polynomials Having Some Concentration on a Lacunary Set

Let $(n_j)_{j\geq 0}$ be a sequence of integers satisfying $n_o = 0$, and $n_{j+1} > 2n_j$, $j \geq 1$. Let $E = \{(n_j)_{j\in\mathbb{N}}\}$. Such a set is called (Hadamard) lacunary. A translate of E, $\zeta_a E$, will be $(n_j + a)_{j\in\mathbb{N}}$.

If $P = \sum_{j\geq 0} a_j x^j$, we put $\Pi_E P = \sum_{j\in E} a_j x^j$. The next Theorem is analogous to Theorem 2.5: instead of having some concentration at a point (that is, a large coefficient), P has some concentration on a lacunary set.

<u>Theorem 1</u>. Let $\delta, \delta' > 0$, $k \in \mathbb{N}$. There exists a constant $\lambda(\delta, \delta'; k) > 0$ such that, for every lacunary set E, all polynomials P and Q satisfying:

$$|\Pi_E P|_1 \geq \delta\ |P|_1 \tag{6}$$

$$|Q|^k|_1 \geq \delta'\ |Q|_1, \tag{7}$$

there is a translate of E, $\zeta_a E$, with

$$|\Pi_{\zeta_a E}(PQ)|_1 \geq \lambda\ |P|_1 \cdot |Q|_1.$$

This theorem can be extended to higher structures for E, if we strengthen the assumption (6) on P. More precisely, if m is an integer, we shall say that $E_m \subset \mathbb{N}$ is an m-mesh on $\{(n_j)_{j\in\mathbb{N}}\}$, if we have:

$$E_m = \{n_{j_1} + \dots + n_{j_m};\ j_1, j_1, \dots, j_m, j_m \in \mathbb{N}\}.$$

(Observe that by the assumption $n_o = 0$, we have $E_{m-1} \subset E_m$). The assumption we now make on P is less general than (6): it still means that P has some concentration on E_m, but also that it disappears for high degrees.

Theorem 2. Let $k, m \in \mathbb{N}$, $\delta, \delta' > 0$. Let $(n_j)_{j\geq 0}$ be a sequence of integers with $n_o = 0$, $n_1 = 1$, and $n_j > 2jn_{j-1}$ $(j \geq 1)$. Let ρ_j be a decreasing sequence of real numbers, with $\lim_{j\to+\infty} \rho_j = 0$.

There exists a constant $\lambda((n_j), (\rho_j); \delta, \delta'; k, m) > 0$ such that, for every polynomial P written under canonical form:

$$P = \sum_\alpha a_\alpha x^{\alpha_1 n_1+\alpha_2 n_2+\ldots+\alpha_N n_N}$$

(with $\alpha_1 n_1 < n_2$, $\alpha_2 n_2 < n_3$, ...) and satisfying, for some integer u:

a) $|\Pi_{\zeta_u E_m} P|_1 \geq \delta |P|_1$

b) $\forall M > 0$, $\sum_{|\alpha|>m+M} |a_\alpha| \leq \rho_M |P|_1$

and for every polynomial Q satisfying (7) as before, we can find an integer ν such that

$$|\Pi_{\zeta_\nu E_m}(PQ)|_1 \geq \lambda |P|_1 |Q|_1.$$

The proof goes by induction on m, and what is proved is actually a simultaneous version of Theorem 2, dealing with a family of P's. The first step of the induction is the simulaneous version of Theroem 2.5.

4. Products of Polynomials in Many Variables

As we already mentioned, the results of this paragraph are due to P. Enflo [2]; we therefore refer the reader to [2] for complete proofs.

We now consider polynomials in many variables:

$$A = \sum_\alpha a_\alpha x_1^{\alpha_1} \ldots x_N^{\alpha_N}, \quad B = \sum_\beta b_\beta x_1^{\beta_1} \ldots x_N^{\beta_N}$$

where $\alpha = (\alpha_1, \ldots, \alpha_N)$, $\beta = (\beta_1, \ldots, \beta_N)$. For such a polynomial, we put $|A|_1 = \sum_\alpha |a_\alpha|$. If $n \in \mathbf{N}$, we define:

$$A_n = \sum_{|\alpha| \le n} a_\alpha x_1^{\alpha_1} \ldots x_N^{\alpha_n}$$

$$A_{[n]} = \sum_{|\alpha| = n} a_\alpha x_1^{\alpha_1} \ldots x_N^{\alpha_N},$$

where $|\alpha| = \alpha_1 + \ldots + \alpha_N$.

The main theorem deals with products of polynomials having concentration at low (total) degrees.

Theorem 1. Let $\delta, \delta' > 0$, $n, m \in \mathbb{N}$. There is a constant $\lambda(\delta, \delta'; m, n)$ such that, for all polynomials A, B satisfying:

$$|A_m|_1 \ge \delta \, |A_{m+n}|_1$$

$$|B_n|_1 \ge \delta' |B_{m+n}|_1,$$

one has $|(AB)_{m+n}|_1 \ge \lambda |A_m|_1 \, |B_n|_1$.

The important thing in this statement is that the constant λ is independent of the number of variables in A and B.

It should be noticed also that the concentration hypothesis involves only A_m and A_{m+n}, not A itself. Of course, it is a fortiori fulfilled if

$$|A_m|_1 \ge \delta \, |A|_1,$$

but this is not necessary. Such a weaker condition did not appear in §2.

Theorem 1 will follow from the corresponding result for homogeneous polynomials.

Theorem 2. Let $m, n \in \mathbb{N}$. There is a constant $\lambda(m, n)$ such that for all homogeneous polynomials A, B, of degree m and n respectively, one has:

$$|AB|_1 \geq \lambda(m, n)\ |A|_1 \cdot |B|_1.$$

Here again, the important thing is that λ is independent of the number of variables. To have a constant depending on this number is easy, by a compactness argument, as we saw in the introduction.

Proof that Theorem 2 implies Theorem 1. Assume, for the sake of simplicity, $n \geq m$. Put $\lambda = \inf_{\substack{m' \leq m \\ n' \leq n}} \lambda(m', n')$.

Take two polynomials A and B, and assume:

$$|A_m|_1 = 1,\ |B_n|_1 = 1,\ |A_{m+n}|_1 \leq \frac{1}{\delta},\ |B_{m+n}|_1 \leq \frac{1}{\delta}.$$

There is a smallest $j_1 \leq m \leq n$ such that

$$|A_{[j_1]}|_1 \geq \frac{1}{2n}$$

and a smallest $j_2 \leq n$ such that

$$|B_{[j_2]}|_1 \geq \frac{1}{2n}.$$

Then look at $|(AB)_{[j_1+j_2]}|$. Two cases can occur:

- either it is $\geq \frac{1}{2}\lambda \frac{1}{4n^2}$, and the result is proved,

- or not. In this latter case, since by Theorem 2:

$$|A_{[j_1]}\ B_{[j_2]}|_1 \geq \frac{\lambda}{4n^2},$$

it means that

$$\sum_{\substack{|\alpha+\beta| = j_1+j_2 \\ |\alpha| \neq j_1, |\beta| \neq j_2}} |a_\alpha\ b_\beta| \geq \frac{1}{2}\lambda \frac{1}{4n^2}.$$

So either
$$\sum_{\substack{|\alpha|+|\beta|=j_1+j_2 \\ |\alpha|<j_1}} |a_\alpha\, b_\beta| \geq \frac{1}{4}\lambda\ \frac{1}{4n^2}$$

or

$$\sum_{\substack{|\alpha|+|\beta|=j_1+j_2 \\ |\beta|<j_2}} |a_\alpha\, b_\beta| \geq \frac{1}{4}\lambda\ \frac{1}{4n^2}\ .$$

In the first case, we get

$$|A_{j_1-1}|_1 \geq \frac{1}{4}\lambda\ \frac{1}{4n^2}\ \delta',$$

and in the second,

$$|B_{j_2-1}|_1 \geq \frac{1}{4}\lambda\ \frac{1}{4n^2}\ \delta.$$

Now consider the smallest j_3 such that

$$|A_{[j_3]}| \geq \frac{1}{2n} \cdot \frac{1}{4}\lambda \cdot \frac{1}{4n^2}\ \delta'$$

and the smallest j_4 such that

$$|B_{[j_4]}| \geq \frac{1}{2n} \cdot \frac{1}{4}\lambda \cdot \frac{1}{4n^2}\ \delta.$$

Then $j_3 + j_4 < j_1 + j_2$.

We begin again on $(AB)_{[j_3+j_4]}$, and we find $j_5 + j_6 < j_3 + j_4$, and so on. This process stops after at most $2n$ steps, and Theorem 1 follows from Theorem 2.

Theorem 2 will follow from a more general result. Let $A_1, -, A_n$ be homogeneous polynomials, in many variables of degrees $d_1, -, d_n$ respectively. Let P be a polynomial in n variables, such that $P(A_1, -, A_n)$ is homogeneous: thus P

satisfies, for some d:

$$P(t^{d_1}z_1, -, t^{d_n}z_n) = t^d P(z_1, -, z_n).$$

We shall study the case when $P(A_1, -, A_n)$ is small.

Theorem 3. Assume we have a sequence $P_j, A_{j,1}, -, A_{j,n}$, with $|A_{j,i}|_1 \le 1\ \forall i, j$,

$$|P_j|_1 \le 1, \forall j$$

and

$$|P_j(A_{j,1}, -, A_{j,n})|_1 \underset{j\to+\infty}{\longrightarrow} 0 \quad .$$

Then there is a number m, depending only on $d_1, -, d_n$ and $d^\circ P$, and, for some subsequence of the j's, polynomials

$$Q_{j,1}, \ldots, Q_{j,n} \text{ and } B_{j,1}, \ldots, B_{j,m},$$

of bounded degrees, such that

1) $A_{j,i} = Q_{j,i}(B_{j,1}, \ldots, B_{j,m}) + o(1) \quad j \to +\infty$

2) $|Q_{j,i}|_1 = O(1)$, $|B_{j,i}|_1 = O(1)$, $\quad j \to +\infty$

3) If we put

$$R_j(t_1, \ldots, t_m) = P_j(Q_{j,1}(t_1, \ldots, t_m), \ldots, Q_{j,n}(t_1,\ldots,t_m)),$$

then $|R_j|_1 \underset{j\to+\infty}{\longrightarrow} 0$.

Moreover, the degrees of B_{ji}, Q_{ji} depend only on $d_1, \ldots, d_n$ and d^0P.

In this statement, we have used the notation "$o(1)$" to denote a polynomial X_j, with $|X_j|_1 \underset{j\to+\infty}{\longrightarrow} 0$, and the notation "$O(1)$" to denote a quantity which is bounded when $j \to +\infty$.

Observe that conversely, if 1), 2), 3) are satisfied, we must have $|P_j(A_{j,1}, \ldots, A_{j,n})|_1 \underset{j\to+\infty}{\longrightarrow} 0$. Indeed, in 3), we have $Q_{j,i}(t_1, \ldots, t_m)$ instead of $A_{j,i}$, that is, up to a term which is $0(1)$, instead of $Q(B_{j,1}, \ldots, B_{j,m})$.

Substituting $B_{j1}, \ldots, B_{j,m}$ to $t_1, \ldots, t_m$ may bring some cancellation, so the norm of the resulting polynomial will be smaller.

Proof that Theorem 3 implies Theorem 2. Here we deal only with $P(z_1, z_2) = z_1 \times z_2$. Assume we have $A_{j,1}, A_{j,2}$ with $|A_{j,1}|_1 = |A_{j,2}|_1 = 1$, and

$$|A_{j,1} \times A_{j,2}| \underset{j\to+\infty}{\longrightarrow} 0 .$$

Then, by 1):

$$A_{j,1} = Q_{j,1} (B_{j,1}, \ldots, B_{j,m}) + o(1)$$

$$A_{j,2} = Q_{j,2} (B_{j,1}, \ldots, B_{j,m}) + o(1),$$

which implies, for j, large enough

$$|Q_{j,1}|_1 \geq \frac{1}{2}, \ |Q_{j,2}|_1 \geq \frac{1}{2} .$$

Moreover, by 3):

$$|Q_{j,1} (t_1, \ldots, t_m) \cdot Q_{j,2}(t_1, \ldots, t_m)|_1 \underset{j\to+\infty}{\longrightarrow} 0 .$$

But this is impossible: the degrees of the $Q_{j,1}, Q_{j,2}$ are bounded, the number of variables is bounded, so this contradicts Theorem 2 for polynomials with given number of variables, which we already saw to be clear.

Let's now give some notation.

First, we shall drop the index j everywhere. Second, we shall write $a >> 0$ to denote a quantity bounded from below; that is, $\liminf_{j\to+\infty} a_j > 0$.

Finally, if $R(t_1, \ldots, t_n)$ is a polynomial, we shall denote by $|R|_1 (|A_1|_1, \ldots, |A_n|_1)$ the positive number obtained by replacing each coefficient in R by its modulus, and substituting $|A_i|_1$ instead of t_i.

If $A_i = \lambda_i B_i$, and if we put $S(t_1, \ldots, t_m) = R(\lambda_1 t_1, \ldots, \lambda_m t_m)$, then clearly

$$|S|_1 (|B_1|_1, \ldots, |B_m|_1) = |R|_1 (|A_1|_1, \ldots, |A_m|_1).$$

We also put $A^{(i)} = \dfrac{\partial A}{\partial z_i}$.

We have:

Lemma 4. If A is homogeneous of degree d,

$$\sum_i |A^{(i)}|_1 = d\, |A|_1 .$$

Lemma 5. Let $(P(A_1, \ldots, A_n))^{(i)} = \sum_k \dfrac{\partial P}{\partial z_k} A_k^{(i)}$

$$= R_i(A_1, \ldots, A_n, A^{(i)}, \ldots, A_n^{(i)}).$$

Then

$$\sum_i |R_i|_1 (|A_1|_1, \ldots, |A_n|_1, |A_1^{(i)}|_1, \ldots, |A_n^{(i)}|_1)$$

$$\leq K\, |P|_1 (|A_1|_1, \ldots, |A_n|_1)$$

where K depends only on $d^\circ P$ and $d^\circ A_1, \ldots d^\circ A_n$.

To prove this lemma, one reduces it to the case where P is a monomial, where the proof is immediate.

Theorem 3 will follow from Theorem 6 below:

Theorem 6. Assume again $|A_1|_1 \leq 1, \ldots, |A_n|_1 \leq 1$, $|P|_1 \leq 1$, $|P(A_1, \ldots, A_n)|_1 = o(1)$.

Suppose $d^\circ A_1$ is maximal among $d^\circ A_i$. Write:

$$P = A_1^r C_0(A_2, \ldots, A_n) + A_1^{r-1} C_1(A_2, \ldots, A_n) + \ldots +$$

$$C_r(A_2, \ldots, A_n).$$

If $|C_0(A_2, \ldots, A_n)|_1 >> 0$, we can find polynomials $B_1, \ldots, B_m$, with $d^\circ B_i < d^\circ A_1$, and a polynomial Q, with $|Q|_1 = 0(1)$, $|B_i|_1 = 0(1)$, such that

$$A_1 = Q(A_2, \ldots, A_n, B_1, \ldots, B_m) + o(1).$$

Furthermore, the degree of each monomial in $Q(A_2, \ldots, A_n, B_1, \ldots, B_n)$ is equal to $d^\circ A_1$.

<u>Proof of Theorem 6</u>. Assume first that $d^\circ C_0 > 0$. Then:

$$(P(A_1, \ldots, A_n))^{(i)} = A_1^r (C_0(A_2, \ldots, A_n))^{(i)} +$$

$$r\, A_1^{r-1} A_1^{(i)}\, C_0 + \ldots$$

$$= R_i(A_1, \ldots, A_n, A_1^{(i)}, \ldots, A_n^{(i)}).$$

We have by lemma 4:

$$\sum_i |(C_0(A_2, \ldots, A_n))^{(i)}|_1 >> 0$$

and by lemma 5:

$$\sum_i |R_i|_1(|A_1|_1, \ldots, |A_n|_1, |A_1^{(i)}|_1, \ldots, |A_n^{(i)}|_1) = 0(1)$$

and

$$\sum_i |(P(A_1, \ldots, A_n))^{(i)}|_1 = o(1).$$

From this, we deduce that, for some i, if we put

$$\alpha_i = |(C_0(A_2, \ldots, A_n))^{(i)}|_1$$

then

$$\left|\frac{1}{\alpha_i} (P(A_1, \ldots, A_n))^{(i)}\right|_1 = o(1)$$

and

$$\frac{1}{\alpha_i} |R_i|_1(|A_1|_1, \ldots, |A_n|_1, |A_1^{(i)}|, \ldots, |A_n^{(i)}|_1) = O(1).$$

For this i, we write $A_k^{(i)} = \lambda_k B_k$, so that $|B_k|_1 = 1$, and put $S = \frac{1}{\alpha} R_i$. So, S can be written:

$$A = A_1^r D(A_2, \ldots, A_n; B_2, \ldots, B_n) + A_1^{r-1} \ldots ,$$

with

$$|S|_1 = O(1),$$

$$|D(A_2, \ldots, A_n, B)|_1 >> 0,$$

and

$$|S(A_1, \ldots, A_n, B)|_1 = o(1)$$

(where B stands for $B_2, \ldots, B_n$).

So S is a polynomial involving more variables than P, but which has total degree which is one unit less than the total degree of P. Repeating this process as many times as needed (at most $d°P$ times), we will obtain a polynomial in many variables, of degree 1, for which the result is clear.

Let us now consider the case when C_0 = constant. Then $(P(A))^{(i)} = A_1^{r-1}(r\, C_0\, A_1^{(i)} + C_1^{(i)}) + A_1^{r-2} \ldots$

- If $|r\, C_0\, A_1 + C_1|_1 = o(1)$, since $C_0 >> 0$, we have

$A_1 = -(r\ C_0)^{-1}\ C_1 + o(1)$, and the result follows since A_1 is then a polynomial in $A_2, \ldots, A_n, B$.

- If $|r\ C_0\ A_1 + C_1|_1 >> 0$, then

$$\sum_i |r\ C_0\ A_1^{(i)} + C_1^{(i)}|_1 >> 0,$$

and the proof goes as before. This proves Theorem 6.

Proof that Theorem 6 implies Theorem 3. Let $P = A_1^r\ C_0(A_2, \ldots, A_n)$ as before.

1) If $|C_0(t_2, \ldots, t_n)|_1 = o(1)$, the first term can be dropped and we have lowered the degree of A_1 in P.

2) If $|C_0(t_2, \ldots, t_n)|_1 >> 0$ and $|C_0\ (A_2, \ldots, A_n)|_1 >> 0$, we make repeated uses of Theorem 6.

3) If $|C_0(t_2, \ldots, t_n)|_1 >> 1$ but $|C_0(A_2, \ldots, A_n)|_1 = o(1)$, making an induction also on the number of variables, we can write for C_0:

$$A_2 = Q_2(B) + o(1), \ldots, A_n = Q_n(B) + o(1),$$

satisfying Theorem 3, and substituting in P, we get

$$A_1^r\ C_0(Q_2(B), \ldots, Q_n(B)) + \ldots .$$

So $|C_0(Q_2(t), \ldots, 0_n(t)| = o(1)$ by Theorem 3 (induction hypothesis), so we can neglect the first term, and we have lowered the $d°$ at which A_1 appears. The theorem follows by induction.

REFERENCES

1. Beauzamy, B., and Enflo, P., "Estimations de produits de polynômes", to appear.

2. Enflo, P., "On the invariant subspace problem in Banach spaces", to appear.

3. Domar, Y., "Harmonic analysis based on certain commutative algebras", Acta Math, 96, 1956, pp. 1-66.

4. Kahane, J. P., "Séries de Fourier absolument convergentes", Springer-Verlag, 1970.

5. Rudin, W., "Real and complex analysis", Tata McGraw-Hill.

6. Waldschmidt, M., "Nombres transcendants", Springer-Verlag, Lecture Notes.

EXCHANGE ALGORITHMS, ERROR ESTIMATIONS AND STRONG UNICITY IN CONVEX PROGRAMMING AND CHEBYSHEV APPROXIMATION

Hans-Peter Blatt

Katholische Universität Eichstätt

In the first section an exchange algorithm is given which is a generalization of the Remez algorithm: Using an idea of Töpfer we consider at each step a finite sequence of finite subproblems, which is called an optimization problem with respect to a *chain of references*. Modifying a strategy of Carasso and Laurent we replace in the exchange theorem the zero-checks by practical δ-checks with $\delta > 0$. This allows us to reduce the numerical calculations of the functional to be minimized, to raise the stability of the algorithm and to introduce multiple exchange techniques known from the Remez algorithm for getting faster convergence.

Using the concept of chain of references a new characterization for strongly unique optimal solutions is given. These chain of references are generalizations of H-sets in the sense of Collatz and lead in a natural way to error estimations in the parameter space. It turns out that the exchange algorithms of the first section are excellent methods for obtaining sharp estimations for optimal solutions.

In section 3 we consider the approximation of a function f on $[a, b]$ by elements of a Haar-subspace. The strong unicity constants are characterized by a certain type of approximation problems. Using this characterization the asymptotic behaviour of the strong unicity constants is discussed for polynomial approximations.

S. P. Singh et al. (eds.), Approximation Theory and Spline Functions, 23–63.

1. A MULTIPLE EXCHANGE ALGORITHM IN CONVEX PROGRAMMING

1.1. The Minimization Problem

Let T be an arbitrary set and

$$a : T \to \mathbb{R}^n, \quad f : T \to \mathbb{R}$$

be two bounded mappings, V a subspace of $\mathbb{R}^n$ with $\dim V = d$, W an affine subspace parallel to V. Denoting by $\langle \cdot, \cdot \rangle$ the scalar product in $\mathbb{R}^n$, we consider the continuous convex functional

$$g(x) := \sup_{t \in T} (\langle a(t), x \rangle - f(t)).$$

The problem consists in *minimizing $g(x)$ with respect to $x \in W$.* We suppose that

$$\alpha := \inf_{x \in W} g(x)$$

is finite.

Furthermore we may assume without loss of generality that the set

$$\tilde{A} := \{(a(t), f(t)) \mid t \in T\} \subset \mathbb{R}^{n+1}$$

is convex and we define $A := \{a(t) \mid t \in T\}$.

Carasso and Laurent [8] proposed a dual method for this problem which uses some generalized notions of Stiefel [24].

Example. Let E be a normed linear space over $\mathbb{C}$ with norm $\|\cdot\|$, $V = \text{span}(v_1, v_2, \ldots, v_n)$ a n-dimensional subspace of E. For a given element $f \in E$ we want to find an element $\overline{v} \in V$ such that

$$\min_{v \in V} \|f - v\| = \|f - \overline{v}\| . \tag{1.1}$$

If S_{E^*} denotes the unit cell of the dual space E^* it is known that

$$\|h\| = \max_{L \in S_{E^*}} \operatorname{Re} L(h).$$

Consider for $v \in V$ the representation

$$v = \sum_{k=1}^{n} (x_k + i x_{k+n}) v_k$$

and define

$$x := (x_1, \ldots, x_{2n})$$

$$a(L) := (\operatorname{Re}L(v_1), \ldots, \operatorname{Re}L(v_n), -\operatorname{Im}L(v_1), \ldots, -\operatorname{Im}L(v_n))$$

$$f(L) := \operatorname{Re}L(f).$$

Then

$$\operatorname{Re}L(v - f) = \langle a(L), x\rangle - f(L)$$

and the problem (1.1) is equivalent to minimize the functional

$$g(x) = \max_{L \in S_{E^*}} \langle a(L), x\rangle - f(L)$$

with respect to $x \in \mathbb{R}^{2n}$.

1.2 References and Chain of References

Definition. A subset $R = \{a(t_i) \mid 1 \leq j \leq k + 1\} \subset A$ is called a *V-reference* if

(α) there exists $\lambda_i > 0 (1 \leq i \leq k + 1)$ such that

$$\sum_{i=1}^{k+1} \lambda_i = 1 \quad \text{and} \quad \sum_{i=1}^{k+1} \lambda_i a(t_i) \in V^{\perp},$$

where $V^{\perp}$ is the orthogonal complement of V in $\mathbb{R}^n$.

(β) R is minimal, i.e. there exists no proper subset of R such that (α) holds.

The numbers λ_i are uniquely determined and called the *characteristic numbers* of the V-reference R.

Definition. $x \in W$ is called *solution of the V-reference R in W*, if

$$\min_{z \in W} \max_{1 \le i \le k+1} (\langle a(t_i), z\rangle - f(t_i)) = \max_{1 \le i \le k+1} (\langle a(t_i), x\rangle - f(t_i)).$$

For constructing such a solution one defines

$$h_R = \sum_{i=1}^{k+1} \lambda_i (\langle a(t_i), z\rangle - f(t_i)) \tag{1.2}$$

for a fixed $z \in W$ and obtains $y \in V$ by solving the linear system

$$\langle a(t_i), y\rangle = f(t_i) - \langle a(t_i), z\rangle + h_R \tag{1.3}$$

for $1 \leqq i \leqq k+1$. Then $x = z + y$ is a solution of the V-reference R in W, in general not unique. h_R is called *deviation of the V-reference* R in W.

Consider now the following construction: Let $R_1 \subset A$ be a V-reference, W_1 the set of solutions of the V-reference R_1 in W, h_1 the deviation of R_1 in W,

$$V_1 := [R_1]^{\perp} \cap V,$$

where we denote for abbreviation by $[R_1]$ the subspace in $\mathbb{R}^n$ spanned by the vectors of R_1.

Then $\dim V_1 = \dim V - k_1$, if R_1 has $k_1 + 1$ elements, and W_1 is parallel to V_1.

We repeat the same construction for V_1, W_1 instead of V, W: Let $R_2 \subset A$ be a V_1-reference, W_2 the set of solutions of the V_1-reference R_2 in W_1, h_2 the deviation of R_2 in W_1,

$$V_2 := [R_2]^{\perp} \cap V_1.$$

Then $\dim V_2 = V - k_1 - k_2$, if R_2 has $k_2 + 1$ elements, and W_2 is parallel to V_2, etc...

In this way we possibly can find a *chain (or string) of references*

$$\mathcal{R} = (R_1, R_2, \ldots, R_s)$$

with corresponding subspaces

$$V_1 \supset V_2 \supset \ldots \supset V_s$$

and parallel affine subspaces

$$W_1 \supset W_2 \supset \ldots \supset W_s,$$

and a *vector of deviations*

$$h = (h_1, h_2, \ldots, h_s)$$

such that

(1) R_i is a V_{i-1}-reference,

(2) $V_i = [R_i]^{\perp} \cap V_{i-1}$,

(3) h_i is the deviation of R_i in W_{i-1},

(4) $V_s = [0]$

$(V_o := V, W_o := W)$.

Such a chain of references determines a unique point $x_{\mathcal{R}} \in W$ $(\{x_{\mathcal{R}}\} = W_s)$, which is the *solution of the chain* $\mathcal{R}$ *in* W.

Definition. $\mathcal{R}$ is called *regular*, if each R_i has at least two elements.

If $\mathcal{R}$ is not regular we get a regular chain by cancelling in $\mathcal{R}$ each R_j having only one element. Thereby the solution of the chain is not changed.

A chain of references does not need to exist in any case. In approximation problems as in the example of 1.1 a chain of references always exist. More general, if we know a priori for a solution x of the minimization problem a bound α for the norm x, then it is possible to define an equivalent problem which has x as a solution and contains a chain of references (Carasso, Laurent [9]).

1.3 The δ-exchange Procedure

Let $\delta \geqq 0$ and

$$\mathcal{R} = (R_1, \ldots, R_j, R_{j+1}, \ldots, R_s)$$

a chain of references with deviation vector

$$h = (h_1, \ldots, h_j, h_{j+1}, \ldots, h_s).$$

Our aim is "to exchange R_{j+1} and R_j" such that we get a new chain of references

$$\tilde{\mathcal{R}} = (R_1, \ldots, R_{j-1}, \tilde{R}_j, \tilde{R}_{j+1}, R_{j+2}, \ldots, R_s)$$

with deviation vector

$$\tilde{h} = (h_1, \ldots, h_{j-1}, \tilde{h}_j, \tilde{h}_{j+1}, \ldots, \tilde{h}_s)$$

satisfying $\tilde{h}_j > h_j$ for $h_{j+1} > h_j$.

We know that there exist characteristic numbers $\lambda_i > 0$ for the V_{j-1}-reference

$$R_j = \{a(t_i) \mid 1 \leqq i \leqq k + 1\}$$

such that $\sum_{i=1}^{k+1} \lambda_i \, a(t_i) \in V_{j-1}^{\perp}$ and $\dim [R_j]/V_{j-1}^{\perp} = k$.

Analogously there exist for the V_j-reference

$$R_{j+1} = \{a(t_i) \mid 1 \leqq i \leqq \ell + 1\}$$

characteristic numbers $\mu_i > 0$ such that

$$z_o := \sum_{i=1}^{\ell+1} \mu_i \, a(\tilde{t}_i) \in V_j^{\perp} \quad \text{and} \quad \dim [R_{j+1}]/V_j^{\perp} = 1. \tag{1}$$

Because of $V_j^{\perp} = [R_j] + V_{j-1}^{\perp}$ we can find real numbers s_i such that

$$z_o + \sum_{i=1}^{k} w_i \, a(t_i) \in V_{j-1}^{\perp}.$$

With $w_{k+1} = 0$ we get for $\theta \in \mathbb{R}$:

$$z_o + \sum_{i=1}^{k+1} (w_i - \theta\lambda_i) \, a(t_i) \in V_{j-1}^{\perp}. \tag{1.4}$$

We choose

$$\theta = \min_{1 \leqq i \leqq k+1} (w_i/\lambda_i)$$

and define

$$\kappa := 1 + \sum_{i=1}^{k+1} (w_i - \theta\lambda_i), \tag{1.5}$$

$$\rho_i := (w_i - \theta\lambda_i)/\kappa \quad \text{for} \quad 1 \leqq i \leqq k+1,$$

$$\tilde{\mu}_i := \mu_i/\kappa \qquad \text{for} \quad 1 \leqq i \leqq \ell+1.$$

Then at least one of the ρ_i is zero,

$$\operatorname{sgn} \rho_i = \operatorname{sgn} \lambda_i > 0 \quad \text{for} \quad \rho_i \neq 0,$$

and

$$\tilde{\mu}_i > 0 \quad \text{for all} \quad 1 \leqq i \leqq \ell+1.$$

In the usual Remez algorithm (with Haar condition) or the method of Carasso and Laurent [9] all $a(t)$ in (1.4) with positive coefficients constitute $\tilde{R}_j$, all $a(t)$ with zero coefficients the reference $\tilde{R}_{j+1}$. This leads to the deviation

$$\tilde{h}_j = \kappa^{-1} h_{j+1} + (1 - \kappa^{-1}) h_j \tag{1.6}$$

of the V_{j-1}-reference $\tilde{R}_j$ in W_{j-1}.

Since in the minimization problem considered, the Haar condition is not satisfied in general, two situations may occur which should be reflected by the algorithm:

(α) For very small κ^{-1} it makes not much sense to exchange R_j with R_{j+1}, if R_{j+1} consists of more than one vector.

(β) It is difficult to recognize small ρ_i: Therefore in any case for practical reasons the 0-checks of ρ_i in [9] have to be replaced by δ-checks with a small, but nevertheless positive δ.

To overcome the difficulty (α) we redefine for $\kappa^{-1} < \delta$:

$$\mathcal{R} := (R_1, \ldots, R_j, \{z_0\}, R_{j+1}, \ldots, R_s)$$

with deviation vector

$$h := (h_1, \ldots, h_j, h_{j+1}, h_{j+1}, \ldots, h_s).$$

Hence we may assume that R_{j+1} consists of one single vector if $\kappa^{-1} < \delta$.

Now we define

$$I(\delta) := \{i \mid 1 \leq i \leq k + 1, \rho_i \leq \delta\},$$

$$\tilde{R}_{j+1} := \{a(t_i) \mid i \in I(\delta)\},$$

$$\hat{\mu}_1 := \tilde{\mu}_1 + \sum_{i \in I(\delta)} \rho_i,$$

$$\tilde{a} := (\tilde{\mu}_1 a(\tilde{t}_1) + \sum_{i \in I(\delta)} \rho_i a(t_i))/\hat{\mu}_1,$$

$$\tilde{f} := (\tilde{\mu}_1 f(\tilde{t}_1) + \sum_{i \in I(\delta)} \rho_i f(t_i))/\hat{\mu}_1.$$

Then there exists a point $\tilde{t} \in T$ such that

$\tilde{a} = a(\tilde{t})$ and $\tilde{f} = f(\tilde{t})$.

Using the same arguments as Carasso, Laurent [9] we see that

$$\tilde{R}_j := (R_j \cup R_{j+1} \cup \{a(\tilde{t})\})\backslash\tilde{R}_{j+1}\backslash\{a(\tilde{t}_1)\}.$$

is a V_{j-1}-reference and $\tilde{R}_{j+1}$ a $\tilde{V}_j$-reference with $\tilde{V}_j = [\tilde{R}_j]^{\perp} \cap V_{j-1}$.

Furthermore we get for the deviation $\tilde{h}_j$ of the V_{j-1}-reference $\tilde{R}_j$ the relation (1.6).

We remark that the deviation $\tilde{h}_j$ does not depend on δ. For $\delta = 0$ we get the exchange theorem of Carasso and Laurent ([8], [9]).

4. The Exchange Algorithm

The algorithm depends on the parameters

$$\varepsilon > 0 \quad \text{and} \quad \frac{1}{2} \leqq \beta < 1.$$

We want to construct an ε-solution of our problem, i.e. $z \in W$ such that

$$g(z) \leqq \alpha + \varepsilon.$$

We assume that we can find a chain of references

$$\mathcal{R}^1 = (R_1^1, \ldots, R_{s_1}^1)$$

with the solution x^1 and the vector of deviations

$$h^1 = (h_1^1, \ldots, h_{s_1}^1).$$

We put

$$M_1 := g(x^1)$$

and describe now *the step* ν *of the algorithm:*

We have a chain of references

$$\mathcal{R}^{\nu} = (R_1^{\nu}, \ldots, R_{s_\nu}^{\nu})$$

with solution x^{ν} and vector of deviations

$$h^{\nu} = (h_1^{\nu}, \ldots, h_{s_\nu}^{\nu}),$$

a number $M_\nu \in \mathbb{R}$ and an index $1 \leqq \mu \leqq \nu$ such that

$$g(x^{\mu}) \leq M_\nu.$$

We distinguish the cases:

(a) $h_1^{\nu} \geqq M_\nu - \varepsilon$:

Then x^{μ} is an ε-solution and the algorithm stops.

(b) $h_1^{\nu} < M_\nu - \varepsilon$:

In this case we have to consider two cases again.

(b1) If there exists a *minimal* index j with $1 \leqq j \leqq s_\nu - 1$ such that

$$h_{j+1}^{\nu} \geqq \beta h_j^{\nu} + (1 - \beta)K_\nu,$$

we determine as in (1.5) the number $\kappa = \kappa_\nu$.

Let $k_\nu + 1$ be the number of points of R_j, we define

$$\delta_\nu := \frac{\varepsilon}{(k_\nu + 1)(2^j(M_\nu - h_j^{\nu}) - \varepsilon)}. \tag{1.7}$$

Then we exchange R_j^{ν} and R_{j+1}^{ν} following the δ_ν-exchange of the preceding section and define $M_{\nu+1} := M_\nu$.

(b2) If such a minimal j as in (b1) does not exist, we form

R^ν to a regular chain R, determine a point $t_o \in T$ such that

$$\langle a(t_o), x^\nu \rangle - f(t_o) > g(x^\nu) - \frac{\varepsilon}{2^{s_\nu}}$$

and put

$$M_{\nu+1} := \min (M_\nu, g(x^\nu)).$$

Now we define the new chain of references by

$$R^{\nu+1} = (R, \{a(t_o)\}).$$

The algorithm is well-defined and does not get into a cycle [4].

Under certain conditions the Remez algorithm for the approximation of functions with respect to n-dimensional Haar subspaces has quadratic convergence if one exchanges not only one point with maximal deviation in the error curve, but $n + 2$ local extreme points of the error curve. We want to suggest such a proceeding to get a faster convergence.

We assume that the set T is a compact metric space. Then we can modify the iteration step of type (b2) in the following way:

($\tilde{b}2$) If a minimal j as in (b1) does not exist, we form R^ν to a regular chain r and determine the relative maxima t of the function

$$\langle a(t), x^\nu \rangle - f(t)$$

in T. Let M be the set of these relative maxima and let $M = \{t_i \mid 1 \leqq i \leqq r\}$ be a finite set, then we define the new chain by

$$R^{\nu+1} = (R, \{a(t_1)\}, \ldots, \{a(t_r)\})$$

and put

$$M_{\nu+1} := \min (M_\nu, g(x^\nu)).$$

In general the vectors h^ν are not monotonically increasing in the lexicographical sense. But if we consider the regular chain $\overline{R}^\nu$ belonging to R^ν with corresponding deviation vector $\overline{h}^\nu$ we can easily see that the sequence $\{\overline{h}^\nu\}$ increases monotonically in the lexicographical sense.

Moreover we have for the convergence of the algorithm the following.

Theorem 1 [4]. After a finite number of steps the algorithm with single or multiple exchanges leads to an ε-solution.

1.5. Implementation of the Algorithm

We denote by $\mathbb{R}_k^m$ the set of real matrices with k rows and m columns (k, m positive integers). Let $e_1, e_2, \ldots, e_n$ be the canonical basis of $\mathbb{R}^n$ with

$$e_i = (\underbrace{0, \ldots, 0}_{i-1}, 1, 0, \ldots, 0)$$

as unit vector for the index i. Moreover, let $P = (p_{ik}) \in \mathbb{R}_n^n$ be a regular matrix and let us consider the new basis

$$\tilde{e}_k = \sum_{i=1}^{n} p_{ik}\, e_i \tag{1.8}$$

for $k = 1, 2, \ldots, n$. If $x \in \mathbb{R}^n$ and $x = (x_1, \ldots, x_n)$ is the vector of coordinates with respect to the canonical basis,

$$x = (x_1, \ldots, x_n) = \sum_{i=1}^{n} x_i\, e_i,$$

then we get the vector $y = (y_1, \ldots, y_n)$ of coordinates with respect to the new basis $\tilde{e}_1, \tilde{e}_2, \ldots, \tilde{e}_n$ by

$$x^T = P \cdot y^T \quad \text{or} \quad y^T = P^{-1} \cdot x^T. \tag{1.9}$$

Therefore we have got

$$\langle a(t), x\rangle = a(t) \cdot x^T = a(t) \cdot P \cdot y^T \\ = \langle a(t) \cdot P, y\rangle . \tag{1.10}$$

As above let V be a supspace of $\mathbb{R}^n$ with $\dim V = d$ and $R = \{a(t_i) \mid 1 \leq i \leq k+1$ a V-reference. We assume

(S): We have a basis $\tilde{e}_1, \tilde{e}_2, \ldots, \tilde{e}_n$ of $\mathbb{R}^n = V^\perp + V$ such

$$V^\perp = \text{span } (\tilde{e}_1, \ldots, \tilde{e}_{n-d})$$

and

$$V = \text{span } (\tilde{e}_{n-d+1}, \ldots, \tilde{e}_n).$$

Let $P \in \mathbb{R}^n_n$ denotes the transformation matrix connecting the basis $\tilde{e}_1, \ldots, \tilde{e}_n$ with the canonical basis as in (1.8) and (1.9). We consider the matrix

$$B = \begin{pmatrix} a(t_1) \cdot P \\ \cdot \\ \cdot \\ \cdot \\ a(t_{k+1}) \cdot P \end{pmatrix} , \tag{1.11}$$

associated with the elements of the V-reference R, and construct a regular matrix

$$P_1 = \left(\begin{array}{c|c} I_1 & \\ \hline 0 & \end{array} \tilde{P}_1 \right) \tag{1.12}$$

where I_1 is the unit matrix in $\mathbb{R}^{n-d}_{n-d}$, such that

$$B\, P_1 = (D_1, C_1, 0) \tag{1.13}$$

with $D_1 \in \mathbb{R}^{n-d}_{k+1}$ and

$$C_1 = \begin{pmatrix} \underline{*} & & & & 0 \\ \cdot & \underline{*} & & & \\ \cdot & \cdot & & & \\ \cdot & \cdot & \cdot & & \underline{*} \\ * & \cdot & \cdot & * & \underline{*} \end{pmatrix} = \begin{pmatrix} c_1 \\ c_2 \\ \cdot \\ \cdot \\ \cdot \\ c_{k+1} \end{pmatrix} \in \mathbb{R}^k_{k+1}. \qquad (1.14)$$

($c_i \in \mathbb{R}^k$ for $i = 1, \ldots, k + 1$).

Because R is a V-reference such a matrix P_1 always exists. For example we can use for P_1 a product of matrices as in Gaussian elimination method. But more appropriate are matrices of Householder transformations with respect to the columns, because in this case for Euclidean norm the condition numbers of the matrices in the systems (1.3) of linear equations are not changed by these matrix multiplications. Moreover D_1 consists of the first $n - d$ columns of B and all elements of C_1, indicated in (1.14) by $\underline{*}$, are different from zero.

Thus we have introduced another basis

$$e_1^*, e_2^*, \ldots, e_n^*$$

associated with the product $P \cdot P_1$, instead of P, as in (1.8) and (1.9). Because of the special structure of P_1 we know that

$$e_1^* = \tilde{e}_1, \ldots, e_{n-d}^* = \tilde{e}_{n-d}.$$

The first step for calculating a solution of the V-reference R in W is the determination of the characteristic numbers of the V-reference R: Using (S) and (1.11) - (1.14), we have only to solve the system of linear equations

$$\sum_{i=1}^{k} \mu_i c_i = -c_{k+1},$$

what can be done recursively because of (1.14). Then the numbers

$$\lambda_i := \frac{\mu_i}{1 + \sum_{i=1}^{k} \mu_i} \qquad (i = 1, \ldots, k + 1)$$

(with $\mu_{k+1} := 1$) are the characteristic numbers of the V-reference R.

For calculating a solution of the V-reference R in W, we consider instead of (1.3) the linear equations:

$$c_i \cdot \begin{pmatrix} \tilde{y}_1 \\ \cdot \\ \cdot \\ \cdot \\ \tilde{y}_k \end{pmatrix} = f(t_i) - \langle a(t_i), z\rangle + h_R \tag{1.15}$$

for $i = 1, \ldots, k + 1$, where z is fixed in W. Then

$$x = z + (\underbrace{0_1, \ldots, 0}_{n=d}, \tilde{y}_1, \ldots, \tilde{y}_k, 0, \ldots, 0) \in \mathbb{R}^n$$

is a solution of the V-reference R in W. Moreover the affine subspace of all solutions is parallel to

$$V_1 := [R_1]^{\perp} \cap V = \text{span}\ (e^*_{n-d+k+1}, \ldots, e^*_n)$$

and

$$\mathbb{R}^n = V_1^{\perp} + V_1$$

with

$$V_1^{\perp} = \text{span}\ (e^*_1, \ldots, e^*_{n-d+k})$$

Hence our assumption (S) is satisfied for V_1 instead of V. Therefore we can use the same method for the further calculations in a whole chain of references.

To summarize the results for a chain of references

$$R = (R_1, \ldots, R_s)$$

with

$$R_j = \{a(t^j_i) \mid 1 \leqq i \leqq k_j + 1\}, \tag{1.16}$$

we assume without loss of generality, that $V = W = \mathbb{R}^n$. Then in any step of the algorithm we have a basis

$$\tilde{e}_1, \tilde{e}_2, \ldots, \tilde{e}_n \tag{1.17}$$

and a regular matrix $P \in \mathbb{R}_n$, such that $\tilde{e}_1, \ldots, \tilde{e}_n$ and the canonical basis $e_1, e_2, \ldots, e_n$ are connected by (1.8) and (1.9).

Moreover if we define

$$B_j = \begin{pmatrix} a(t_1^{\,j}) \\ \cdot \\ \cdot \\ \cdot \\ a(^{\,j}_{k_j+1}) \end{pmatrix}$$

we have got

$$B_j \cdot P = (D_j, C_j, 0),$$

where $D_j \in \mathbb{R}^{n-d_j}_{k_j+1}$, $d_j: = k_1 + k_2 + \ldots + k_{j-1}$, $(k_o: = 0)$ and

$C_j \in \mathbb{R}^{k_j}_{k_j+1}$ has the same structure as in (1.14).

With

$$U_i: = \text{span}\ (\tilde{e}_{d_i+1}, \ldots, \tilde{e}_{d_i+k_i}) \tag{1.19}$$

we have got a decomposition of $\mathbf{R}^n$

$$\mathbb{R}^n = U_1 \oplus U_2 \oplus \ldots \oplus U_s, \tag{1.20}$$

into orthogonal subspaces.

We want to mention that in a step of the algorithm, exchanging R_j and R_{j+1}, only the elements of the basis belonging to

U_j and U_{j+1} have to be transformed. All other elements remain automatically unchanged by using Householder matrices or Gaussian elimination for the columns.

A lot of numerical problems in real and complex Chebyshev-approximation and L_1-approximation was treated by these algorithms in [4], [16], [21], [26].

Naturally the speed of convergence depends on the problem to be treated. In any case, if it is possible to introduce multiple exchange techniques, we have observed that the multiple exchange is superior to the single exchange. Furthermore it was possible to get suitable approximations for starting faster local methods such as Newton's method, even if the solution of the problem was extremely bad conditioned. Here bad condition means that the solution, as in the L_1-case, is characterized by a trivial reference consisting only of one element.

2. ERROR ESTIMATION AND STRONG UNICITY

2.1 Error Estimations for Near-Best Solutions

The algorithm of the first section is based on the notion of ε-solution z, which means that for z the inequality

$$g(z) \leqq \alpha + \varepsilon$$

has to be satisfied. But nevertheless the norm of the difference $z - x^*$ can be much larger than ε for any solution x^* of our minimization problem. The aim of this section is to get realistic estimations for the difference $x_{\mathcal{R}} - x^*$, where $x_{\mathcal{R}}$ is the solution of the chain $\mathcal{R}$.

We consider a step of the algorithm above:
We have a regular chain of references

$$\mathcal{R} = (R_1, R_2, \ldots, R_s) \tag{2.1}$$

with

$$R_j = \{a(t_i^{\,j}) \mid 1 \leqq i \leqq k_j + 1\} \tag{2.2}$$

for $j = 1, \ldots, s$. Assuming again that $V = W = \mathbb{R}^n$, we have a decomposition of $\mathbf{R}^n$ as in (1.20) in orthogonal subspaces:

$$\mathbb{R}^n = U_1 \oplus U_2 \oplus \ldots \oplus U_s. \tag{2.3}$$

For a fixed $x \in \mathbb{R}^n$ and an optimal solution x^* we consider the decompositions with respect to (2.3):

$$x = u_1 + u_2 + \ldots + u_s,$$

$$x^* = u_1^* + u_2^* + \ldots + u_s^*$$

with u_i, $u_i^* \in U_i$ for $i = 1, 2, \ldots, s$.

Let us denote by $\lambda_1^j, \lambda_2^j, \ldots, \lambda_{k_j+1}^j$ the characteristic numbers of R_j associated with $a(t_1^j), a(t_2^j), \ldots, a(t_{k_j+1}^j)$. Then we have

Theorem 2. If $x \in \mathbb{R}^n$ and x^* is an optimal solution of the minimization problem, then for $1 \leqq j \leqq s$ the inequality

$$g(x^*) \geqq \min_{1 \leqq i \leqq k_j+1} \left(\langle a(t_i^j), x + \sum_{i=1}^{j-1} (u_i^* - u_i)\rangle - f(t_i^j)\right)$$
$$+ \gamma^j \cdot \max_{1 \leqq i \leqq k_j+1} |\langle a(t_i^j), u_j^* - u_j\rangle| \tag{2.4}$$

holds with $\gamma^j = \dfrac{\lambda^j}{1 - \lambda^j}$ and $\lambda^j = \min_{1 \leqq i \leqq k_j+1} \lambda_i^j$.

Proof. For each $1 \leq i \leq k_j + 1$

$$g(x^*) \geqq \langle a(t_i^j), x + \sum_{i=1}^{j-1} (u_i^* - u_i)\rangle - f(t_i^j)$$
$$+ \langle a(t_i^j), \sum_{i=j}^{s} (u_i^* - u_i)\rangle.$$

Since $\langle a(t_i^j), u_i^* - u_i\rangle = 0$ for $i = j + 1, j + 2, \ldots, s$, we get

$$g(x^*) \geqq \min_{1 \leqq i \leqq k_j+1} (\langle a(t_i^j), x + \sum_{i=1}^{j-1} (u_i^* - u_i)\rangle - f(t_i^j)) \tag{2.5}$$
$$+ \max_{1 \leqq i \leqq k_j+1} \langle a(t_i^j), u_j^* - u_j\rangle .$$

We define

$$K := \max_{1 \leqq i \leqq k_j+1} |\langle a(t_i^j), u_j^* - u_j\rangle| .$$

If there exists an index i with

$$K = \langle a(t_i^j), u_j^* - u_j\rangle ,$$

then the inequality (2.4) holds with $\gamma^j = 1$. Otherwise there exists an index i_o with

$$K = -\langle a(t_{i_o}^j), u_j^* - u_j\rangle .$$

Since

$$0 = \langle \sum_{i=1}^{k_j+1} \lambda_i^j a(t_i^j), u_j^* - u_j\rangle$$
$$= -\lambda_{i_o}^j \cdot K + \sum_{i \neq i_o} \lambda_i^j \langle a(t_i^j), u_j^* - u_j\rangle$$

we have got

$$K = \frac{1}{\lambda_{i_o}^j} \sum_{i \neq i_o} \lambda_i^j \langle a(t_i^j), u_j^* - u_j\rangle$$
$$\leqq \frac{1 - \lambda^j}{\lambda^j} \max_{1 \leqq i \leqq k_j+1} \langle a(t_i^j), u_j^* - u_j\rangle$$

and the theorem is proved.

The expression

$$\max_{1 \leq i \leq k_j+1} |\langle a(t_i^j), u\rangle|$$

with $u \in U_j$ is a norm in the subspace U_j. Hence we can get from (2.4) immediately error bounds for the coefficients of the difference

$$u_j^* - u_j$$

with respect to the elements

$$\tilde{e}_{d_j+1}, \ldots, \tilde{e}_{d_j+k_j}$$

of the basis of U_j (compare (1.19), (1.20)), if we know already estimations for

$$u_1^* - u_1, u_2^* - u_2, \ldots, u_{j-1}^* - u_{j-1}.$$

Therefore we apply Theorem 2 successively: For $j = 1$ we get estimations for $u_1^* - u_1$, then for $j = 2$ we can estimate $u_2^* - u_2$, etc.

In the algorithm presented above in each step ν only one characteristic number in each of the references R_j $(1 \leq j \leq s)$ can be smaller than δ_ν, where the parameter δ_ν controls the exchange procedure in the step ν. Moreover from (1.7) follows that there exists a real number $\beta(\varepsilon) > 0$ with

$$\frac{1}{2} > \delta_\nu > \beta(\varepsilon) > 0$$

where $\beta(\varepsilon)$ depends only on the number ε controlling the desired precision, in the sense that we construct an ε-solution z satisfying

$$g(z) \leq \inf_{x \in W} g(x) + \varepsilon.$$

So it is natural to sharpen Theorem 2 in the following way:

Theorem 3. Under the conditions of Theorem 1 we have

$$g(x^*) \geqq \min_{1 \leqq i \leqq k_j+1} \langle a(t_i^{\,j}),\ x + \sum_{i=1}^{j-1} u^*_i - u_i \rangle \qquad (2.6)$$

$$+\ \tilde{\gamma}^j \max_{1 \leqq i \leqq k_j} |\langle a(t_i^{\,j}),\ u^*_j - u_j \rangle|$$

with

$$\tilde{\gamma}^j = \frac{\lambda^j}{1 - \lambda^j} \quad \text{and} \quad \tilde{\lambda}^j = \min_{1 \leqq i \leqq k_j} \lambda_i^{\,j}.$$

<u>Proof.</u> We define

$$K := \max_{1 \leqq i \leqq k_j} |\langle a(t_i^{\,j}),\ u^*_j - u_j \rangle| \ .$$

If there exists an index i with $1 \leqq i \leqq k_j$ such that

$$K = |\langle a(t_i^{\,j}),\ u_j^* - u_j \rangle|$$

we get from (2.5) the inequality (2.6) with $\tilde{\gamma}^j = 1$. Otherwise there exists an index i_o with $1 \leqq i_o \leqq k_j$ such that

$$K = -\langle a(t_{i_o}^{j}),\ u_j^* - u_j \rangle \ .$$

Following the same lines as in the proof above we get

$$K = \frac{1}{\lambda_{i_o}^j} \sum_{\substack{i=1 \\ i \neq i_o}}^{k_j+1} \lambda_i^{\,j} \langle a(t_i^{\,j}),\ u_j^* - u_j \rangle$$

$$\leqq \frac{1 - \tilde{\lambda}^j}{\tilde{\lambda}^j} \max_{1 \leqq i \leqq k_j+1} \langle a(t_i^{\,j}),\ u_j^* - u_j \rangle$$

and hence (2.6).

In applications we arrange R_j in such a way that

$$\lambda^j_{k_j+1} = \min_{1 \leq i \leq k_j+1} \lambda_i^j .$$

Then we get for the number $\tilde{\gamma}^j$ of Theorem 3:

$$\tilde{\gamma}^j > \frac{\beta(\varepsilon)}{1 - \beta(\varepsilon)} .$$

Therefore $\tilde{\gamma}^j$ is not only larger than the constant γ^j of Theorem 2, but $\tilde{\gamma}^j$ is not arbitrary small.

Nevertheless it should be mentioned that this advantage can be destroyed by worse norm properties of

$$\max_{1 \leq i \leq k_j} |\langle a(t_i^j), u_j^* - u_j \rangle| .$$

Another possibility for obtaining error estimations for near-best approximations is the use of H-sets in the sense of Collatz [10], if the Haar condition is satisfied, or generally H_s-sets ([3], [22]). But we believe that the concept of chain of references considered above is more appropriate to the general case, because this concept leads to a new characterization of strong unicity.

For numerical examples using H_s-sets and chain of references see [3].

We remark, that in Theorem 2 and Theorem 3 it is not necessary, that x* is an optimal solution. Both theorems are true for any fixed x and x*. We have chosen the formulation above because we are interested in error estimations for optimal solutions.

2.2 Strong Unicity of Optimal Solutions

We consider again the minimization problem of section 1 and assume furthermore, that T is a compact set and $a: T \to \mathbb{R}^n$ and $f: T \to \mathbb{R}$ are continuous mappings. We denote by $\|\cdot\|$ a norm in $\mathbb{R}^n$.

Definition. A solution x* of the minimization problem is

called *strongly unique* if there exists a constant $\gamma > 0$ such that

$$g(x) \geqq g(x^*) + \gamma \|x - x^*\|$$

for all $x \in W$.

Let us define the set E by

$$E: = \{a(t) \in A \mid g(x^*) = \langle a(t), x^*\rangle - f(t)\} .$$

Then we can characterize strong unicity by the following theorem.

Theorem 4. A solution x^* is strongly unique if and only if the set E contains a chain R of references.

Interchanging the role of x and x^* in Theorem 2, we get immediately that x^* is strongly unique if E contains a chain R of references. For a detailed complete proof see [6].

3. STRONG UNICITY CONSTANTS IN LINEAR CHEBYSHEV APPROXIMATION

Let $C[a, b]$ be the collection of all continuous real-valued functions on the interval $[a, b]$ endowed with the uniform norm $\|\cdot\|$. We denote by V_n a Haar subspace of dimension n of $C[a, b]$. For a given $f \in C[a, b]$ let $T_n(f)$ be the best uniform approximation of f with respect to V_n. Then $T_n(f)$ is strongly unique, i.e. there exists a positive real number γ such that

$$\|f - v\| \geqq \|f - T_n(f)\| + \gamma \|v - T_n(f)\| \tag{3.1}$$

for all $v \in V_n$ ([18]).

The largest constant with (3.1) is denoted by $\gamma_n(f)$ and is called *strong unicity constant*.

Let

$$E_n(f): = \{x \in [a, b] \mid |f(x) - T_n(f)(x)| = \|f - T_n(f)\| \} \tag{3.2}$$

and

$$\sigma(x) = \text{sgn } (f - T_n(f))(x) \tag{3.3}$$

then Bartelt and McLaughlin [1] gave the following characterization of $\gamma_n(f)$:

$$\gamma_n(f) = \min_{\substack{v \in V_n \\ \|v\| = 1}} \max_{x \in E_n(f)} \sigma(x) \cdot v(x). \tag{3.4}$$

Poreda [20] raised the question to describe the asymptotic behaviour of $\gamma_n(f)$ for fixed f and $n \to \infty$, if $V_n = \Pi_{n-1}$, where Π_n denotes the set of all algebraic polynomials of degree n or less. Later Henry and Roulier [11] conjectured, that

$$\underline{\lim}_{n\to\infty} \gamma_n(f) = 0 \quad \text{or} \quad f \text{ is a polynomial,}$$

if $V_n = \Pi_{n-1}$ $(n = 1, 2, \ldots)$.

Recently quite a number of papers ([2], [11], [12], [13], [17], [19], [23]) used the structure of the extremal point sets $E_n(f)$ to get results on the asymptotic behaviour of $\gamma_n(f)$ for $n \to \infty$ and $V_n = \Pi_{n-1}$.

3.1 Characterization of $\gamma_n(f)$

The crucial point for characterizing $\gamma_n(f)$ is the point set $E_n(f)$. For $f \notin V_n$ we decompose $E_n(f)$ into a finite number of *sign components*

$$E_n^{\,0} < E_n^{\,1} < \ldots < E_n^{m}$$

such that

(i) $\sigma(x)$ is constant on $E_n^{\,i}$,

(ii) m is minimal.

Now we consider for $0 \le i \le m$ special approximation problems:

<u>Problem (A_i)</u>. We fixe a point $y_i \in E_n^{\,i}$, then the approximation problem (A_i) consists in determining a function $p_i \in V_n$, such that

$$p_i(y_i) = -\sigma(y_i), \tag{3.6}$$

$$\sigma(x) \cdot p_i(x) \leqq h_i \quad \text{for all} \quad x \in E_n(f), \tag{3.7}$$

$$h_i \quad \text{is minimal,} \tag{3.8}$$

Lemma 1. p_i is a solution of (A_i), if and only if there exist points $x_k \in E_n(f)$,

$$x_o < x_1 < \ldots < x_{j-1} < y_i < x_{j+1} < \ldots < x_n, \tag{3.9}$$

such that

$$\sigma(x_k) = (-1)^{j-k} \sigma(y_i) \quad \text{for} \quad k \neq j, \tag{3.10}$$

$$\sigma(x_k) \cdot p_i(x_k) = h_i \quad \text{for} \quad k \neq j, \tag{3.11}$$

$$\sigma(x) \cdot p_i(x) \leqq h_i \quad \text{for all} \quad x \in E_n(f). \tag{3.12}$$

Moreover the solution p_i is unique.

Proof. For a point set

$$R = \{x_o, x_1, \ldots, x_{j-1}, y_i, x_{j+1}, \ldots, x_n\}$$

with the properties (3.9) and (3.10) we consider the approximation problem (B_R):
Minimize $\max_{x \in R} |p(x)|$ with respect to $p \in V_n$ satisfying the interpolation condition (3.6).

Then the solution q of this problem satisfies

$$\sigma(x_k) \cdot q(x_k) = h \quad \text{for all} \quad k \neq j$$

with a number $h > 0$ (see proof of Lemma 1 in Blatt, Klotz [5]). If $\sigma(x) \cdot q(x) \leqq h$ for all $x \in E_n(f)$, then q has the properties (3.9) - (3.12). Otherwise we determine a point $\xi \in E_n(f)$ such that

$$\max_{x \in E_n(f)} \sigma(x)\, q(x) = \sigma(\xi)\, q(\xi)$$

and exchange ξ with a point x_k, using the exchange theorem as in the Remez algorithm for uniform approximation problems with respect to Haar subspaces and one fixed interpolation condition. For the new point set R' we consider the problem $(B_{R'})$ and use the same method as above for R' instead of R, etc.. This iterative method leads (as in the proof for the usual Remez method) to a point set

$$\tilde{R} = \{z_0 < z_1 < \ldots < z_{j-1} < y_i < z_{j+1} < \ldots < z_n\}$$

and a function $\tilde{q} \in V_n$, such that $\tilde{R}$ and $\tilde{q}$ satisfy (3.9) - (3.12). On the other hand if we have a polynomial $p_i \in V_n$ satisfying (3.9) - (3.12), then let us assume that there exists $q \in V_n$ such that

(i) $q(y_i) = -\sigma(y_i)$,

(ii) $\sigma(x)\, q(x) \leqq h$ for all $x \in E_n(f)$,

(iii) $h < h_i$.

Then

$$\sigma(x_k)\, [p_i(x_k) - q(x_k)] > 0 \quad \text{for} \quad k \neq j$$

and

$$p_i(y_i) = q(y_i) = 0.$$

Consequently $p_i - q$ has n zeros counting multiplicity, a contradiction to the Haar condition of V_n. Analogously the unicity of the solution of (A_i) is proved.

We remark that the solution p_i of (A_i) depends on the choice of $y_i \in E_n^i$. But if we fix another point $\tilde{y}_i \in E_n^i$, then the solution of (A_i) is multiplied only by a positive factor. The same is true if we fix another point $z \in [a, b]$ with

$$E_n^{i-1} < z < E_n^{i+1}$$

satisfying the interpolation condition

$$\sigma(z) = -\sigma(E_n^i),$$

where $\sigma(E_n^i)$ is the sign of the error function $f - T_n(f)$ over the set E_n^i. $(E_n^{-1} := a - 1,\ E_n^{m+2} := b + 1)$.

Lemma 1 immediately leads to

Lemma 2.

$$\gamma_n(f) \leqq \tilde{\gamma}_n(f) := \min_{0 \leqq i \leqq m} \frac{h_i}{\|p_i\|},$$

where p_i is the solution of (A_i) for $i = 0, \dots, m$.

For $m = n$ this result is due to Bartelt and Schmidt [2]. But in Lemma 2 the estimation is connected in any case with a finite number of approximation problems of type (A_i), contrary to the characterization given in [2] using all possible interpolation problems satisfying (3.11). The number of these interpolation problems is infinite if the number of points in $E_n(f)$ is infinite.

For characterizing $\gamma_n(f)$ it is necessary to construct another kind of polynomials: Let $z \in \operatorname{conv}(E_n^i) \setminus E_n^i$, where $\operatorname{conv}(E_n^i)$ denotes the convex hull of E_n^i. We consider

Problem (C_z). We want to construct a polynomial $p \in \pi_n$, such that

$$p(z) = \sigma(E_n^i), \tag{3.13}$$

$$\sigma(x)\, p(x) \leqq h_z \quad \text{for all } x \in E_n(f), \tag{3.14}$$

$$h_z \text{ is minimal.} \tag{3.15}$$

Following the same lines of the proof of Lemma 1 we get

Lemma 3. p_z is a solution of (C_z) if and only if there exist points $x_k \in E_n(f)$ with (3.9), where y_i is replaced by z, and

$$\sigma(x_k) = (-1)^{j-k+1}\,\sigma(z) \quad \text{for } k \neq j, \tag{3.16}$$

$$\sigma(x_k)\, p_z(x_k) = h_z \quad \text{for } k \neq j \tag{3.17}$$

$$\sigma(x)\, p_z(x) \leqq h_z \quad \text{for all } x \in E_n(f). \tag{3.18}$$

Theorem 5. Let p_i $(i = 0, \ldots, m)$ denote the polynomials of Lemma 1 and p_z the polynomials of Lemma 3, then

$$\gamma_n(f) = \min\,(\tilde{\gamma}_n(f), \min_{z \in Z} h_z)$$

with

$$Z = \bigcup_{i=0}^{m} (\operatorname{conv}(E_n^{\,i}) \setminus E_n^{\,i}).$$

Proof. Let $p \in \Pi_n$ with $\|p\| = 1$ and

$$\gamma_n(f) = \max_{x \in E_n(f)} \sigma(x)\, p(x).$$

We consider a point $\xi \in [a, b]$ such that $p(\xi) = 1$. Then there exists an index i with $0 \leqq i \leqq m$ such that

$$E_n^{\,i-1} < \xi < E_n^{\,i+1}.$$

We distinguish the two cases:

(a) $p(\xi) = -\sigma(E_n^{\,i})$: Then $p = \alpha \cdot p_i$ with a factor $\alpha > 0$. Hence $\gamma_n(f) = \tilde{\gamma}_n(f)$. (b) $p(\xi) = \sigma(E_n^{\,i})$: If $\xi \in E_n^{\,i}$, then $\gamma_n(f) = 1 = \tilde{\gamma}_n(f)$. If $\xi \in \operatorname{conv}(E_n^{\,i}) \setminus E_n^{\,i}$, then $p = p_\xi$ as in Lemma 3 and therefore

$$\gamma_n(f) \leq h_\xi.$$

If $\xi \notin \text{conv}(E_n^{\,i})$, then we have either $\xi < E_n^{\,i}$ or $\xi > E_n^{\,i}$. then either

$$E_n^{\,i-1} < \xi < E_n^{\,i}$$

or

$$E_n^{\,i} < \xi < E_n^{\,i+1}.$$

In any case we have the situation considered in (a).

It can be shown by examples, that $\gamma_n(f)$ is different from $\tilde{\gamma}_n(f)$. Therefore it seems to be very interesting that it is possible to derive a lower bound for $\gamma_n(f)$ using only the number $\tilde{\gamma}_n(f)$.

Theorem 6.

$$[\tilde{\gamma}_n(f)]^2 \leqq \gamma_n(f) \leqq \tilde{\gamma}_n(f).$$

Proof. Because of Theorem 5 we have only to consider the case, that $\gamma_n(f) = h_z$ for some $z \in \text{conv}(E_n^{\,i}) \setminus E_n^{\,i}$ and $0 \leqq i \leqq m$. Let p_z be the solution of the associated problem (C_z) and let us consider the polynomial p_i of problem (A_i) with deviation h_i. We define

$$\tilde{p}_i : = \frac{p_i}{\|p_i\|} .$$

Then there exists a point set

$$x_o < x_1 < \ldots < x_{j-1} < y_i < x_{j+1} < \ldots < x_n$$

with

$$\sigma(x_k) \cdot \tilde{p}_i\,(x_k) = \frac{h_i}{\|p_i\|} \quad \text{for} \quad k \neq j.$$

Let $p = \tilde{p}_i + p_z$ and $\varepsilon := \sigma(E_n^i)$. Then $\operatorname{sgn} p(z) = \sigma(E_n^i) = \varepsilon$. We distinguish :

(a) $\sigma(x_k)\, p(x_k) \leqq 0$ for all $k \neq j$:

Because of $\sigma(x_k) = (-1)^{j-k}\, \sigma(E_n^i) = (-1)^{j-k} \cdot \varepsilon$, the function p has at least n zeros counting multiplicity, hence $\tilde{p}_i = -p_z$. Then

$$\gamma_n(f) = h_z = \max_{x \in E_n(f)} \sigma(x)\, p_z(x)$$

$$= -\frac{1}{\|p_i\|} \min_{x \in E_n(f)} \sigma(x)\, p_i(x)$$

$$\geqq \frac{1}{\|p_i\|} \geqq \tilde{\gamma}_n(f).$$

(b) $\sigma(x_{k_0})\, p(x_{k_0}) < 0$:

Then

$$\begin{aligned} \sigma(x_{k_0})\, p_z(x_{k_0}) &= -\sigma(x_{k_0})\, \tilde{p}_i(x_{k_0}) \\ &= -\frac{h_i}{\|p_i\|}\,. \end{aligned} \tag{3.19}$$

We consider the solution $\tilde{p}_{k_0}$ of the problem (A_{k_0}) such that

$$\tilde{p}_{k_0}(x_{k_0}) = -\sigma(x_{k_0}),$$

$$\sigma(x)\, \tilde{p}_{k_0}(x) \leqq \tilde{h}_{k_0} \quad \text{for all } x \in E_n(f),$$

and $\tilde{h}_{k_0}$ is minimal. Then from (3.19) we get

$$\tilde{h}_{k_0} \leqq \frac{\|p_i\|}{h_i} \max_{x \in E_n(f)} \sigma(x)\, p_z(x)$$

$$= \frac{\|p_i\|}{h_i} \gamma_n(f).$$

Because of $\|\tilde{p}_{k_o}\| \geq 1$ we have got

$$\gamma_n(f) \geqq \frac{h_i}{\|p_i\|} \tilde{h}_{k_o} = \frac{h_i}{\|p_i\|} \cdot \frac{\tilde{h}_{k_o}}{\|\tilde{p}_{k_o}\|} \cdot \|\tilde{p}_{k_o}\|$$

$$\geqq [\tilde{\gamma}_n(f)]^2.$$

3.2 Upper Bounds for $\gamma_n(f)$

Theorem 7. Suppose that $E_n(f)$ consists of exactly $n + 1$ points, then

$$\gamma_n(f) \leqq \frac{1}{n}.$$

Proof. Let

$$E_n(f): x_o < x_1 < \ldots < x_n.$$

We consider the functions p_i of the problem (A_i) as in Lemma 1 and assert that

$$\min_{0 \leqq i \leqq n} h_i \leqq \frac{1}{n}. \tag{3.20}$$

Let us assume that (3.20) is false. Then we define

$$p: = \sum_{i=0}^{n} p_i.$$

Since $p_j(x_j) = -\sigma(x_j)$ and $\sigma(x_j)\, p_i(x_j) = h_i$ for $i \neq j$, we get

$$\sigma(x_j)\, p(x_j) = -1 + \sum_{\substack{i=0 \\ i \neq j}}^{n} h_i > 0$$

for $j = 0, \ldots, n$. Therefore $p \not\equiv 0$ and p has at least n zeros in (a, b). This contradicts the Haar condition. Hence (3.20) and the theorem is proved.

Corollary (Schmidt [2]). Let $V_n = \Pi_{n-1}$ for $n = 1, 2, \ldots$ and suppose that there exists a subsequence $\{n_k\}$ such that $E_{n_k}(f)$ consists of exactly $n_k + 1$ points for $k = 1, 2, \ldots$, then

$$\lim_{n \to \infty} \gamma_n(f) = 0,$$

For generalizing this result to a larger class of functions f, let $l(n)$ denote the number of sign components E_n^i in (3.5) which have more than one point.

Theorem 8. Let $f \in C[a, b]$ and $V_n = \Pi_{n-1}$ $(n = 1, 2, \ldots)$. If $\lim\limits_{n \to \infty} \frac{l(n)}{\log n} = 0$, then $\lim\limits_{n \to \infty} \gamma_n(f) = 0$.

We remark that $\lim\limits_{n \to \infty} \frac{l(n)}{\log n} = 0$ if $\lim\limits_{n \to \infty} l(n) = l_o \in \mathbb{N}_o$. As a special case, $l_o \leq 2$, we get a result of Bartelt and Schmidt [2].

For convenience we suppose that $[a, b] = [-1, 1]$. Then we transform the interval $[-1, 1]$ into $[0, \pi]$ by

$$\psi: [-1, 1] \to [0, \pi]$$

$$x \to \arccos x.$$

Hence we have reduced the approximation of f with respect to Π_{n-1} to the approximation of $f(\cos \phi) = g(\phi)$ with respect to

$$\tilde{\Pi}_{n-1} := \operatorname{span} \{1, \cos \phi, \ldots, \cos(n-1)\phi\}. \tag{3.21}$$

In the following let $E_n(g)$ be the set of extremal points of the optimal error function $(g - T_n(g))(\phi)$, if we consider the uniform approximation of g with respect to $\tilde{\Pi}_{n-1}$. Just as in (3.5) we decompose $E_n(g)$ into different sign components:

$$E_n(g):\ E_n^o < E_n^1 < \ldots < E_n^m. \tag{3.22}$$

If $T_n(g) \neq T_{n+1}(g)$ we know that $m = n$. Moreover we need a result of Bartelt and Schmidt [2]:

Let

$$\delta_n := \max_{0 \leq i \leq m} \{\min E_n^{i+1} - \max E_n^{i-1}\}$$

with

$$E_n^{-1} := 0 \quad \text{and} \quad E_n^{m+1} := \pi,$$

then

$$\lim_{n \to \infty} \gamma_n(f) = 0, \quad \text{if} \quad \overline{\lim_{n \to \infty}}\ \delta_n \cdot n = \infty.$$

Proof of Theorem 8. Because of the above mentioned result of Bartelt and Schmidt, we have only to consider the case that $\overline{\lim_{n \to \infty}}\ \delta_n \cdot n \leq \alpha \in \mathbb{R}$ or equivalently, that

$$\delta_n \leq \frac{\alpha + 1}{n} \quad \text{for all} \quad n \geq \tilde{n}.$$

We consider only indices n mit $T_n(g) \neq T_{n+1}(g)$, which means that in (3.22) we have $m = n$.

Let

$$r := \left[\frac{\alpha + 1}{\varepsilon}\right]$$

for fixed $0 < \varepsilon < 1$. Then r is independent of n. We decompose any interval

$$J_\nu = [\min E_n^\nu, \max E_n^\nu]$$

with more than one point into $r + 1$ subintervals of equal length by the points

$$\min E_n^\nu = z_o^\nu < z_1^\nu < \ldots < z_{r+1}^\nu = \max E_n^\nu.$$

Then we define

$$I_n := \{i \mid 0 \leq i \leq n, |E_n^{\,i}| = 1\},$$

$$\tilde{E}_n^{\,\nu} := \{z_0^{\,\nu}, \ldots, z_{r+1}^{\nu}\} \quad \text{for} \quad k \notin I_n, \ \tilde{E}_n^{\nu} := E_n^{\nu} \text{ for } k \in I_n,$$

$$\tilde{E}_n(g) := \bigcup_{k=0}^{n} \tilde{E}_n^{\,\nu}.$$

Let us consider for $i \in I_n$ the problem (A_i) with $V_n = \tilde{\Pi}_{n-1}$ and $f, E_n(f), E_n^{\,i}$ replaced by $g, \tilde{E}_n(g), \tilde{E}_n^{\,i}$. The solution of this problem is characterized as in Lemma 1 by a point set of type (3.9) with deviation h_i. The number of different point sets of type (3.9) is bounded by the number $(r + 2)^{l(n)}$. Since

$$\lim_{n \to \infty} \frac{l(n)}{\log n} \cdot \log (r + 2) = 0,$$

there exists a subsequence $\{n_k\}$ such that $T_{n_k}(g) \neq T_{n_k+1}(g)$ and

$$\lim_{k \to \infty} \frac{l(n_k)}{\log n_k} \cdot \log (r + 2) = 0,$$

or

$$(r + 2)^{l(n)} = o(n_k) \quad \text{for} \quad k \to \infty.$$

Because of

$$|I_{n_k}| = n_k + 1 - l(n_k)$$

we know that $n_k = O(|I_{n_k}|)$ for $k \to \infty$. Now for each $i \in I_{n_k}$ we have a solution p_i of (A_i) and an associated point set of type (3.9). Then we denote by B the point set of type (3.9) which occurs in the characterization of (A_i) with $i \in I_{n_k}$

most frequently and define

$$\tilde{I}_{n_k} := \{i \in I_{n_k} \mid p_i \text{ is characterized by } B\}.$$

Since we have only $o(n_k)$ different point sets of type (3.9) we get from $n_k = O(|I_{n_k}|)$ that

$$|\tilde{I}_{n_k}| \to \infty \quad \text{for} \quad k \to \infty.$$

We assert now that

$$\lim_{k\to\infty} \min_{i \in \tilde{I}_{n_k}} h_i = 0. \tag{3.23}$$

Let us assume that contrary to (3.23) there exists a number $\beta > 0$ such that

$$\min_{i \in \tilde{I}_{n_k}} h_i > \beta$$

for a subsequence of $\{n_k\}$. Then we consider as in the proof of Theorem 7 the even trigonometric polynomial

$$p = \sum_{i \in \tilde{I}_{n_k}} p_i \in \tilde{\Pi}_{n_k-1}.$$

There exists $k_o > 0$ such that $p \not\equiv 0$ for $k \geqq k_o$, and p has at each point of B the same sign as the error function $g - T_{n_k}(g)$. But this is a contradiction, since $\tilde{\Pi}_{n_k-1}$ is a Haar subspace. Therefore (3.23) is true.

Let $i_o \in \tilde{I}_{n_k}$ with $h_{i_o} = \min_{i \in \tilde{I}_{n_k}} h_i$ and define

$$\tilde{p} := \frac{p_{i_o}}{\|p_{i_o}\|}$$

For each $z \in E_{n_k}(g)$ there exists a point $\tilde{z} \in \tilde{E}_{n_k}(g)$ with $|z - \tilde{z}| \leqq \frac{\varepsilon}{n_k}$. Using Bernstein's inequality we get

$$|\tilde{p}(z) - \tilde{p}(\tilde{z})| \leqq n_k \cdot \frac{\varepsilon}{n_k} = \varepsilon.$$

Therefore we have got for all $z \in E_{n_k}(g)$ that

$$\sigma(z)\,\tilde{p}(z) \leqq \frac{h_{i_o}}{\|p_{i_o}\|} + \varepsilon$$

with $\sigma(z) = \operatorname{sgn}(g - T_{n_k}(g))(z)$. Since $\|p_{i_o}\| \geqq 1$, it follows that

$$\lim_{k \to \infty} \gamma_{n_k}(f) = 0$$

and the theorem is proved.

3.3 Lower Bounds For $\gamma_n(f)$

For convenience we consider in this section the uniform approximation of $f \in C[0, \pi]$ with respect to $\tilde{\Pi}_{n-1}$, the set of even trigonometric polynomials of degree $n - 1$ or less. $\gamma_n(f)$ denotes in the following the strong unicity constant for the approximation of f with respect to $\tilde{\Pi}_{n-1}$.

The first results on lower bounds for $\gamma_n(f)$ were obtained by Kroo [17].

Theorem 9 [17]. Let $f \in C[0, \pi]$ and $f \notin \tilde{\Pi}_{n-1}$ for any $n = 1, 2, \ldots$. Then

$$\overline{\lim_{n \to \infty}} \sqrt[n]{\gamma_n(f)} \, (\log n)^{\beta} = \infty \tag{3.24}$$

for arbitrary $\beta > 2$. Moreover, if $f \in \text{Lip}\,\alpha$ for some $0 < \alpha \leqq 1$, then

$$\overline{\lim_{n \to \infty}} \sqrt[n]{\gamma_n(f)} > 0. \tag{3.25}$$

Kroo [17] showed that $\overline{\lim\limits_{n \to \infty}}$ in Theorem 9 cannot be strengthened to $\lim\limits_{n \to \infty}$. Our aim is to show that (3.25) holds not only for $f \in \text{Lip}\ \alpha$, but for all $f \in C[a, b]$.

For $f \in C[0, \pi]$ let $T_n(f)$ denote again the best uniform approximation of f with respect to $\tilde{\Pi}_{n-1}$. Then there exist $n + 1$ points

$$0 \leqq x_0^{(n)} < x_1^{(n)} < \ldots < x_n^{(n)} \leqq \pi, \tag{3.26}$$

such that the error function $f - T_n(f)$ takes its maximum modulus value at $x_k^{(n)}$ with alternating signs. If we introduce

$$\xi_k^{(n)} = \frac{\pi k}{n} \quad \text{for} \quad k = 0, \ldots, n,$$

then $\xi_k^{(n)}$ are the extremal points of $\cos nx$.

Defining

$$\Delta_n := \max_{0 \leqq k \leqq n} |x_k^{(n)} - \xi_k^{(n)}|,$$

then a theorem of Kadec [15] shows that

$$\underline{\lim}_{n \to \infty} \Delta_n n^{1/2-\varepsilon} = 0 \tag{3.27}$$

for any $\varepsilon > 0$. Kroo [15] used in the proof of Theorem 9 results about the *minimal distance* d_n between two consecutive alternation points:

$$d_n := \min_{0 \leqq k \leqq n} (x_k^{(n)} - x_{k-1}^{(n)}), \tag{3.28}$$

where $x_{-1}^{(n)} := -x_1^{(n)}$.

It is important, that it is possible to combine the numbers Δ_n and d_n:

Theorem 10. For every $\delta > 0$ there exists a number $\alpha > 0$ and a subsequence $\{n_j\}$ such that

$$\Delta_n \leqq \alpha \frac{\log n}{\sqrt{n}} \tag{3.29}$$

and

$$d_n \geqq \frac{2}{(n + 1)^{2+\delta}} \tag{3.30}$$

for $n = n_j$ and $j = 1, 2, \ldots$.

We want to emphasize, that (3.29) is already sharper than Kadec's result (3.27). For a proof see [7].

With Theorem 10 and the estimation

$$[\tilde{\gamma}_n(f)]^2 \leqq \gamma_n(f)$$

of Theorem 6, it is possible to prove

Theorem 11 [7]. There exists a positive number $\beta > 0$, such that

$$\overline{\lim_{n \to \infty}} \sqrt[n]{\gamma_n(f)} > \beta$$

and this number β is independent of f.

REFERENCES

1. Bartelt, M. W., and McLaughlin, H. W., "Characterizations of strong unicity in approximation theory", J. Approximation Theory 9, 1973, pp. 255-266.

2. Bartelt, M. W., and Schmidt, D., "On Poreda's problem for strong unicity constants", J. Approximation Theory 33, 1981, pp. 69-79.

3. Blatt, H.-P., "Strenge Eindeutigkeitskonstanten und Fehlerabschätzungen bei linearer Tschebyscheff-Approximation", appeared in L. Collatz, G. Meinardus, H. Werner (ed.): Numerische Methoden der Approximationstheorie, Birkhäuser, Basel, 1982, pp. 9-25.

4. Blatt, H.-P, Kaiser, U., and Ruffer-Beedgen, B., "A multiple exchange algorithm in convex programming", appeared in J. B. Hiriart-Urruty, W. Oettli, J. Stoer (ed.): Optimization: Theory and Algorithms, Marcel Dekker, New York, 1983, pp. 113-130.

5. Blatt, H.-P., and Klotz, V., "Zur Anzahl der Interpolationspunkte polynomialer Tschebyscheff-Approximationen im Einheitskreis", appeared in L. Collatz, G. Meinardus, H. Werner (ed.): Numerische Methoden der Approximations theore, ISNM 42, Birkhäuser, Basel, 1978, pp. 61-77.

6. Blatt, H.-P., "Strong unicity in semi-infinite optimization", to appear in Proceedings of the conference on Parametrische Optimierung und Approximation, Oberwolfach 1983.

7. Blatt, H.-P., "Lower bounds for strong unicity constants", preprint.

8. Carasso, C., and Laurent, P. J., "Un algorithme de minimisation en chaine en optimisation convexe", SIAM J. Control and Optimization 16, 1978, pp. 209-235.

9. Carasso, C., and Laurent, P. J., "An algorithm of successive minimization in convex programming", R.A.I.R.O., Analyse numérique, Numerical Analysis, 1978, pp. 377-400.

10. Collatz, L., "Approximation von Funktionen bei einer und bei mehreren unabhängigen Veränderlichen", Z. Angew. Math. Mech. 36, 1956, pp. 198-211.

11. Henry, M. S., and Roulier, J. A., "Lipschitz and strong unicity constants for changing dimension", J. Approximation Theory 22, 1978, pp. 85-94.

12. Henry, M. S., Swetits, J. J., and Weinstein, S., "Orders of strong unicity constants", J. Approximation Theory 31, 1981, pp. 175-187.

13. Henry, M. S., and Swetits, J. J., "Precise Orders of strong unicity constants for a class of rational functions", J. Approximation Theory 32, 1981, pp. 292-305.

14. Henry, M. S., Swetits, J. J., and Weinstein, S., "On extremal sets and strong unicity constants for certain C^{∞}-functions", J. Approximation Theory 37, 1983, pp. 155-174.

15. Kadec, M. I., "On the distribution of points of maximum deviation in the approximation of continuous functions by polynomials", Amer. Math. Soc. Transl. (2), 1963, pp. 231-234.

16. Kaiser, U., "Ein Aufstiegsverfahren für konvexe Optimierungsprobleme mit Nebenbedingungen", Diplomarbeit, Universität Mannheim, 1981.

17. Kroo, A., "The Lipschitz constant of the operator of best approximation", Acta Math. Acad. Sci. Hungar. 35, 1980, pp. 279-292.

18. Newman, D. J., and Shapiro, H. S., "Some theorems on Chebyshev approximation", Duke Math. J. 30, 1963, pp. 673-682.

19. Nürnberger, G., "Strong unicity constants for spline functions", Num. Funct. Anal. and Optim. 5, 1982-83, pp. 319-347.

20. Poreda, S. J., "Counterexamples in best approximation", Proc. Amer. Math. Soc. 56, 1976, pp. 167-171.

21. Ruffer-Beedgen, B., "Der Rémèz-Algorithmus ohne Haarsche Bedingung", Diplomarbeit, Universität Mannheim, 1981.

22. Schaback, R., "Bemerkungen zur Fehlerabschätzung bei linearer Tschebyscheff-Approximation", appeared in L. Collatz, G. Meinardus, H. Werner (ed.): Numerische Methoden der Approximationstheorie, ISNM 52, Birkhäuser, Basel, 1980, pp. 255-276.

23. Schmidt, D., "On an unboundedness conjecture for strong unicity constants", J. Approximation Theory 24, 1978, pp. 216-223.

24. Stiefel, E., "über diskrete und lineare Tschebyscheff-Approximation, Num. Math. 1, 1959, pp. 1-20.

25. Töpfer, H. J., "Tschebyscheff-Approximation und Austauschverfahren bei nicht erfüllter Haarscher Bedingung", appeared in L. Collatz, G. Meinardus, H. Werner (ed.): Numerische Methoden der Approximationstheore, ISNM 7, Birkhäuser, Basel, 1967, pp. 71-89.

26. Witte, J., "Remes- und Newton-Verfahren bei L_1-Approximation", Staatsexamensarbeit, Universität Mannheim, 1981.

FOUR LECTURES ON MULTIVARIATE APPROXIMATION

E. W. Cheney

The University of Texas

1. INTRODUCTION

From the abstract viewpoint of Banach space theory, we need not draw any distinction between univariate (one-variable) approximation and multivariate (many-variable) approximation. In either case, the central problem of best approximation can be stated thus: a Banach space X and a subspace Y of X are prescribed, and for a particular $x \in X$ we seek a best approximation of x in Y. If that exists, it is an element $y^* \in Y$ such that

$$\|x - y^*\| \leq \|x - y\| \qquad (y \in Y).$$

Whether the elements of X are functions is immaterial at this level of discussion. However, powerful and interesting results are largely concerned with function spaces, and at that stage it is useful to distinguish between functions of one or several variables.

In comparison to the well-developed univariate approximation theory, the multivariate theory exhibits a number of distinctive features. The remainder of this introductory section will elaborate upon these.

The problem of interpolation undergoes a sudden change as we proceed from one to several variables. Let S be a set, and x a real-valued function defined on S. Let Y be an n-dimensional vector space of functions defined on S. If points $s_1, \ldots, s_n$ are given in S, we would like to determine an

S. P. Singh et al. (eds.), Approximation Theory and Spline Functions, 65–87.

element $y \in Y$ which interpolates x at the given points. This means that

$$y(s_i) = x(s_i) \qquad (1 \leq i \leq n).$$

The property of Y which makes this possible for all functions x and for all choices of distinct points $s_1, \ldots, s_n$ is the Haar property. We say that the n-dimensional space Y has the Haar property if each set of n point-functionals $\{\hat{s}_1, \ldots, \hat{s}_n\}$ is linearly independent over Y. A point-functional $\hat{s}$ corresponding to a point $s \in S$ is defined for any function x by the equation $\hat{s}(x) = x(s)$. The polynomial subspace π_{n-1} is a Haar subspace of dimension n on any set $S \subset R$, provided only that S contains at least n points.

In general, polynomials in several variables do not enjoy the Haar property. Indeed, the Haar property is essentially a univariate concept, by the theorem of Mairhauber [23], which states that if S is a compact Hausdorff space and if $C(S)$ contains a Haar subspace of dimension 2 or greater, then S must be homeomorphic to a subset of the real line. Here, and in what follows, $C(S)$ will always denote the Banach space of all real-valued continuous functions defined on the compact Hausdorff space S. The norm in $C(S)$ is always understood to be the "supremum norm":

$$\|x\| = \sup\{|x(s)| : s \in S\} \qquad (x \in C(S)).$$

It should be emphasized that the Mairhauber theorem warns of difficulties to be expected in interpolation by linear subspaces, but does not rule out the possibility of interpolation at selected points, nor does it rule out more general interpolation processes in which the dimension exceeds the number of interpolation points. Examples given later will illustrate some positive results on interpolation. For a simple case in which certain interpolation points are not feasible, consider interpolation by Π_1 in two variables. The interpolating functions are then of the form

$$y(s, t) = a + bs + ct.$$

Interpolation to arbitrary data at three points (s_i, t_i) is possible if and only if the points are not co-linear, for it is only in this case that the linear system from which a, b, and c are determined is non-singular.

For multivariate interpolation at arbitrary points, a number of methods are available in which the subspace of interpolating functions depends upon the nodes of interpolation. One such method of wide applicability can be introduced on an arbitrary metric space (S, d). Let the nodes for interpolation be denoted by $t_1, \ldots, t_n$. Define then

$$y_i(s) = \prod_{\substack{j=1 \\ j \neq i}}^{n} d(s, t_j)/d(t_i, t_j) \qquad (1 \leq i \leq n).$$

This formula imitates the one in Lagrange interpolation on the real line, and, as in that familiar situation, one can see immediately that

$$y_i(t_j) = \delta_{ij} \quad \text{(Kronecker delta)} \qquad (1 \leq i, j \leq n).$$

Therefore an interpolating function for an arbitrary function x is given by

$$Lx = \sum_{i=1}^{n} x(t_i) y_i \qquad (x\colon S \to R).$$

The linear operator L which this equation defines is a projection $(L^2 = L)$, and is moreover, <u>positive</u>:

$$x \geq 0 \Rightarrow Lx \geq 0.$$

A modification of this process given by Shepherd [34] defines

$$a_i(s) = \prod_{\substack{j=1 \\ j \neq i}}^{n} d(s, t_j)$$

$$b(s) = \sum_{i=1}^{n} a_i(s)$$

$$y_i(s) = a_i(s)/b(s).$$

The operator L now has, in addition to the linearity, positivity, idempotency, and interpolatory properties, the property of leaving constant functions invariant. From this one sees that if $\alpha \leq x(s) \leq \beta$ then $\alpha \leq (Lx)(s) \leq \beta$. It follows that $\|L\| = 1$, which is favorable for approximation processes, as we will point out in Section 3. Although the space spanned by $y_1, \ldots, y_n$ has been defined in order to interpolate efficiently at one set of nodes, it will be unsatisfactory for other sets of nodes unless S is 1-dimensional (homeomorphic to a subset of R).

Another feature that distinguishes multivariabe approximation from univariate approximation is the natural occurrence of infinite-dimensional approximating subspaces. Thus, for example, the problem of approximating one continuous bivariate function by a sum of two continuous univariate functions is immediately seen to involve an infinite-dimensional subspace of functions

$$(s, t) \mapsto x(s) + y(t) \qquad x \in C(S),\ y \in C(T).$$

In situations such as this the existence of best approximations is usually not deduced from general theorems involving compactness or weak compactness; rather, ad hoc arguments must be used. As we shall see later, some important spaces which are useful in practical approximation have not yet been proved to be proximanal.

The next distinguishing feature of multivariate approximation to which we wish to draw attention is the natural occurrence of unusual norms. A single example, drawn from the theory of integral equations, will illustrate.

A linear Fredholm integral equation of the first kind is of the form

$$x(s) = \int_S K(s, t)x(t)dt + u(s).$$

If the kernel is separable, i.e. of the form

$$K(s, t) = \sum_{i=1}^{n} a_i(s)b_i(t)$$

then the integral equation is readily solved for the unknown function x. If the kernel is not separable, one can proceed by first approximating it by another kernel K' which is separable. This should be done in such a way that the resulting linear operators are close to each other in the operator norm. The appropriate norm is

$$\|L\| = \sup\{\|Lx\| : x \in C(S), \|x\| \le 1\}.$$

If $Lx = \int_S k(\cdot, t)x(t)dt$ then a short calculation reveals that

$$\|L\| = \sup_s \int_S |K(s, t)|dt.$$

This norm on the bivariate function K is a mixture of an L_∞-norm and an L_1-norm. Approximation problems with this norm are discussed in [24].

A fourth distinguishing feature of multivariate approximation problems is that the geometry of the domain plays a more important role. Here again a single example will illustrate this feature. We have already mentioned the problem of approximating a continuous bivariate function by the sum of two continuous univariate functions. If $z \in C(S \times T)$ then there exists a pair (x, y) with $x \in C(S)$, $y \in C(T)$ and

$$\|z - x - y\| = \sup|z(s, t) - x(s) - y(t)| = \min.$$

If the domain of z is a subset Q of $S \times T$, this theorem is no longer valid. A theorem of Ofman [25] is pertinent here:

<u>Theorem 1</u>. Let $S = T = [0, 1]$ and $Q \subset S \times T$. If Q contains a point (s_0, t_0) such that (s, t_0) and (s_0, t) belong to Q whenever (s, t) belongs to Q, then each Lipschitz continuous function in $C(Q)$ possesses a best approximation on Q by a function of the form $x + y$, $x \in C(S)$, $y \in C(T)$.

The final distinctive feature of multivariate approximation that we call attention to is that univariate functions occur as the "building-blocks" for multivariate functions and are combined in various ways for this purpose. Thus, through multiplication and addition we can build bivariate functions of the form

$$z(s, t) = \sum_{i=1}^{n} a_i(s)b_i(t).$$

By addition and composition we can build functions of the form

$$z(s, t) = \sum_{i=1}^{n} \phi_i(a_i(s) + b_i(t)).$$

In both of these cases there are powerful theorems which tell us that these forms go a long way towards generating "all" bivariate functions. Here are two such theorems.

Theorem 2. If S and T are compact then the set of functions of the form $\sum_{i=1}^{n} a_i(s)b_i(t)$ with $n \in N$, $a_i \in C(S)$, and $b_i \in C(T)$ is dense in $C(S \times T)$.

Theorem 3. (Kolmogoroff) If $S = T = [0, 1]$ then each function in $C(S \times T)$ is of the form

$$\sum_{i=1}^{5} \phi(a_i(s) + b_i(t))$$

where $a_i \in C(S)$, $b_i \in C(T)$, $\phi \in C(R)$. The functions a_i and b_i can be fixed once and for all.

The first of these theorems is a consequence of the Stone-Weierstrass Theorem [29]. The second theorem can be conveniently found in [22] or [33].

The approximation problems suggested by these two theorems are quite difficult. Here are two open problems:

(1) Devise an algorithm which, for a given $z \in C(S \times T)$, will find ϕ, a, b for which $\phi(a(s) + b(t))$ is a best approximation to z.

(2) Devise an algorithm which, for a given $z \in C(S \times T)$ and for a given n, will find x_i, y_i for which $\sum_{i=1}^{n} x_i(s)y_i(t)$ is a best approximation to z.

The second of these problems can be reformulated as a problem of "n-widths". Here is an outline of how that can be done. Define the (nonlinear) manifold M_n in $C(S \times T)$ by

$$M_n = \{ \sum_{i=1}^{n} x_i(s)y_i(t) : x_i \in C(S),\ y_i \in C(T)\}.$$

Now the distance of z to M_n is

$$\begin{aligned} \operatorname{dist}(z, M_n) &= \inf_{w \in M_n} \|z - w\| \\ &= \inf_{x_i} \inf_{y_i} \|z - \sum_1^n x_i y_i\| \\ &= \inf_{\dim G = n} \operatorname{dist}(z, G \otimes C(T)) \\ &= \inf_{\dim G = n} \sup_t \operatorname{dist}(z^t, G) \\ &= d_n(\{z^t : t \in T\}). \end{aligned}$$

In this calculation, we use the notation $z^t(s) = z(s, t)$. Then $z^t \in C(S)$ and $\{z^t : t \in T\}$ is a compact subset of $C(S)$. The formula

$$\operatorname{dist}(z, G \otimes C(T)) = \sup_t \operatorname{dist}(z^t, G)$$

is proved in [11]. The n-width of a set A in a Banach space X is defined by

$$d_n(A) = \inf_{\dim G = n} \sup_{x \in A} \operatorname{dist}(x, G).$$

For an introduction to the theory of n-widths, see [22] or the forthcoming [26].

The L^2-analogue of this approximation problem was solved in 1905 by Schmidt [31]. See also [13]. Here is an outline of the solution. We want to choose L^2 functions x_i and y_i to minimize the expression

$$\int_S \int_T [z(s, t) - \sum_{i=1}^n x_i(s) y_i(t)]^2 \, ds \, dt.$$

Define a kernel K by

$$K(s, \sigma) = \int_T z(s, t)z(\sigma, t)dt.$$

Define an operator T on $L^2(S)$ by

$$Tx = \int_S K(\cdot, \sigma)x(\sigma)d\sigma.$$

The x_i sought should be eigenfunctions corresponding to the n largest eigenvalues of T (each eigenvalue repeated a number of times equal to the dimension of its eigenspace).

II. TENSOR PRODUCTS IN APPROXIMATION

In this section we will discuss the tensor product of Banach spaces and show how this concept is used in approximation theory.

In forming the tensor product of two Banach spaces, there are several steps. If X and Y are the given spaces, we begin by forming the set $X \otimes Y$ of all formal "objects"

$$\sum_{i=1}^{n} x_i \otimes y_i \qquad (n \in N,\ x_i \in X,\ y_i \in Y).$$

With each object $\sum_{i=1}^{n} x_i \otimes y_i$ we can associate a linear operator L from X^* into Y. This operator L is defined on the conjugate space X^* by the equation

$$L\phi = \sum_{i=1}^{n} \phi(x_i)y_i \qquad (\phi \in X^*).$$

We want to regard two objects, $\sum_{i=1}^{n} x_i \otimes y_i$ and $\sum_{i=1}^{n'} x_i' \otimes y_i'$, as being equal if they produce the same linear operators. Therefore we define equality by the condition

$$\sum_{i=1}^{n} \phi(x_i)y_i = \sum_{i=1}^{n'} \phi(x_i')y_i' \qquad (\text{all } \phi \in X^*).$$

Formally, this amounts to introducing an equivalence relation in the set $X \otimes Y$.

In $X \otimes Y$, addition and multiplication by scalars are defined so as to be consistent with our interpretation of "objects" as linear operators. Thus we put

$$\sum_{i=1}^{n} x_i \otimes y_i + \sum_{i=n+1}^{m} x_i \otimes y_i = \sum_{i=1}^{m} x_i \otimes y_i$$

$$\lambda \sum_{i=1}^{n} x_i \otimes y_i = \sum_{i=1}^{n} \lambda x_i \otimes y_i = \sum_{i=1}^{n} x_i \otimes \lambda y_i .$$

At this juncture, we have a linear space, $X \otimes Y$. It can be regarded as a subspace (not necessarily closed) in $L(X^*, Y)$. In order to make this into a Banach space, we define a norm on $X \otimes Y$ and then take its completion. One of the natural norms that can be put on $X \otimes Y$ is the norm it receives automatically as a subspace of $L(X^*, Y)$. This norm is denoted by λ:

$$\lambda\Big(\sum_{i=1}^{n} x_i \otimes y_i \Big) = \sup_{\substack{\phi \in X^* \\ \|\phi\| \le 1}} \Big\| \sum_{i=1}^{n} \phi(x_i) y_i \Big\| .$$

Not only is this norm "natural" when we are interpreting $\sum x_i \otimes y_i$ as a linear operator, but it is "natural" for the space $C(S) \otimes C(T)$. Let us examine this important case.

Let $X = C(S)$ and $Y = C(S)$. Then the objects $\sum x_i \otimes y_i$ can be interpreted as elements in $C(S \times T)$. In other words, $x \otimes y$ can be interpreted as an ordinary product of two continuous functions. If we compute the λ-norm of an object the result is

$$\lambda(\sum x_i \otimes y_i) = \sup_{\substack{\phi \in X^* \\ \|\phi\| = 1}} \|\sum \phi(x_i) y_i\|$$

$$= \sup_{\phi} \sup_{t} |\sum \phi(x_i) y_i(t)| = \sup \sup |\phi(\sum y_i(t) x_i)|$$

$$= \sup_{t} \|\sum y_i(t) x_i\| = \sup_{t} \sup_{s} |\sum y_i(t) x_i(s)|$$

$$= \|\sum x_i y_i\| .$$

The last norm in this equation is a sup-norm in $C(S \times T)$. The completion of $X \otimes Y$ in the λ-norm is just its ordinary closure as a subspace of $C(S \times T)$. Since the functions $\sum_{i=1}^{n} x_i(s)y_i(t)$ are dense in $C(S \times T)$ (by the Stone-Weierstrass Theorem), we have proved:

Theorem 4. If S and T are compact Hausdorff spaces, then $C(S) \otimes_\lambda C(T) \simeq C(S \times T)$, where $\simeq$ denotes a natural isometry.

A more general theorem of Grothendieck states the following. See [32, page 357].

Theorem 5. If S is a compact Hausdorff space and Y is a Banach space, then $C(S) \otimes_\lambda Y \simeq C(S, Y)$.

In this theorem, $C(S, Y)$ is the Banach space of all continuous maps $f: S \to Y$, normed by defining

$$\|f\| = \sup_{s \in S} \|f(s)\| .$$

If $\sum x_i \otimes y_i$ is an element of $C(S) \otimes Y$, it induces a map $f \in C(S, Y)$ by the definition $f(s) = \sum x_i(s)y_i$. This relationship is a linear isometry.

Now we return to the general theory of forming a tensor product of two Banach spaces. The construction $X \otimes Y$ is an algebraic entity, and many different norms can be introduced in it. In order for the general theory to be applicable we assume that the norm α used in $X \otimes Y$ is a "reasonable" one. This means that in addition to the usual axioms for a norm, we require

(1) $\alpha(x \otimes y) = \|x\| \, \|y\|$

(2) $\alpha(\sum Ax_i \otimes By_i) \le \|A\| \, \|B\| \, \alpha(\sum x_i \otimes y_i)$

(3) $\alpha \ge \lambda$.

In (2), A and B are arbitrary linear operators on X and on Y respectively. This property is the "uniform" property of the norm α. Property (1) is the "cross-norm" property.

For information on constructing tensor products, the reader is referred to the books of Day [5], Schatten [30], Horvath [15], Diestel and Uhl [6], Kelley and Namioka [20], Semadeni [32], Robertson and Robertson [28] and to the article of Gilbert and Leih [12]. The nomenclature is not completely standard. In particular, some authors insist that a "reasonable" norm shall be one which can be defined on $X \otimes Y$ for arbitrary Banach spaces, without taking into account the nature of the elements. Schatten refers to such norms as being of "general character" [30, p. 39]. The norm λ is certainly of this type. Properties (1) and (2) for λ are easily verified. Because of property (3), λ is the "least" of the reasonable norms.

Another reasonable norm of general character is the "greatest" reasonable norm, γ. It is the norm of $\sum x_i \otimes y_i$ when this object is interpreted as a functional on $L(X, Y^*)$. Hence the definition of γ must be

$$\gamma(\sum x_i \otimes y_i) = \sup\{\sum [Ax_i, y_i]: A \in L(X, Y^*), \|A\| \le 1\}.$$

It is not immediately clear how to compute γ, but it can be shown [6] that $\gamma(\sum x_i \otimes y_i)$ is also equal to the infimum of $\sum \|u_i\| \|v_i\|$ taken over the set of all objects $\sum u_i \otimes v_i$, equal to $\sum x_i \otimes y_i$. That γ is the greatest cross norm is shown as follows. If $t \in X \otimes Y$ and if α is a cross norm, we write

$$\alpha(t) = \alpha(\sum x_i \otimes y_i) \le \sum \alpha(x_i \otimes y_i) = \sum \|x_i\| \|y_i\| .$$

By taking an infimum over all possible representations of t, we obtain $\alpha(t) \le \gamma(t)$.

Not only is the norm γ a natural one for our interpretation of $\sum x_i \otimes y_i$ as a functional on $L(X, Y^*)$, but it is also natural for the space $L_1(S) \otimes L_1(T)$. See [32, p. 477] for the following theorem, due to Dunford and Schatten.

Theorem 6. Let S and T be locally compact spaces furnished with finite regular Borel measures. Then there is a natural isometry

$$L_1(S) \otimes_\gamma L_1(T) \simeq L_1(S \times T).$$

Here $S \times T$ is given the product measure from S and T.

Another important result which plays a role in approximation theory is this one due to Schatten [30, p. 47]. See also [6, p. 230].

Theorem 7. There is a natural isometry $(X, Y^*) \simeq (X \otimes_\gamma Y)^*$.

The identification of an operator $A \in L(X, Y^*)$ with a functional ϕ on $X \otimes Y$ is through the equation

$$[\phi, \sum x_i \otimes y_i] = \sum [Ax_i, y_i].$$

Applications of this theorem to approximation problems exploit the fact that $L(X, Y^*)$ is a conjugate Banach space, and therefore has a weak*-topology. This makes available weak* compactness arguments for proving proximinality theorems. A notable success of this method of proof is in establishing the following result [27, Theorem 20].

Theorem 8. Let G and H be finite-dimensional subspaces in conjugate spaces X^* and Y^* respectively. Then the subspace

$$G \otimes Y^* + X^* \otimes H$$

is proximinal in $X^* \otimes_\lambda Y^*$.

Another approximation-theoretic result proved with similar techniques is this one due to Isbell and Semadeni [16].

Theorem 9. If Y is a complemented linear subspace of a normed space X and if Y is isometric to a dual Banach space, then there exists a projection of minimal norm from X onto Y.

Since approximation theory deals mainly with function spaces, let us return to the spaces of type $C(S) \otimes_\lambda X$, where S is a compact Hausdorff space and X is a Banach space. If G is a subspace of X then $C(S) \otimes_\lambda G$ is a subspace of $C(S) \otimes_\lambda X$.

We ask: Is $C(S) \otimes_\lambda G$ proximinal in $C(S) \otimes_\lambda X$? I.e., does each element of $C(S) \otimes_\lambda X$ have a best approximation in $C(S) \otimes_\lambda G$? The answer is generally "no", even if G is proximinal in X. The following theorem from [11] addresses this question.

Theorem 10. If G is a subspace of X having a continuous proximity map, then $C(S) \otimes_\lambda G$ is proximinal in $C(S) \otimes_\lambda X$.

A proximity map is a (nonlinear) operator $A: X \to G$ such that $\|x - Ax\| = \text{dist}(x, G)$ for all $x \in X$. The analogous question for L_1-spaces is answered by this theorem from [21].

Theorem 11. Let S and T be two σ-finite measure spaces. If G is a finite-dimensional subspace in $L_1(T)$, then $L_1(S) \otimes G$ is proximinal in $L_1(S \times T)$.

In a space $L_\infty(S \times T)$ a better result can be proved. See [21].

Theorem 12. Let S and T be σ-finite measure spaces. If G and H are finite-dimensional subspaces in $L_\infty(S)$ and $L_\infty(T)$ respectively, then $L_\infty(S) \otimes H + G \otimes L_\infty(T)$ is proximinal in $L_\infty(S \times T)$. (It is also complemented and weak*-closed.)

III. PROJECTION OPERATORS

A projection of a normed linear space X onto a subspace Y is a linear map $P: X \twoheadrightarrow Y$ such that $P^2 = P$, or equivalently $Py = y$ for all $y \in Y$. (The double arrow notation indicates a surjection.) Such operators are used to provide approximations in Y. The elementary inequality

$$\|x - Px\| \le \|I - P\| \, \text{dist}(x, Y)$$

shows that if $\|I - P\|$ is close to 1, then Px is nearly a best approximation to x. The projections in common use are Lagrange interpolation projections into polynomials, interpolation projections into spline subspaces, and orthogonal projections associated with various function spaces bearing inner products.

The structure of a projection having a finite-dimensional range is elementary.

Theorem 13. Let $\dim(Y) = n$ and let $P: X \twoheadrightarrow Y$ be a projection. If $\{y_1, \ldots, y_n\}$ is a basis for Y then there exist functionals $\phi_1, \ldots, \phi_n \in X^*$ such that $Px = \sum_{i=1}^{n} [x, \phi_i]y_i$.

Since $Py_i = y_i$ for $1 \le i \le n$, the functionals and the basis in the theorem have the biorthogonality property: $[y_i, \phi_j] = \delta_{ij}$.

If X and Y are fixed, the projections of X onto Y which are close to I are of greatest importance for approximation. Of course, $\|I - P\| \le 1 + \|P\|$, and so projections of small norm are useful too. In some cases these criteria are the same.

Daugavet's Theorem 14. If S is a compact Hausdorff space without isolated points and if L is a compact operator on $C(S)$ then $\|I - L\| = 1 + \|L\|$.

Since projections with finite-dimensional range are compact, we see that minimizing $\|I - P\|$ is the same as minimizing $\|P\|$ if P ranges over all projections of $C(S)$ onto a prescribed finite-dimensional subspace.

It has recently been proved [1] that Daugavet's theorem is also true in spaces of type $L^1(S)$. The Theorem in $C(S)$ occurs in [4].

Corrollary. If S is as in Daugavet's theorem then no finite-dimensional subspace in $C(S)$ can have a linear proximity map ("metric selection").

The minimal projection problem is that of determining a projection of least norm from a space X onto a prescribed subspace Y. The relative projection constant of Y as a subspace of X is defined by

$$\lambda(Y, X) = \inf\{\|P\| : P \in \mathcal{L}(X, Y),\ P(X) = Y,\ P^2 = P\}.$$

If there is no projection of X onto Y, we define $\lambda(Y, X)$ to be $+\infty$. If $\dim(Y) = n$ then $\lambda(Y, X) \le \sqrt{n}$ (Theorem of Kadec and Snobar [19]). It is only in rare instances that $\lambda(Y, X)$ is known precisely. Thus, for example we do not know the projection constants of the polynomial subspaces Π_n in $C[a, b]$.

It often happens that two subspaces Y_1 and Y_2 are given in a Banach space X, and we would like to construct a projection onto the vector sum, $Y_1 + Y_2$. If $P_i: X \twoheadrightarrow Y_i$ are projections $(i = 1, 2)$ then we define their Boolean sum by the equation

$$P_1 \oplus P_2 = P_1 + P_2 - P_1P_2.$$

Theorem 15. If $P_1P_2P_1 = P_2P_1$ then $P_1 \oplus P_2$ is a projection of X onto $Y_1 + Y_2$.

Proof. Clearly $P_1 \oplus P_2$ is a bounded linear map from X into $Y_1 + Y_2$. In order to see that $P_1 \oplus P_2$ leaves invariant each element of Y_1 we compute

$$(P_1 \oplus P_2)P_1 = P_1^2 + P_2P_1 - P_1P_2P_1 = P_1.$$

Similarly $P_1 \oplus P_2$ leaves invariant the elements of Y_2 since

$$(P_1 \oplus P_2)P_2 = P_1P_2 + P_2^2 - P_1P_2^2 = P_2.$$

Hence $P_1 \oplus P_2$ leaves invariant each element of $Y_1 + Y_2$. □ (See [14].)

Next we develop a little of the theory of projections in tensor product spaces. For the following theorem, refer to [27].

Theorem 16. If G and H are complemented subspaces in Banach spaces X and Y respectively, then

$$G \otimes_\alpha Y + X \otimes_\alpha H$$

is complemented in $X \otimes_\alpha Y$, for any reasonable norm α.

In the proof of this, one starts with two projections $P: X \twoheadrightarrow G$ and $Q: Y \twoheadrightarrow H$. A standard construction is used to form $P \otimes_\alpha I$ and $I \otimes_\alpha Q$. For example, $(P \otimes_\alpha I)(x \otimes y) = Px \otimes y$. By linearity and continuity, the map $P \otimes_\alpha I$ thus defined is extended to $X \otimes_\alpha Y$ in a unique manner. It is a projection of $X \otimes_\alpha Y$ onto $G \otimes_\alpha Y$. Similar remarks are valid for $I \otimes_\alpha Q$.

Finally, the Boolean sum of these two projections is a projection of $X \otimes_\alpha Y$ onto the desired subspace. That $I \otimes_\alpha Q$ commutes with $P \otimes_\alpha I$ is automatic.

An important observation concerning the construction just described is that if the projections P and Q are also proximity maps (which means $\|I - P\| = \|I - Q\| = 1$) then $P \otimes_\alpha I$, $I \otimes_\alpha Q$, and $(P \otimes_\alpha I) + (I \otimes_\alpha Q)$ are also linear proximity maps. These matters are discussed in [27]. Their application to orthogonal projections is particularly important. Here is a summary.

Theorem 17. Let G and H be subspaces of $L^2(S)$ and $L^2(T)$ respectively, with corresponding orthogonal projections P and Q. Then $(P \otimes I) \oplus (I \otimes Q)$ is the orthogonal projection of $L^2(S \times T)$ onto $G \otimes L^2(T) + L^2(S) \otimes H$.

For practical approximations, even in continuous function spaces, one often utilizes orthogonal projections. Think of the Fourier-Chebyshev projections in $C[-1, 1]$ for an outstanding example. Thus, the preceding theorem supplies a rich variety of useful projections in $C(S \times T)$; one only has to introduce suitable inner products defined with weight functions. Of course, interpolating projections also exist in great profusion; rearranging the distribution of interpolation nodes corresponds roughly to changing the weight function in an orthogonal projection. Incidentally, there exists a coherent theory of projections in which the concept of "orthogonal" projection is extended so as to encompass interpolating projections. This class of projections has been studied in [7]; it has many nice properties.

How much is known about projection constants of subspaces in $C(S \times T)$? We ask this only of the type of subspace mentioned previously. One result is the following, from [8]:

Theorem 18. If G and H are finite-dimensional subspaces of $C(S)$ and $C(T)$ respectively, then the relative projection constant of $G \otimes H$ in $C(S \times T)$ is the product of the relative projection constants of G and H.

The preceding result is easily established with the aid of a difficult theorem of Tomczak-Jaegermann [35]: For any two finite-dimensional Banach spaces, X and Y, we have $\lambda(X \otimes_\lambda Y) = \lambda(X) \cdot \lambda(Y)$.

For the more versatile approximating subspace

$$W = C(S) \otimes H + G \otimes C(T)$$

with $H \subset C(T)$ and $G \subset C(S)$, we have the following theorem from [8].

Theorem 19. If S and T are infinite sets and if G and H are finite-dimensional subspaces containing the constants, then W has relative projection constant at least 3 in $C(S \times T)$.

The proof of this theorem uses techniques which Jameson and Pinkus had used earlier [17] to establish that $C(S) + C(T)$ has relative projection constant 3 in $C(S \times T)$. For this subspace, a minimal projection is obtained in the form $(P \otimes I) \oplus (I \otimes Q)$, where $P: C(S) \to \Pi_0(S)$, $Q: C(T) \to \Pi_0(T)$, $Px = x(s_0)$, and $Qy = y(t_0)$ for arbitrary fixed $s_0 \in S$ and $t_0 \in T$.

IV. SIMULTANEOUS APPROXIMATION

Problems in multivariate approximation often lead to univariate simultaneous approximation problems. In simultaenous approximation we attempt to approximate a set of functions by one function. To make this idea precise, suppose that a bounded set A has been prescibed in a Banach space X. At first, we may look for a single element $x \in X$ which is a good approximation to all the elements of A. We therefore define

$$r(A) = \inf_{x \in X} \sup_{a \in A} \|a - x\| .$$

This is called the Chebyshev radius of the set A. The points of interest to us are the elements of

$$E(A) = \{x \in X: \sup_{a \in A} \|a - x\| = r(A)\}.$$

This set is the Chebyshev center of A; it may be empty. One of the goals of the theory is to establish conditions on X and A such that $E(A)$ is nonempty.

The problem of simultaneous approximation arises naturally when a function is not known precisely because of experimental errors, for example. It can be interpreted as a set, which then is to be represented by a single best element.

One of the most elegant results in this subject can now be stated.

Theorem 20. Every bounded set in $C(S)$ has a nonempty Chebyshev center.

In this theorem S can be an arbitrary topological space, and $C(S)$ consists of all bounded real-valued functions. The first theorem of this type was proved by Kadets and Zamyatin [18], and the form given is to be found in [9].

Now, usually a set A is not to be approximated by an element x which is free in the Banach space X. Rather, we expect a subspace Y of "approximants" to be specified. In this case we define the restricted Chebyshev radius of A by putting

$$r_Y(A) = \inf_{y \in Y} \sup_{a \in A} \|a - y\|.$$

Then the restricted center is

$$E_Y(A) = \{y \in Y: \sup_{a \in A} \|a - y\| = r_Y(A)\}.$$

Here is a multivariate approximation problem which leads directly to a problem of Chebyshev centers: a function $z \in C(S \times T)$ is given, and we wish to approximate it as well as possible by an element of $C(S)$. If $x \in C(S)$, then its deviation from z is

$$\|z - x\| = \sup_t \sup_s |z(s, t) - x(s)|$$

$$= \sup_t \|z^t - x\| .$$

Here we have used the t-sections of z defined by $z^t(s) = z(s, t)$. It is clear now that x should be chosen as any element in the Chebyshev center of

$$\{z^t: t \in T\}.$$

This set is compact in $C(S)$, and so its center is nonempty. Actually, an explicit solution for x is given by the formula

$$x(s) = \frac{1}{2} \max_{t \in T} z(s, t) + \frac{1}{2} \min_{t \in T} z(s, t).$$

We have seen how a straightforward problem in multivariate approximation leads to a problem of simultaneous approximation. The reverse is also true. Suppose that $A \subset X$ and $Y \subset X$. We seek an element $y \in Y$ for which the expression $\sup_{a \in A} \|a - y\|$ is as small as possible. This can be couched as an ordinary approximation problem in the space $C(A, X)$ of all bounded continuous maps from A into X, normed by $\|f\| = \sup_{a \in A} \|f(a)\|$.

To do this, define $e \in C(A, X)$ by $e(a) = a$ for all $a \in A$. Then define $\bar{y} \in C(A, X)$ for each $y \in Y$ by putting $\bar{y}(a) = y$ for all $a \in A$. Then let $\bar{Y} = \{\bar{y}: y \in Y\}$. For $y \in Y$ we have

$$\|e - \bar{y}\| = \sup_{a \in A} \|e(a) - \bar{y}(a)\| = \sup_{a \in A} \|a - y\|.$$

Here we are looking for a best approximation of e in the set $\bar{Y}$. Incidentally, the space $C(A, X)$ is isometric to the tensor product $C(A) \otimes X$ if A is locally compact, and $\bar{y} \in C(A, Y) = C(A) \otimes_\lambda Y$ in this case.

Since simultaneous approximation can be reduced to ordinary Chebyshev approximation in a space $C(A, X)$, it is to be expected that characterizations of the solutions and duality theorems can be easily obtained. Here is a sample, taken from [9].

Theorem 21. If A is a compact set in a Banach space X and if Y is a subspace of X then the restricted Chebyshev radius of A satisfies the equation

$$r_Y(A) = \max_{\phi \perp Y} \inf_{\phi(z)=0} \max_{a \in A} \|a - z\|.$$

In this equation, $\phi \in X^*$ and $(y) = 0$ for all $y \in Y$. As with all duality theorems, this one provides an easy method for obtaining a lower bound on the extreme value.

An open problem in this area of study is whether a proximinal set in $C(S)$ remains proximinal when embedded in $C(S \times T)$. We state this as a conjecture, using different but equivalent language.

Conjecture. If Y is a proximinal set in $C(S)$ then $E_Y(A) \neq \phi$ for all compacts sets A in $C(S)$.

In order to see that this statement of the open problem

is equivalent to the earlier statement, one needs the following lemma.

Lemma. Every compact set in $C(S)$ is the set of t-sections of an element of $C(S \times T)$, where T is a compact metric space.

Proof. Let K be a compact set in $C(S)$. Define $f \in C(S \times K)$ by the equation $f(s, k) = k(s)$. Then $K = \{f^k : k \in K\}$. □

One result which relates to the above conjecture is the following, taken from [10].

Theorem 22. Let Y be a proximinal set in $C(S)$, S being compact Hausdorff. In order that $E_Y(A)$ be nonempty for all compact $A \subset C(S)$ it is necessary and sufficient that $\mathrm{dist}(x, Y)$ attain its infimum on every order interval $[u, v]$ such that $u \in C(S)$, $v \in C(S)$, and $\max[u(s) - v(s)] = 0$.

These theorems can be used to obtain various proximinality theorems. Here, for example, is a result related to an old theorem of Mazur. See [10, Theorem 13].

Theorem 23. Let S and T be compact Hausdorff spaces, and let $f: S \twoheadrightarrow T$ be open and continuous. Let Y be a proximinal subspace in $C(T)$. If $E_Y(A)$ is nonempty for each compact set A in $C(T)$ then the subspace $\{v \circ f : v \in V\}$ is proximinal in $C(S)$.

REFERENCES

1. Babenko, V. F., and Pichugov, S. A., "A property of compact operators in the space of integrable functions", Ukrain. Mat. Zhurnal 33, 1981, pp. 491-492. Translation, Plenum Publishing Corp. MR82m: 47018.

2. Cheney, E. W., and Price, K. H., "Minimal projections", pp. 261-289 in Approximation Theory, ed. by A. Talbot. Academic Press, New York, 1970.

3. Cheney, E. W., "Projection operators in approximation theory", pp. 50-80 in Studies in Functional Analysis, ed. by R. G. Bartle. Mathematical Association of America, 1980.

4. Daugavet, I. K., "A property of compact operators in the space C", Uspehi Mat. Nauk SSSR 18, 1963, pp. 157-158. MR28-431.

5. Day, M. M., "Normed linear spaces", Academic Press, New York, 1962.

6. Diestel, J., and Uhl, J. J., "Vector measures", Amer. Math. Soc., Providence, 1977.

7. Franchetti, C., and Cheney, E. W., "Orthogonal projections in continuous functions spaces", J. Math. Analysis and Applications, 63, 1978, pp. 253-264.

8. Franchetti, C., and Cheney, E. W., "Minimal projections in tensor-product spaces", CNA Report 184, University of Texas, July 1983. To appear, J. Approximation Theory.

9. Franchetti, C., and Cheney, E. W., "Simultaneous approximation and restricted Chebyshev centers in function spaces", in Approximation Theory and Applications, ed. by Z. Ziegler. Academic Press, New York, 1981, pp. 65-88. MR82f: 41042.

10. Franchetti, C., and Cheney, E. W., "The embedding of proximinal sets", CNA Report 183, June 1983, University of Texas.

11. Franchetti, C., and Cheney, E. W., "Best approximation problems for multivariate functions", Boll. Unione Mat. Italia 18, 1981, pp. 1003-1015.

12. Gilbert, J. E., and Leih, T. J., "Factorization, tensor products, and bilinear forms in Banach space theory", in Notes in Banach Spaces, H. E. Lacey, ed. University of Texas Press, Austin 1980, pp. 182-305.

13. Golomb, M., "Approximation by functions of fewer variables", in On Numerical Approximation, ed. by R. E. Langer, University of Wisconsin Press, 1959, pp. 275-327.

14. Gordon, W. J., and Cheney, E. W., "Bivariate and multivariate interpolation with noncommutative projectors", in Linear Spaces and Approximation, ed. by P. L. Butzer and B. Sz.-Nagy, Birkhauser, Basel, 1978, pp. 381-388.

15. Horvath, J., "Topological Vector Spaces and Distribution", Addison-Wesley Publ. Co., London, 1966.

16. Isbell, J. R., and Semadeni, Z., "Projection constants and spaces of continuous functions", Trans. Amer. Math. Soc., 107, 1963, pp. 38-48.

17. Jameson, G. J. O., and Pinkus, A., "Positive and minimal projections in functions spaces", J. Approximation Theory 37, 1983, pp. 182-195.

18. Kadets, M., and Zamyatin, V. N., "Chebyshev centers in the space C[a, b]", Teo. Funk. Funkcion. Anal. Pril. 7, 1968, pp. 20-26. MR42#3480.

19. Kadec, M. I., and Snobar, M. G., "Some functionals over a compact Minkowski space", Math. Notes USSR 10, 1971, pp. 694-696.

20. Kelley, J., and Namioka, I., "Linear Topological Spaces", D. Van Nostrand Co., New York, 1963.

21. Light, W. A., and Cheney, E. W., "Some best approximation theorems in tensor-product spaces", Math. Proc. Cambridge Phil. Soc., 89, 1981, pp. 385-390.

22. Lorentz, G. G., "Approximation of Functions", Holt-Reinhart-Winston, New York, 1966.

23. Mairhuber, J. C., "On Haar's theorem concerning Chebyshev approximation problems having unique solutions", Proc. Amer. Math. Soc., 7, 1956, pp. 609-615.

24. McGabe, J. H., Phillips, G. M., and Cheney, E. W., "A mixed-norm bivariate approximation problem with application to Lewanowicz operators", in Multivariate Approximation, ed. by D. C. Handscomb, Academic Press, New York, 1978, pp. 315-323.

25. Ofman, Ju. P., "Best approximation of functions of two variables by functions of the form $\phi(x) + \psi(y)$", Amer. Math. Soc. Transl. Series 2, vol. 44, 1965, pp. 12-28.

26. Pinkus, A., "Theory of n-Widths", Lecture Notes in Math., Springer-Verlag. To appear.

27. Respess, J. R., and Cheney, E. W., "Best approximation problems in tensor product spaces", Pacific J. Math., 102, 1982, pp. 437-446.

28. Robertson, A. P., and Robertson, W., "Topological Vector Spaces", Cambridge University Press, 1964.

29. Rudin, W., "Functional Analysis", McGraw Hill Book Co., New York, 1973.

30. Schatten, R., "A Theory of Cross-Spaces", Annals of Math. Studies, No. 26, Princeton University Press, 1950.

31. Schmidt, E., "Zur Theorie der linearen und nichtlinearen Integral-gleichungen", Math. Annalem 63, 1907, pp. 433-476.

32. Semadeni, Z., "Banach spaces of continuous functions", Polish Scientific Publishers, Warsaw, 1971.

33. Shapiro, H. S., "Topics in Approximation Theory", Lecture Notes in Mathematics, vol. 187, Springer, New York.

34. Shepard, D., "A two-dimensional interpolation function for irregularly spaced data", Proc. ACM Nat. Conf., 1964, 517-524.

35. Tomczak-Jaegermann, N., "Finite-dimensional operator ideals and the Banach-Maxur theorem", Lecture Notes in Mathematics, Springer-Verlag. To appear.

36. Treves, F., "Topological vector spaces, distributions and kernels", Academic Press, New York, 1967.

THE APPROXIMATION OF CERTAIN FUNCTIONS BY COMPOUND MEANS

D. M. E. Foster and G. M. Phillips*

University of St. Andrews

1. INTRODUCTION

We begin with the double recurrence relation

$$u_{n+1} = \frac{1}{2}(u_n + v_n), \tag{1a}$$

$$v_{n+1} = (u_{n+1}\, v_n)^{1/2}, \tag{1b}$$

where u_0, v_0 are given. We will refer to (1) as the Schwab-Borchardt algorithm following Schoenberg [10, 11], who has made a careful study of its origins. Hitherto this process has been more commonly identified in the literature solely with the name of Borchardt. (See, for example, Carlson [2], Todd [13].)

If we write $u_j = 1/a_j$, $v_j = 1/b_j$ in (1), we obtain

$$1/a_{n+1} = \frac{1}{2}(1/a_n + 1/b_n), \tag{2a}$$

$$b_{n+1} = (a_{n+1}\, b_n)^{1/2}, \tag{2b}$$

which we will call, for convenience rather than for entirely convincing historical reasons, the Archimedean algorithm. The special case of (2) where $a_0 = 3\sqrt{3}$, $b_0 = \frac{1}{2}\, 3\sqrt{3}$ generates sequences (a_n) and (b_n) which have a simple geometrical

S. P. Singh et al. (eds.), Approximation Theory and Spline Functions, 89–95.

interpretation: a_n and b_n are the semi-perimeters of respectively the escribed and inscribed regular polygons of 3×2^n sides of the unit circle. In the third century B.C., Archimedes is said to have computed a_n and b_n for $0 \le n \le 5$ to obtain his famous inequalities for π:

$$3\frac{10}{71} < \pi < 3\frac{1}{7}.$$

According to Dijksterhuis [3], Archimedes generated his values of a_n and b_n from separate recurrence relations and not from the recurrence relation (2). Incidentally, in computing his values of a_n, which are all upper bounds for π, Archimedes consistently rounded up and in computing the lower bounds b_n, he always rounded down. (See Miel [7].) Phillips [8] shows how, by using repeated extrapolation to the limit, the three figure accuracy for π obtained by Archimedes can be stretched to give an approximation which is accurate to about 19 decimal places.

Explicit expressions for the sequences (a_n) and (b_n), and therefore for (u_n) and (v_n), are well known in the literature. If $a_0 > b_0 > 0$,

$$a_n = 2^n \lambda \tan(\theta/2^n), \quad b_n = 2^n \lambda \sin(\theta/2^n), \tag{3}$$

where

$$\lambda = (1/b_0^2 - 1/a_0^2)^{-1/2}, \quad \theta = \cos^{-1}(b_0/a_0). \tag{4}$$

It is clear from (3) that the sequences (a_n) and (b_n) have the common limit

$$\lambda\theta = (1/b_0^2 - 1/a_0^2)^{-1/2} \cos^{-1}(b_0/a_0).$$

If $b_0 > a_0 > 0$, then

$$a_n = 2^n \lambda \tanh(\theta/2^n), \quad b_n = 2^n \lambda \sinh(\theta/2^n), \tag{5}$$

so that the circular functions in (3) are replaced by the corresponding hyperbolic functions. In this case

$$\lambda = (1/a_0^2 - 1/b_0^2)^{-1/2}, \quad \theta = \cosh^{-1}(b_0/a_0) \tag{6}$$

and the common limit of (a_n) and (b_n) is

$$\lambda\theta = (1/a_0^2 - 1/b_0^2)^{-1/2} \cosh^{-1}(b_0/a_0). \tag{7}$$

This latter form also shows how we may compute the logarithm function. For if we take

$$a_0 = (t^2 - 1)/(t^2 + 1), \quad b_0 = (t^2 - 1)/2t, \quad t > 1,$$

we may deduce from (7) that the Archimedean process converges to log t. See, for example, Schoenberg [10] or Phillips [8]. Schoenberg's account [10] is explicitly concerned with the Schwab-Borchardt algorithm. When 'translated' by taking reciprocals as described above, Schoenberg attributes a beautiful geometrical proof of the circular function case (3) to Schwab (dated 1813) and credits Pfaff with the hyperbolic function case (5), also circa 1800.

Given arbitrary positive 'starting values' a_0 and b_0, we see from (2) that the sequences (a_n) and (b_n) are both monotonic and converge to a common limit. These conclusions follow from elementary analysis and do not require the use of the explicit forms for a_n and b_n cited above.

Phillips [9] has given an analysis of the Archimedean process when a_0 and b_0 are complex and this generalizes the above treatment where a_0 and b_0 are positive. In the complex case the square root in (2b) is chosen so that the vector b_{n+1} lies (midway) within the smaller angle between a_{n+1} and b_n. It is shown that the two sequences of vectors (a_n) and (b_n) rotate 'monotonically' towards, and converge to, a common limiting vector. Also, at least for all n sufficiently large, the moduli of the sequence of vectors (a_n) and (b_n) also converge monotonically to the modulus of the common limiting vector.

2. GENERALIZATION OF THE ARCHIMEDEAN PROCESS

Foster and Phillips [4] have generalized (1) and (2) as follows. First they define a mean:

Definition. A mean M is a continuous mapping from $\mathbb{R}^+ \times \mathbb{R}^+$ into $\mathbb{R}^+$ such that

(a) $0 \le a \le b \Rightarrow a \le M(a, b) \le b$,

(b) $M(a, b) = M(b, a)$,

(c) $a = M(a, b) \Rightarrow a = b$.

Such means include the arithmetic, geometric and harmonic means and the Minkowski means

$$M(a, b) = ((a^p + b^p)/2)^{1/p}, \quad 1 \le p < \infty.$$

Foster and Phillips give further examples of such means and discuss the following generalization of (1) and (2):

$$\begin{aligned} a_{n+1} &= M(a_n, b_n), \\ b_{n+1} &= M'(a_{n+1}, b_n), \end{aligned} \tag{8}$$

where M and M' are means in the sense defined above. It then follows very readily that, given any $a_0, b_0 > 0$, the process (8) generates two sequences (a_n) and (b_n) which converge monotonically to a common limit. It is also interesting that the rate of convergence of the process (8) is independent of the choice of means M and M', provided these mappings are sufficiently smooth. The rate of convergence is always linear. Specifically, provided the partial derivatives of M and M' are continuous up to second order, then

$$\lim_{n\to\infty} \frac{a_{n+1} - \alpha}{a_n - \alpha} = \lim_{n\to\infty} \frac{b_{n+1} - \alpha}{b_n - \alpha} = \frac{1}{4}, \tag{9}$$

where α denotes the common limit of (a_n) and (b_n). It is trivial to verify (9) for the Archimedean algorithm by using (3) or (5); the general case (9) is established in the absence of explicit formulas for a_n and b_n. It is also shown that if

$$c_n = (a_n + 2b_n)/3, \tag{10}$$

then (c_n) converges to α faster than (a_n) or (b_n). The significance of c_n in the special case of the Archimedean algorithm is easily seen from the Maclaurin series for the circular and hyperbolic functions in (3) and (5). (It is worth noting here that, in the early seventeenth century, Snell [12] effectively used (10) to compute π to 34 decimal places. Of course, Snell did not have access to the series for the sine and tangent, which make the derivation of (10) from (3) so easy.) In the general case, again for sufficiently smooth M and M', it is shown that

$$\lim_{n\to\infty} \frac{c_{n+1} - \alpha}{c_n - \alpha} = 1/16, \tag{11}$$

provided $4M_{xx}(\alpha, \alpha) + M'_{xx}(\alpha, \alpha) \neq 0$.

Foster and Phillips [5] have also discussed some other specific examples of the double-mean process (8). In particular, they have explored the arithmetic-harmonic mean in some detail.

3. THE ARITHMETIC-GEOMETRIC MEAN

It is interesting to see what happens when we make an apparently trivial change to (8) and write

$$\begin{aligned} a_{n+1} &= M(a_n, b_n), \\ b_{n+1} &= M'(a_n, b_n), \end{aligned} \tag{12}$$

where M and M' are means of the type defined in the last section. The celebrated case where M and M' are respectively the arithmetic and geometric means was investigated by Gauss circa 1800. It can be shown that the integral

$$I(a_n, b_n) = \int_0^{\pi/2} \frac{d\theta}{\sqrt{(a_n^2 \cos^2\theta + b_n^2 \sin^2\theta)}} \tag{13}$$

is constant for all n, where a_n and b_n are generated by the arithmetic-geometric case of (12). Assuming convergence of (a_n) and (b_n) to α, it follows that

$$I(a_n, b_n) = I(\alpha, \alpha) = \pi/(2\alpha)$$

and, if we write $b_0 = k\, a_0$, (13) is seen to be the elliptic integral

$$\frac{1}{a_0} \int_0^{\pi/2} \frac{d\theta}{\sqrt{(1 - (1 - k^2)\sin^2\theta)}} . \tag{14}$$

Thus the arithmetic-geometric process provides a method of computing the integral (14). See, for example, Todd [13]. A discussion of the arithmetic-geometric mean process is given by the brothers Borwein [1].

Foster and Phillips [4] have shown that the general process (12) always converges quadratically and special cases of (12) have also been discussed by Lehmer [6]. If

$$\kappa = M_{xx}(\alpha, \alpha) - M'_{xx}(\alpha, \alpha), \tag{15}$$

it can be shown that, for all n sufficiently large,

$$\begin{aligned} \delta_{n+1} &= \kappa\, \delta_n^2 + O(\delta_n^3), \\ \varepsilon_{n+1} &= -\kappa\, \varepsilon_n^2 + O(\varepsilon_n^3), \end{aligned} \tag{16}$$

where δ_n and ε_n are the errors at each stage, defined by

$$a_n = \alpha + \delta_n, \quad b_n = \alpha + \varepsilon_n .$$

Note that, provided $\kappa \neq 0$, (16) implies that the errors δ_n and ε_n are ultimately of opposite sign.

ACKNOWLEDGEMENT

We are most grateful to Professor J. M. Borwein for providing the reference to the work of Snell.

REFERENCES

1. Borwein, J. M. and P. B., "The arithmetic-geometric mean and fast computation of elementary functions", preprint.

2. Carlson, B. C., "Algorithms involving arithmetic and geometric means", Amer. Math. Monthly, Vol. 78, 1971, 496.

3. Dijksterhuis, E. J., "Archimedes", Ejnar Munksgaard, Copenhagen, 1956.

4. Foster, D. M. E. and Phillips, G. M., "A generalization of the Archimedean double sequence", J. Math. Anal. and Applics., to appear.

5. Foster, D. M. E. and Phillips, G. M., "The arithmetic-harmonic mean", Math. of Comp., to appear.

6. Lehmer, D. H., "On the compounding of certain means", J. Math. Anal. and Applics., Vol. 36, 1977, 183.

7. George Miel, "Of calculations past and present: the Archimedean algorithm", Amer. Math. Monthly, Vol. 90, 1983, 17.

8. Phillips, G. M., "Archimedes the numerical analyst", Amer. Math. Monthly, Vol. 88, 1981, 165.

9. Phillips, G. M., "Archimedes and the complex plane", Amer. Math. Monthly, to appear.

10. Schoenberg, I. J., "On the arithmetic-geometric mean", Delta, Vol. 7, 1978, 49.

11. Schoenberg, I. J., "On the arithmetic-geometric mean and similar iterative algorithms", Chapter 12 of 'Mathematical Time Exposures', to appear.

12. Snell, Willebrord, "Cyclometricus", Leyden, 1621.

13. Todd, J., "Basic Numerical Mathematics", Vol. 1, Birkhäuser, Basel, 1979.

---

* Paper presented by this author.

A PRACTICAL METHOD FOR OBTAINING A PRIORI ERROR BOUNDS IN POINTWISE AND MEAN-SQUARE APPROXIMATION PROBLEMS

Jean Meinguet

Université Catholique de Louvain

ABSTRACT

Unlike optimal error bounds, realistic quantitative results can often be obtained at reasonable cost, by manipulating suitable explicit representations of any function as a polynomial plus a remainder term. This is demonstrated here as regards pointwise or uniform approximation in $C^m(\overline{\Omega})$, and pointwise and mean-square approximation in $H^m(\Omega)$. Two concrete examples of application are considered in detail: multivariate numerical integration with nonnegative weights, bivariate affine interpolation at the vertices of a triangle (finite element called Courant's triangle).

1. INTRODUCTION

Error *estimates* often involve unknown constants, in which case they are naturally ill-suited to *quantitative* studies. The present lectures are based essentially on some fairly recent research work whose primary purpose was precisely to cover this shortcoming by devising a *method of practical value* for obtaining *realistic a priori upper bounds of approximation errors* in a *wide variety of situations*. As explained in detail in [14, 15, 16, 20], such a method can be elaborated from a *structural analysis of error coefficients* (see Section 2.2), which evolves quite naturally in the setting of operator theory in normed linear spaces while referring to such classical tools as the *Peano kernel theorem* and its *qualitative* generalization known as the

S. P. Singh et al. (eds.), Approximation Theory and Spline Functions, 97–125.

Bramble-Hilbert lemma. As exemplified in the above references and also in [1, 12], that method can yield at *reasonable cost* explicit upper bounds for those *generic constants* which tend to pervade the modern literature on error estimation (typically in spline analysis, in connection with the rate of convergence of the finite element method, see e.g. [6] and [7], Theorems 2, 4, 5, 6).

We will review and summarize here some of the significant results we have obtained so far in matter of *quantitative estimation* of interpolation and (more generally) approximation errors. Two wide classes of (multivariate) applications need to be considered in the first instance, to wit:

(a) *pointwise or uniform approximation in* $C^m(\bar{\Omega})$ (see Section 2).

(b) *pointwise and mean-square approximation in* $H^m(\Omega)$ (see Section 3).

Our approach basically amounts to manipulating (most carefully!) suitable *explicit representations* of any element v of such spaces as a polynomial Pv (where P denotes a linear projector) plus a remainder term (expressed in integral form, in terms of partial derivatives of v), in order to get *key estimates* of the general form

$$|v-Pv|_I \le e|v|_{II},$$

where $|\cdot|_I$ and $|\cdot|_{II}$ denote seminorms involving appropriately related subsets of (possibly generalized) partial derivatives of v, and e, the so-called *error coefficient*, is a numerical constant (with respect to v).

Needless to say, such inequalities are formally related to *variations of the well known Bramble-Hilbert lemma* (see e.g. [4, 5]) and could be proved quite similarly, that is in a strictly nonconstructive way, by making use of Morrey-like results (see [21], pp. 85-86) and generalizations thereof. Our proof, which is somewhat reminiscent of Sobolev's approach to imbedding theorems (see e.g. [25], pp. 50-56), is completely different; essentially constructive throughout, it can yield directly (i.e., without passing through the analysis of "change of scale" effects) realistic upper bounds for the best possible error coefficient e_0; of course, these results depend on the specific representation formula at hand: the (truncated) *Taylor series* (see e.g. [14, 15, 20]), the *Kowalewski-Ciarlet-Wagschal formula* (see e.g. [1, 8, 12]), the *Sard-Taylor expansion* (see e.g. [3, 23]), the *averaged Taylor series* [15, 16, 19] and more generally the *Sobolev representation formula* (see [25] and [10, 11]), etc. It should be noted that a systematic, constructive approach to the important topic of (multivariate) *representation formulas* can be found in the recent papers [17, 18] whose main concern is different from the present one, however; it is shown there, in particular, how easily *convolutions* (or *Fourier transforms*) can be used, together with *basic integral inequalities*, to find in a unified way "appropriate" expressions of distributions (or functions) in terms of a prescribed subset of partial derivatives; such representation formulas can provide appropriate substitutes for the truncated Taylor series and for its integral remainder term, such as requested for extending the Peano kernel theorem to the representation of errors in intrinsically multivariate situations.

It turns out that *optimal error coefficients* e_o can often be characterized, at least in the Sobolev case (for an example, see Section 3.1.2). Though such characterizations are essentially theoretical results (they require indeed the solution of unduly complicated boundary value or eigenvalue problems), they can occasionally be treated further, in order to yield upper bounds for e_o (see specially [24]); however, there are apparently simple situations (see e.g. [22]) for which this alternative to the more elementary and general approach presented here leads to quite unrealistic results. As regards the determination of *lower bounds* for e_o, this is comparatively a trivial problem (see e.g. [14] for the class (a), and [19, 22] for the class (b)).

By way of *concrete applications* of the above material, we will consider here two significant, non-trivial approximation problems, to wit:

(i) *multivariate numerical integration with nonnegative weights* (see Section 2.1).

(ii) *bivariate Lagrange interpolation over a triangulated domain* (see Section 2.3 for the class (a), and Section 3.1.3 for the class (b)):

owing to the inherent versatility of the method, it proves surprisingly easy to derive *quantitative* error bounds, whether the geometry of finite elements satisfies the *classical minimum angle condition* or the refined *maximal angle condition* (see e.g. [2, 13] and, as mentioned on p. 140 in [26], the "historical" reference [27], p. 211).

For simplicity, all functions and vector spaces considered

in this paper are real.

2. POINTWISE (OR UNIFORM) APPROXIMATION IN $C^m(\overline{\Omega})$

2.1. A Motivating Example of Application: Multivariate Approximate Integration

2.1.1. The problem (J. Descloux [9]): Find a (realistic) upper bound for the scalar $|Rv|$, with

$$Rv: = \int_\Omega v(x)dx - \sum_i w_i v(x^i), \tag{1}$$

under the *natural* assumptions:

(a) Ω is a *bounded* open subset of $\mathbb{R}^n$ whose closure $\overline{\Omega}$ is *star-shaped* with respect to (at least) one of its points, say a.

(b) v is any element of $C^m(\overline{\Omega})$ (i.e., the set of all functions that are *uniformly continuous* in Ω together with all their partial derivatives of order $\le m$), with m *integer* ≥ 1.

(c) the *abscissas* x^i are distinct points of $\overline{\Omega}$ (their number is finite but need not be specified any further).

(d) the *weights* w_i are nonnegative.

(e) the *remainder* functional R vanishes on the space P_{m-1} of all polynomials of degree $\le m-1$.

2.1.2. Our approach to solutions.

(i) In view of assumptions (a) and (b), the well known *Taylor formula of the m-th order* (with integral expression of the remainder) about any of the points a with respect to which $\overline{\Omega}$ is star-shaped, viz.,

$$v(x) = \sum_{k=0}^{m-1} \frac{D^k v(a)\cdot(x-a)^k}{k!} + \int_0^1 \frac{(1-t)^{m-1}}{(m-1)!} D^m v(a+t(x-a))\cdot(x-a)^m dt,$$

$$\forall v \in C^m(\overline{\Omega}),\ \forall x \in \overline{\Omega}, \tag{2a}$$

with the following definitions of the coordinate free notations (analyzed in detail in [14], Section 3):

$$\frac{D^k v(a)\cdot(x-a)^k}{k!} := \sum_{|\alpha|=k} \frac{\partial^\alpha v(a)(x-a)^\alpha}{\alpha!} = \sum_{i_1,\ldots,i_k=1}^{n}$$

$$\frac{\partial_{i_1\ldots i_k} v(a) \prod_{j=1}^{k} (x_{i_j}-a_{i_j})}{k!}, \tag{2b}$$

can serve for representing explicitly over $\bar{\Omega}$ any "admissible" v as a *polynomial* $P_{m-1}v \in P_{m-1}$ plus a *remainder* term $V_m(D^m v)$ depending linearly on v (as readily verified, P_{m-1} is a *linear projector onto* P_{m-1}, while V_m is a *linear right inverse of* D^m). The concretization (2a) of the structural decomposition formula (see Section 2.2.2)

$$v = P_{m-1}v + V_m D^m v, \quad \text{for every admissible } v, \tag{3}$$

is clearly equivalent to the *direct sum* decomposition

$$C^m(\bar{\Omega}) = P_{m-1} \oplus C^m_a(\bar{\Omega}),$$

where

$$C^m_a(\bar{\Omega}) := \{v \in C^m(\bar{\Omega}) : \partial^\alpha v(a) = 0, \text{ for all } \alpha \in \mathbb{N}^n \text{ such that } |\alpha| \le m-1\}.$$

(ii) In view of assumptions (c) and (e),

$$Rv = R(v-P_{m-1}v) = \int_\Omega (V_m D^m v)(x)\,dx - \sum_i w_i (V_m D^m v)(x^i),$$

which immediately yields, by assumption (d), the "sharp" upper bound

$$|Rv| \le \int_\Omega |V_m D^m v(x)|\,dx + \sum_i w_i |V_m D^m v(x^i)|, \ \forall v \in C^m(\bar{\Omega}).$$

(iii) But $V_m D^m v(x)$, for any $x \in \bar{\Omega}$, is nothing else than the remainder term in (2a), so that we immediately get, by applying Hölder's inequality (for sums) under the integral sign, the "sharp" bounds

$$|V_m D^m v(x)| \le \frac{\|x-a\|_1^m}{m!} |v|_m \le \frac{\|x-a\|^m n^{m/2}}{m!} |v|_m, \ \forall v \in C^m(\overline{\Omega}),$$

$$\forall x \in \overline{\Omega},$$

where $\|\cdot\|_1$ (resp. $\|\cdot\|$) denotes the usual 1-norm (resp. Euclidean norm) over $\mathbb{R}^n$ ($\|\cdot\|_1 \le n^{m/2} \|\cdot\|$ by the Cauchy-Schwarz inequality) and

$$|v|_m := \max_{|\alpha|=m} \max_{x \in \overline{\Omega}} |\partial^\alpha v(x)|$$

is the *Chebyshev* m-seminorm over $C^m(\overline{\Omega})$.

Hence the conclusion: by combining the bounds obtained in (ii) and (iii), and since $\int_\Omega dx = \sum_i w_i$ in view of assumptions (d) and (e), it readily follows that, for example,

$$|Rv| \le 2Sn^{m/2}(h^m|v|_m/m!), \ \forall v \in C^m(\overline{\Omega}), \tag{4}$$

where h and S denote the Euclidean diameter and the Lebesgue measure of Ω, respectively.

It must be emphasized that this quite satisfactory result was achieved by resorting only to most elementary means!

2.1.3. Hints for obtaining sharper bounds.

(i) Among various *possible refinements* of (4), we will mention here the following two ones:

$$|Rv| \le \frac{m+2n}{m+n} Sn^{m/2} \frac{\sup_{z \in \partial\Omega} \|z-a\|^m}{h^m} \left(\frac{h^m |v|_m}{m!}\right),$$

which actually follows from the remarkable (though hardly known!) identity (proved simply by changing to spherical polar coordinates)

$$\int_\Omega \|x-a\|^m dx = (m+n)^{-1} \int_{\partial\Omega} \|z-a\|^{m+n} dw_a(z). \tag{5a}$$

and its specialization (for m=0)

$$S = n^{-1} \int_{\partial\Omega} \|z-a\|^n \, dw_a(z), \tag{5b}$$

where $dw_a(z)$ stands for the elementary solid angle or element of hypersurface area on the unit $(n-1)$-dimensional sphere centered at the point a.

- provided Ω is *convex*, then

$$|Rv| \le \frac{Sh_1^m}{2^{m-1} m!} \, |v|_m \le \frac{Sn^{m/2}}{2^{m-1}} \left(\frac{h^m |v|_m}{m!}\right), \ \forall v \in C^m(\bar{\Omega}), \tag{6}$$

where $h_1 := \max \|x-y\|_1$ for $x, y \in \bar{\Omega}$; this follows from the elementary inequalities

$$\|x-a\| \le \|x-a\|_1 \le h_1/2 \le n^{1/2} h/2, \ \forall x \in \bar{\Omega}, \tag{7}$$

which indeed hold valid provided the reference point a is fixed at the center of the smallest parallelepiped that is circumscribed to $\bar{\Omega}$ and whose faces are parallel to those of the unit 1-ball of $\mathbb{R}^n$ (this point a necessarily belongs to $\bar{\Omega}$, since Ω is convex, and is "optimal" in the sense that it minimizes the expression $\max\|x-a\|_1$ for $x \in \bar{\Omega}$).

(ii) An *alternative direct sum decomposition* of $C^m(\bar{\Omega})$ is particularly worth mentioning here, namely the one corresponding to the definition of the term $P_{m-1}v$ in (3) as the Lagrange P_{m-1}-*interpolant of* $v \in C^m(\bar{\Omega})$ *on* a P_{m-1}-unisolvent subset of $\bar{\Omega}$ (consisting of $\binom{m+n-1}{n}$ distinct points with respect to which $\bar{\Omega}$ is star-shaped). In the one-dimensional case (i.e., $n=1$ and $\bar{\Omega}$ is a compact interval, say $[\alpha, \beta]$), Descloux [9] proved (in 1973!) that

$$\sup_{\substack{v \in C^m(\bar{\Omega}) \\ |v|_m = 1}} \ \inf_{p \in P_{m-1}} \|v-p\|_\infty = (\beta-\alpha)^m/(2^{2m-1} m!),$$

the supremun being attained for the (suitably normalized) Chebyshev polynomial $T_m[(2x-\alpha-\beta)/(\beta-\alpha)]$ whose zeros can therefore be regarded as "optimal" interpolation nodes. Since this elegant result still holds for v ranging over the set of "plane wave" polynomials (parallel to faces of the unit 1-ball of $\mathbb{R}^n$), it proves surprisingly easy to obtain lower bounds of practical value (see [14], p. 102) in the convex multivariate case. As a matter of fact, the final conclusion for Ω *convex* is the two-sided estimate

$$\frac{h^m}{2^{2m-1}m!} \le \frac{h_1^m}{2^{2m-1}m!} \le \sup_{\substack{v\in C^m(\bar{\Omega})\\ |v|_m=1}} \inf_{p\in P_{m-1}} \|v-p\|_\infty$$

$$\le \frac{h_1^m}{2^m m!} \le \frac{n^{m/2}h^m}{2^m m!},$$

from which it follows (for example) that the first upper bound in (6) is realistic in the sense that it cannot be improved by a factor 2^{m-1}.

2.2 A Structural Description of Our Method For Quantitative Error Estimation

2.2.1. The abstract problem. As analyzed at length in [14, 15, 16, 20], we are basically concerned in this matter with *inequalities* of the form

$$\|Rv\|_Z \le c\|Uv\|_Y, \ \forall v \in X, \tag{8}$$

where X, Y and Z are given linear spaces (over the real or complex field), $\|\cdot\|_Y$ and $\|\cdot\|_Z$ denote norms given respectively on Y and Z, U: X → Y is a given *linear surjection* (i.e., Y: = U(X)) and R: X → Z is a given *linear mapping* verifying the assumption

(H) R is continuous with respect to U.

This condition is natural: it simply amounts to requiring that (8) hold with c denoting some *finite* constant, so as to yield some genuine appraisal of Rv (regarded as *unknown*) in terms of Uv (supposed to be *known*). From the ensuing inclusion relation:

kernel U ⊂ kernel R,

it follows (by a classical *factorization theorem*) that

$$R = QU,$$

where $Q: Y \to Z$ denotes some uniquely defined linear mapping. The problem here is to find *realistic upper bounds* c for the *theoretical error coefficient*

$$c_o := \|Q\| = \sup_{\substack{v \in X \\ Uv \neq 0}} \|Rv\|_Z / \|Uv\|_Y = \min \{c \text{ verifying } (8)\};$$

this has clearly a great practical interest, specially perhaps in (linear) numerical analysis and approximation theory.

2.2.2. The abstract method. It consists of the following three stages:

(i) Select a representation formula in X of the structural form

$$v = Pv + VUv, \ \forall v \in X,$$

where $P: X \to X$ is a linear projector of X onto kernel U, or equivalently, $V: Y \to X$ is a linear right inverse of the linear surjection U. We are thus led eventually to the following *commutative diagram:*

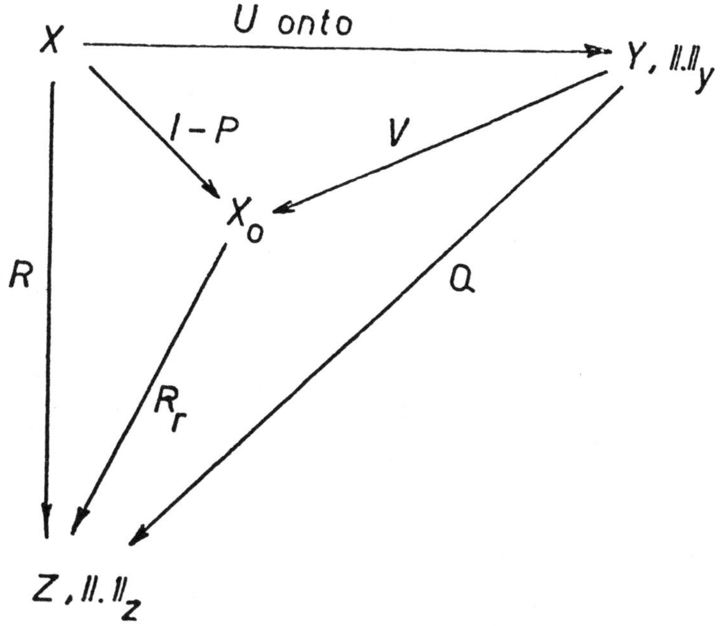

Figure 1

where I is the identity mapping and X_o denotes the linear subspace $VU(X)$ of X, so that

$$X = \text{kernel } U \oplus X_o ,$$

while R_r is the *restriction* of R to X_o, so that

$$W = R_r V;$$

strictly speaking, the notation V means here the bijection canonically associated with the selected right inverse of U, and similarly, $I-P$ is interpreted as a surjection of X onto X_o.

(ii) Decompose the linear mapping $R_r : X_o \to Z$ into a finite sum of *standard* linear mappings $S_j : X_o \to Z$ such that *each composite* $S_j V : Y \to Z$ *is continuous* (typically, S_j is a certain partial derivative, possibly composed with pointwise multiplication by given functions or integration with respect to given measures). There may exist several possibilities in matter of "interesting" decompositions of R_r, in which cases stage (iii) leads naturally to different upper bounds for c_o; as illustrated in Section 2.3, this potential versatility of the method may prove surprisingly useful.

(iii) Determine a (realistic) upper bound d_j for every $\|S_j V\|$, either directly or by making use of suitable *key estimates*.

The ensuing quantitative result

$$\|Rv\|_Z \le (\sum_j d_j) \|Uv\|_Y , \; \forall v \in X, \tag{9}$$

is thus finally achieved without ever trying to evaluate directly theoretical quantities such as $\|R\|$ (defined only if X is equipped with a norm!) or $\|RV\|$ (which is nothing else than c_o, since $R = QU$ and $UV = I$ necessarily imply $Q = RV$). This essential simplification should look surprising to the readers used to the Bramble-Hilbert type presentation of the requested inequality (8) as a consequence of a prior inequality, to wit: $\|Rv\|_Z \le C\|v\|_X$, $\forall v \in X$ (with the norm $\|\cdot\|_X$).

2.2.3. Typical examples of key estimates in $C^m(\overline{\Omega})$.
Particularly significant are the following key estimates:

$$|\partial^\alpha(V_m D^m v)(x) \leq \frac{\|x-a\|_1^{m-|\alpha|}}{(m-|\alpha|)!} |v|_m, \ \forall v \in C^m(\overline{\Omega}), \ \forall x \in \overline{\Omega}, \quad (10a)$$

$$\|\partial^\alpha(V_m D^m v)\|_\infty \leq |V_m D^m v|_{|\alpha|}$$

$$\leq h_1^{m-|\alpha|} |v|_m/(m-|\alpha|)!, \ \forall v \in C^m(\overline{\Omega}), \quad (10b)$$

which hold for every multi-index $\alpha := (\alpha_1, \ldots, \alpha_n)$ or order $\leq m$, provided only that $\overline{\Omega}$ is bounded and star-shaped with respect to the point a. These results readily follow (by applying again Hölder's inequality for sums under the integral sign) from the representation formula in the vector space $\partial^\alpha(C_a^m(\overline{\Omega}))$ for $|\alpha| \leq m-1$:

$$\partial^\alpha(V_m D^m v)(x) = \int_0^1 \frac{(1-t)^{m-|\alpha|-1}}{(m-|\alpha|-1)!} D^{m-|\alpha|}\partial^\alpha v(a+t(x-a)) \quad (11a)$$

$$\cdot (x-a)^{m-|\alpha|} dt,$$

which itself can be regarded as a simple corollary of the well known *commutativity relations*

$$\partial^\alpha(P_{m-1}v) = P_{m-|\alpha|-1}(\partial^\alpha v) \text{ for every } \alpha \text{ such that } |\alpha| \leq m-1, \quad (11b)$$

which hold for the Taylor polynomial $P_{m-1}v$ (i.e., the first term on the right in (2a)).

2.3 Versatility Can Prove Most Useful!

2.3.1. <u>A very significant illustration: Courant's triangle.</u>
Let $\Omega \subset \mathbb{R}^2$ denote the (interior of a nondegenerate) triangle of vertices x^i (i=1,2,3), with angles

$$\alpha_1 \leq \alpha_2 \leq \alpha_3$$

(so that always $\alpha_1 \leq \pi/3$, $\alpha_2 < \pi/2$, $\alpha_3 \geq \pi/3$). Let Tv denote the P_1-interpolant of $v \in C^2(\overline{\Omega})$ at the vertices, i.e. the affine function of x_1, x_2:

$$(Tv)(x) := \sum_{i=1}^{3} v(x^i)p_i(x), \ \forall x = (x_1,x_2) \in \mathbb{R}^2, \tag{12}$$

where $p_i(x)$ denotes the i-th barycentric coordinate of x with respect to the three points x^i.

As regards the corresponding interpolation error $v-Tv$, various realistic upper bounds can be obtained by applying the method described above. For brevity, we will concentrate here on the following two (novel) results:

$$|v-Tv|_1 \leq \begin{cases} (0.908h^2/r)|v|_2, \forall v \in C^2(\bar{\Omega}), \\ (0.908h/tg(\alpha_1/2))|v|_2, \forall v \in C^2(\bar{\Omega}), \\ (1.37h/\sin \alpha_1)|v|_2, \forall v \in C^2(\bar{\Omega}), \end{cases} \tag{13}$$

$$|v-Tv|_1 \leq (2h/\cos(\alpha_3/2))|v|_2, \forall v \in C^2(\bar{\Omega}), \tag{14}$$

where h denotes the greatest side (or Euclidean diameter) of the triangle $\bar{\Omega}$ and r is the Euclidean diameter of the inscribed circle of $\bar{\Omega}$. It must be emphasized here that these two results are essentially different. As a matter of fact, it trivially follows from (13) that

$$|v-Tv|_1 \to 0, \quad \text{as} \quad h \to 0,$$

whenever the *classical minimum angle condition* (due to Zlámal, 1968), viz.,

$$\alpha_1 \text{ bounded above } 0, \tag{15}$$

is satisfied; however, in view of (14), it is clear (though perhaps not sufficiently known) that *Zlámal's condition is not necessary for that convergence to hold;* indeed, what is essential here is the surprisingly different condition:

$$\alpha_3 \text{ bounded below } \pi, \tag{16}$$

or equivalently, α_2 bounded above 0, that is *Synge's maximal angle condition* (1957!).

Estimation problems of this type arise most naturally in connection with the *nodal finite element method*, where indeed piecewise polynomial interpolants over triangulated domains play a leading role. Of course, for the sake of simplicity, we will consider here only the finite element called *Courant's triangle*; in this context, the above comments on (13) and (14) can be reformulated as follows: whereas *arbitrarily thin triangles* always violate Zlámal's condition, they may still be used without spoiling the rate of convergence provided their second smallest angle exceed some positive constant (the "historical" reference here is [27, see p. 211); this may prove useful in practice for the local (i.e., elementwise) approximation of functions changing more rapidly in one direction than in another direction. In actual fact, this type of result concerning the dependence of the interpolation error estimates upon the geometry of the finite element can be extended to various more sophisticated situations (see specially [13] and [2]).

2.3.2. Proof of (13). From the trivial identity

$$v-Tv = (V_2D^2v) - \sum_{i=1}^{3} (V_2D^2v)(x^i)p_i, \tag{17}$$

which holds since $v-Tv = 0$ whenever $v \in P_1$, it follows that (for j=1,2)

$$|\partial_j(v-Tv)(x)| \le |\partial_j(V_2D^2v)(x)| + \sum_{i=1}^{3} |\partial_j p_i(x)||V_2D^2v(x^i)|,$$

$$\forall v \in C^2(\overline{\Omega}),\ \forall x \in \overline{\Omega}.$$

The above *key estimates* (10a) then yield

$$|\partial_j(v-Tv)(x)| \le \{\|x-a\|_1 + (\max_{i=1,2,3} \|x^i-a\|_1^2/2)$$

$$\sum_{i=1}^{3} |\partial_j p_i(x)|\}|v|_2,$$

and since (by elementary geometry!)

$$|\partial_j p_i(x)| \le \|D^1p_i(x)\| = 1/(\text{height of } x^i)$$

with

$$\sum_{i=1}^{3} (\text{height of } x^i)^{-1} = 2/r,$$

it follows that, *for any point* a *fixed in the triangle* $\bar{\Omega}$,

$$|\partial_j(v-Tv)(x)| \leq \{\|x-a\|_1 + \max_{i=1,2,3} \|x^i-a\|_1^2/r\}|v|_2,$$

$$\forall v \in C^2(\bar{\Omega}),\ \forall x \in \bar{\Omega};$$

this is simply a concrete example of (9) (more precisely: $Rv: = \partial_j(v-Tv)$, $U: = D^2$, $\sum_j d_j: = \{\cdot\}$, ...). Now by choosing the reference point a in the "optimal" way described in Section 2.1.3 (at the end of (i)), and by using the inequalities (7) which are then valid, we immediately get the (uniform) error bound:

$$|v-Tv|_1 \leq [h/2^{1/2} + h^2/(2r)]|v|_2,\ \forall v \in C^2(\bar{\Omega}). \tag{18}$$

But it can be proved (by classical trigonometry!) that, for any nondegenerate triangle, the following inequalities hold:

$$3^{1/2} \leq \frac{1+\sin(\alpha_1/2)}{\sin \alpha_1} \leq \frac{h}{r} \leq \operatorname{cotg}(\alpha_1/2); \tag{19}$$

unlike formerly used inequalities (see e.g. [7], p. 185), the present ones are sharp: equality holds indeed, in the left inequality if and only if the triangle is equilateral, in the middle inequality if and only if $\alpha_2 = \alpha_3$, in the right inequality if and only if $\alpha_1 = \alpha_2$. To complete the proof of (13), it remains only to combine (18) and (19), and to note that $(1/2 + 1/6^{1/2}) = 0.908...$ and $(3^{1/2} + 1)/2 = 1.366...$ Compared with our former results, (13) represents a quite *significant improvement*: indeed, as regards the first error bound in (13), the error coefficient given in [14, 20] was only $(1/2 + 1/2^{1/2}) = 1.207...$

2.3.3. Proof of (14). In view of the well known *geometrical interpretation of barycentric* (or "areal") *coordinates*, it is clear that

$$\partial_{\ell_1} p_1(x) = 0,$$

$$\partial_{\ell_1} p_2(x) = -\partial_{\ell_1} p_3(x) = -1/\,\|x^3-x^2\|\,,$$

where

$$\partial_{\ell_1} := D^1\cdot\ell_1 = (\cos\phi_1)\partial_1 + (\sin\phi_1)\partial_2$$

denotes the *directional derivative along the* (oriented) *edge opposite to the vertex* x^1 or, equivalently, the orthogonal projection of grad on that side (of unit tangential vector $\ell_1 := (\cos\phi_1, \sin\phi_1)$). Needless to say, similar results hold for ∂_{ℓ_2} and ∂_{ℓ_3}.

Then the trivial identity (17) immediately yields

$$\partial_{\ell_1}(v-Tv)(x) = \partial_{\ell_1}(V_2D^2v)(x) - \frac{V_2D^2v(x^3) - V_2D^2v(x^2)}{\|x^3-x^2\|},$$

where the last term can actually be rewritten in the form:

$$\frac{V_2D^2v(x^3) - V_2D^2v(x^2)}{\|x^3-x^2\|} = \int_0^1 \partial_{\ell_1}(V_2D^2v)(x^2+s(x^3-x^2))ds.$$

Now the (Taylor-like) representation formula in the vector space $\partial_{\ell_1}(C_a^2(\overline{\Omega}))$, viz.,

$$\partial_{\ell_1}(V_2D^2v)(x) = \int_0^1 D^1\partial_{\ell_1} v(a+t(x-a))\cdot(x-a)dt,$$

which like (11a) simply follows from the commutativity relations (11b), yields in turn the nice *key estimate*

$$|\partial_{\ell_1}(V_2D^2v)(x)| \le \|x-a\|_1\, |\partial_{\ell_1} v|_1,\ \forall v \in C^2(\overline{\Omega}),\ \forall x \in \overline{\Omega},$$

and accordingly (by the triangle inequality)

$$|\partial_{\ell_1}(V_2 D^2 v)(x^2+s(x^3-x^2))| \leq \{(1-s)\ ||x^2-a||_1 + s\ ||x^3-a||_1\}\ |\partial_{\ell_1} v|_1 \quad \text{for} \quad 0 \leq s \leq 1.$$

Hence it follows that, *for any point* a *fixed in* $\overline{\Omega}$,

$$|\partial_{\ell_1}(v-Tv)(x)| \leq \{\ ||x-a||_1 + (\ ||x^2-a||_1 + ||x^3-a||_1)/2\}|\partial_{\ell_1} v|_1,$$
$$\forall v \in C^2(\overline{\Omega}),\ \forall x \in \overline{\Omega},$$

which again is a concrete example of (9). This clearly suggests to choose the reference point a in the same *optimal* way as above, in which case we get the elegant (uniform) error bounds (j=1,2,3)

$$|\partial_{\ell_j}(v-Tv)|_0 \leq h_1|\partial_{\ell_j} v|_1 \leq 2^{1/2}\ ||\ell_j||_1\, h|v|_2 \leq 2h|v|_2, \quad \forall v \in C^2(\overline{\Omega}). \tag{20}$$

As a matter of fact, the following two properties of Courant's triangle:

(a) $Tv = v$ for all $v \in P_1$,

(b) $\partial_{\ell_j}(Tv)(x) = 0$ for all $x \in \overline{\Omega}$ if $\partial_{\ell_j} v(x) = 0$ for all $x \in \overline{\Omega}$, with $v \in C^2(\overline{\Omega})$ and $j=1,2,3$,

provide a concrete illustration of the general conditions found by Jamet (see e.g. [13]) for the replacement of the classical regularity conditions (of the Zlámal type) by less stringent ones (of the Synge type). It should be noted that the special property (b) trivially follows from the first inequality in (20).

But, from the very definition of the directional derivatives ∂_{ℓ_j} (j=1,2,3), it follows that

$$\partial_1 = \frac{(\sin\phi_2)\partial_{\ell_1} - (\sin\phi_1)\partial_{\ell_2}}{\sin\alpha_3},$$

since indeed $\phi_2 - \phi_1 = \pi - \alpha_3 \pmod{2\pi}$, plus two similar expressions (with analogues for ∂_2), so that (by using (20) and Hölder's inequality for sums)

$$|\partial_1(v-Tv)|_0 \le \frac{|\sin\phi_2| + |\sin\phi_1|}{\sin\alpha_3} 2h|v|_2 \le (4/\sin\alpha_3)h|v|_2$$

(with analogues for ∂_2 and for the other two angles α_1, α_2). In view of the remarkable identity

$$\max_{j=1,2,3} (\sin\alpha_j) = \sin(\max_{j=1,2,3} \alpha_j),$$

which actually holds for every triangle (whether $\alpha_3 \le \pi/2$ or not!), we are finally led to the (relatively rough) result

$$|v-Tv|_1 \le (4/\sin\alpha_3)h|v|_2, \forall v \in C^2(\overline{\Omega}). \tag{21}$$

The proof of the sharper result (14) apparently requires to take into account the whole information at hand, namely the fact (implied by (20)) that

$$|D^1(v-Tv)(x) \cdot \ell_j| \le 2h|v|_2, \forall v \in C^2(\overline{\Omega}), \forall x \in \overline{\Omega}, \tag{22}$$

holds valid for all j, *and not only (as is suggested in Jamet's work) for* 2 *values of* j. By setting

$$D^1(v-Tv)(x) = \|D^1(v-Tv)(x)\| \;(\cos\psi, \sin\psi),$$

the constraints (22) can be rewritten in the form

$$\|D^1(v-Tv)(x)\| \;|\cos(\psi-\phi_j)| \le 2h|v|_2, \forall\psi \text{ and } j=1,2,3,$$

or equivalently,

$$\|D^1(v-Tv)(x)\| \le \frac{2}{\min_\psi \max_{j=1,2,3} |\cos(\psi-\phi_j)|} h|v|_2, \tag{23}$$
$$\forall v \in C^2(\overline{\Omega}), \forall x \in \overline{\Omega}.$$

It turns out that the denominator on the right in (23) reduces to $\cos(\alpha_3/2)$, which actually completes the proof of (14).

Although this nice result can be established analytically, Synge's geometrical approach (see [27], p. 211) is by far simpler. This most ingenious proof amounts essentially to interpreting (20) as a bound for the Euclidean length of $\operatorname{grad}(v-Tv)(x)$ projected orthogonally onto the side opposite to the vertex x^j, and this for $j=1,2,3$ and for any x fixed in the triangle $\overline{\Omega}$; then the extremity of the vector $\operatorname{grad}(v-Tv)(x)$ drawn from the center of a circle of radius $2h|v|_2$ cannot lie outside the circumscribed hexagon whose sides are perpendicular to the sides of $\overline{\Omega}$; but it is readily verified (for example, by drawing a picture) that the vertices of the hexagon are at distances $2h|v|_2/\cos(\alpha_j/2)$ from the center, so that (14) finally follows. It should be realized that, strictly speaking, Synge's result (see formula (3.815) in [27]) is actually the following:

$$|v-Tv|_1 \le (1/\cos(\alpha_3/2))h \max_{x\in\overline{\Omega}} \|D^2v(x)\|, \ \forall v \in C^2(\overline{\Omega}); \quad (24)$$

since the elementary inequality

$$\max_{x\in\overline{\Omega}} \|D^2v(x)\| \le 2|v|_2, \ \forall v \in C^2(\overline{\Omega}),$$

is sharp, (14) is just as realistic as (24).

3. POINTWISE AND MEAN-SQUARE APPROXIMATION IN $H^m(\Omega)$

We will summarize here (without proof) the main results of the relevant research work which is presented in detail in [19].

3.1 A Typical Example of Application: Courant's Triangle

3.1.1. <u>The problem</u>. Find a (realistic) estimate of the *theoretical error coefficients* $c_0(x)$ and d_0 associated with the respective inequalities

$$|(v-Tv)(x)| \le c_0(x)|v|_2, \ \forall v \in H^2(\Omega), \ \forall x \in \overline{\Omega}, \quad (25)$$

$$|v-Tv|_1 \le d_0|v|_2, \ \forall v \in H^2(\Omega), \quad (26)$$

it being understood that:

- just as in Section 2.3, $\Omega \subset \mathbb{R}^2$ is a triangular domain and Tv, for any v in the classical Sobolev space $H^2(\Omega) \subset C^0(\overline{\Omega})$, is the P_1-interpolant of v at the vertices x^i $(i=1,2,3)$.

- throughout Section 3, a notation like $|\cdot|_j$ is to be interpreted as the (rotation invariant) *Sobolev* j-seminorm over $H^m(\Omega)$ for $0 \leq j \leq m$ and $\Omega \subset \mathbb{R}^n$, viz., .

$$|v|_j: = \{ \sum_{i_1,\ldots,i_j=1}^{n} \int_\Omega |\partial^j v(x)/\partial x_{i_1} \ldots \partial x_{i_j}|^2 dx\}^{1/2}, \quad (27)$$

of kernel P_{j-1} (if $j \geq 1$ and if Ω is connected).

3.1.2. Theoretical results. By using the *Sobolev imbedding theorem* and the *Rellich compactness lemma*, it is easy to prove that the vector space

$$V: = (I-T)(H^2(\Omega)) = \{v \in H^2(\Omega): v(x^i) = 0 \quad \text{for} \quad i=1,2,3\}, \quad (28)$$

with the norm $|\cdot|_2$, is a *functional Hilbert subspace of* $C^0(\overline{\Omega})$ (equipped with the topology of uniform convergence). Hence the basic *representation formula in* $H^2(\Omega)$:

$$v(x) = (Tv)(x) + (v,k_x)_2, \forall v \in H^2(\Omega), \forall x \in \Omega, \quad (29)$$

where $k_x \in V$ denotes the (unique) *Fréchet-Riesz representer of the 2-dimensional Dirac measure* δ_x *on* V. It turns out that k_x, for each $x \in \Omega$, can be interpreted as the (unique) *classical* solution in $H^2(\Omega)$ of the (distributional) *partial differential equation*

$$\Delta^2 k_x = \delta_x \quad \text{over} \quad \Omega,$$

that satisfies the *essential boundary conditions*

$$k_x(x^i) = 0 \quad \text{for} \quad i=1,2,3,$$

and the *natural boundary conditions*

$$\partial_\nu(\Delta k_x) + \partial_{\nu\tau\tau}k_x = 0 \quad \text{on} \quad \partial\Omega - \{x^1,x^2,x^3\},$$

$$\partial_{\nu\nu}k_x = 0 \quad \text{on} \quad \partial\Omega - \{x^1,x^2,x^3\},$$

where τ (resp. ν) denotes the unit tangential vector (resp. unit outer normal vector) existing everywhere (except at the vertices) along $\partial\Omega$ oriented in the usual way. Since the *reproducing kernel* of V, which is the function k on $\Omega \times \Omega$ defined by $k(x,y) := k_x(y) = (k_x,k_y)_2$, is uniformly continuous on $\Omega \times \Omega$, it can be extended by continuity to $\overline{\Omega} \times \overline{\Omega}$; hence the final result

$$c_o(x) = |k_x|_2 = [k(x,x)]^{1/2}, \ \forall x \in \overline{\Omega}, \tag{30}$$

which immediately follows from (29), by Schwarz's inequality.

As regards the error coefficient d_o, it is clear from (26) that

$$d_o := \sup_{v\in V, |v|_2\leq 1} |v|_1. \tag{31}$$

It can be proved here (from the Rellich compactness lemma and the *Banach-Alaoglu theorem*, together with the fact that V is a Hilbert space) that there must exist in V a function of unit norm $|\cdot|_2$ for which the supremum is attained. Finding d_o thus amounts to solving a variational problem of Rayleigh-Ritz type (or equivalently, of Galerkin type), $1/d_o^2$ being in fact the *smallest positive eigenvalue* for the (self-adjoint, positive definite) *eigenvalue problem* (to be interpreted in the *classical* sense):

$$\Delta^2 v + \lambda\Delta v = 0 \quad \text{over} \quad \Omega,$$

$$v \in V,$$

$$\partial_\nu(\Delta v) + \partial_{\nu\tau\tau}v + \lambda\partial_\nu v = 0 \quad \text{on} \quad \partial\Omega - \{x^1,x^2,x^3\},$$

$$\partial_{\nu\nu}v = 0 \quad \text{on} \quad \partial\Omega - \{x^1,x^2,x^3\},$$

which, incidentally, is the *buckling problem* for a triangular plate fixed only at the vertices.

In spite of the extreme simplicity of Courant's triangle among the potential applications, these complete characterizations of the theoretical error coefficients $c_o(x)$ and d_o, from the respective solutions of "classical" boundary value and eigenvalue problems, prove to be so complicated that there is no hope whatever to derive for them closed form expressions. Hence a strongly motivated need for methods of truly practical value for finding realistic (upper and lower) bounds for such theoretical quantities.

3.1.3. Practical results.

(a) Finding lower bounds is easy! Indeed, from the definition (31) of d_o, it is clear that the *Rayleigh quotient* $|v-Tv|_1/|v|_2$, for any particular $v \in H^2(\Omega)$ such that $|v|_2 \neq 0$, must yield a lower bound for d_o. Consider, by way of example, the *standard rectangular triangle* (of vertices (0,0), (1,0) and (0,1)). It is not very difficult to prove (by explicit computation) that

$$\sup_{v \in P_2, |v|_2 \neq 0} |v-Tv|_1^2/|v|_2^2 = (3+5^{1/2})/24 = 0.218...,$$

the supremum being attained for

$$v(x) := x_1 x_2 - ((5^{1/2}-1)/4)(x_1^2+x_2^2);$$

hence it follows that

$$d_o^2 > 0.218, \quad d_o > 0.467.$$

Needless to say, a lower bound for $c_o(x)$ can be obtained similarly from the alternative definition

$$c_o(x) = \sup_{v \in H^2(\Omega), |v|_2 \neq 0} |(v-Tv)(x)|/|v|_2 \quad \text{for each} \quad x \in \bar{\Omega}.$$

(b) Finding upper bounds is quite feasible! From the trivial identity (17), which holds here for all $v \in H^2(\Omega)$, it immediately follows that, for example,

$$|(v-Tv)(x)| \leq |V_2D^2v(x)| + \sum_{i=1}^{3} |V_2D^2v(x^i)||p_i(x)|, \tag{32a}$$

$$\forall v \in H^2(\Omega),\ \forall x \in \overline{\Omega},$$

$$|v-Tv|_1 \leq |V_2D^2v|_1 + \sum_{i=1}^{3} |V_2D^2v(x^i)||p_i|_1,\ \forall v \in H^2(\Omega). \tag{32b}$$

The following *key estimates*

$$|V_2D^2v(x)| \leq 12^{-1/2}S^{-1/2} \max_{i=1,2,3} \|x-x^i\|^2\, |v|_2,\ \forall x \in \overline{\Omega}, \tag{33a}$$

where S denotes the Lebesgue measure of the triangle $\overline{\Omega}$, and

$$|V_2D^2v|_1 \leq h|v|_2, \tag{33b}$$

where h denotes the greatest side of $\overline{\Omega}$, are known to hold for all $v \in H^2(\Omega)$ (see Theorem 2, formulas (38a,b), and Theorem 3, formulas (40b,c), respectively). Now the three barycentric coordinates $p_i(x)$ of any $x \in \overline{\Omega}$ are nonnegative numbers of sum identically equal to 1, so that we immediately get from (32a,33a) the final result:

$$|(v-Tv)(x)| \leq 3^{-1/2}S^{-1/2}\{1 + \frac{\max_{i=1,2,3} \|x-x^i\|^2}{h^2}\} \frac{h^2|v|_2}{2!},$$

$$\forall v \in H^2(\Omega),\ \forall x \in \overline{\Omega},$$

whose comparison with (25) yields an upper bound of practical value for $c_o(x)$. On the other hand, it can be proved (by elementary geometry!) that

$$\sum_{i=1}^{3} |p_i|_1 = 2S^{1/2}/r,$$

where r denotes the Euclidean diameter of the inscribed sphere of $\overline{\Omega}$, so that (32b), (33a,b) and (19) finally yield the result

$$|v-Tv|_1 \le (3^{-1/2}+r/h)(h^2|v|_2/r) \le 2/3^{1/2}(h^2|v|_2/r), \qquad \forall v \in H^2(\Omega), \tag{34}$$

whose comparison with (26) gives an upper bound of practical value for d_0. Compared with former results, (34) represents a *significant improvement*: indeed, $2/3^{1/2} = 1.1547\ldots$, whereas the similar error coefficients given in [1], [15,16] and [19] were $3, 2^{3/2} = 2.8284\ldots$ and $(5/2)3^{-1/2} = 1.4433\ldots$, respectively. It may be interesting to note that, for the theoretical error coefficient associated with (34), the greatest possible lower bound that can be obtained by using only polynomials of exact degree 2 and for arbitrary isosceles triangles is rigorously 1/6.

3.2 Sharp Key Estimates in $H^m(\Omega)$ for Pointwise Approximation

3.2.1. A suitable representation formula in Sobolev spaces. Unlike other mathematicians also interested in *quantitative* error analysis, we have adopted once and for all a specially simple variant of *averaged Taylor series* as standard representation formula in $H^m(\Omega)$. Except in complicated geometrical situations, this rigid choice proves amply justified for reasons of practical convenience and compares favorably with more popular alternatives such as the *Kowalewski-Ciarlet-Wagschal formula* (see e.g. [1,8, 12]) and the more general *Sobolev representation formula* (advocated, for example, in [10,11]).

Theorem 1. *Let* Ω *be a bounded open convex set in* $\mathbf{R}^n$ *with Lebesgue measure* S. *If* $v \in H^m(\Omega)$ *(with* m *integer* ≥ 1*) and* $x \in \bar{\Omega}$, *then every distributional derivative of order* $< m$ *of* v, *say* $\partial^\alpha v$ *for* $\alpha := (\alpha_1,\ldots,\alpha_n) \in \mathbf{N}^n$ *and* $|\alpha| := \alpha_1 + \ldots + \alpha_n < m$, *can be expressed in the form*

$$\partial^\alpha v(x) \overset{\text{a.e.}}{=} P_{m-|\alpha|-1}(\partial^\alpha v)(x) + V_{m-|\alpha|}D^{m-|\alpha|}(\partial^\alpha v)(x), \tag{35a}$$

where

$$P_{m-|\alpha|-1}(\partial^\alpha v)(x) := S^{-1} \sum_{|\beta|=0}^{m-|\alpha|-1} \int_\Omega \frac{\partial^{\alpha+\beta} v(a)(x-a)^\beta}{\beta!}\, da \tag{35b}$$

is a polynomial (in n *variables) of degree* $\leq m-|\alpha|-1$ *and*

$$V_{m-|\alpha|}D^{m-|\alpha|}(\partial^\alpha v)(x) \tag{35c}$$

$$\overset{\text{a.e.}}{:=} \frac{S^{-1}}{(m-|\alpha|-1)!}\int_\Omega\int_0^1 (1-t)^{m-|\alpha|-1}D^{m-|\alpha|}(\partial^\alpha v)(a+t(x-a))$$

$$\cdot\ (x-a)^{m-|\alpha|}dt\ da.$$

Moreover, if

$$|\alpha| < m-n/2, \tag{36}$$

then $\partial^\alpha v$ *is uniformly continuous on* $\overline{\Omega}$ *and* (35a,c) *holds accordingly everywhere in* $\overline{\Omega}$.

This theorem essentially follows from the Taylor formula of the $(m-|\alpha|)$-th order (with integral expression of the remainder) about the point $a \in \overline{\Omega}$ and for every $\partial^\alpha v$ with $v \in C^m(\Omega)$, which can be readily deduced from (2a), by integration over with respect to the Lebesgue measure da. Indeed, $C^m(\overline{\Omega})$ is known to be dense in $H^m(\Omega)$ (equipped with the norm $\|\cdot\|_m := (|\cdot|_0^2 + \ldots + |\cdot|_m^2)^{1/2}$ for $|\cdot|_j$ defined by (27) and, according to Sobolev's imbedding theorem (see e.g. [6], p. 114), $H^m(\Omega)$ is contained (with a continuous injection) in $C^o(\overline{\Omega})$ if (36) is satisfied, in the Lebesgue space $L^1(\Omega)$ otherwise.

3.2.2. <u>The main result</u>. It consists of the following *key estimates*, whose detailed proof (given only in [19], pp. 268-270) shows that they are even *optimal* (in some *relative* sense).

<u>Theorem 2</u>. *Let* Ω *be a bounded open set in* $\mathbb{R}^n$, *with Lebesgue measure* S *and Euclidean diameter* h, *such that* $\overline{\Omega}$ *is star-shaped with respect to every point in a subset* X. *If* $v \in H^m(\Omega)$ *(with* m *integer* ≥ 1) *and* $x \in X$, *then we have, for every multi-index* $\alpha := (\alpha_1,\ldots,\alpha_n)$ *such that* $|\alpha| := \alpha_1 + \ldots + \alpha_n < m-n/2$, *the "sharp" bound:*

$$|\partial^\alpha(V_mD^mv)(x)| \tag{37}$$

$$\leq \frac{S^{-1}}{(m-|\alpha|)!} \{\frac{m-|\alpha|}{4(m-|\alpha|)^2-n^2} \int_{\partial\Omega} ||x-z||^{2(m-|\alpha|)+n} dw_x(z)\}^{1/2}$$

$$|\partial^\alpha v|_{m-|\alpha|} ,$$

where $dw_x(z)$ *means the elementary solid angle with vertex* x *and in the direction specified by the variable point* $z \in \partial\Omega$. *A reasonable simplification leads to the result of more practical value:*

$$|\partial^\alpha (V_m D^m v)(x)| \leq c(x) \frac{h^{m-j} |\partial^\alpha v|_{m-j}}{(m-j)!} , \tag{38a}$$

where $j := |\alpha| < m-n/2$ *and*

$$c(x) := \{\frac{(m-j)nS^{-1}}{4(m-j)^2-n^2}\}^{1/2} \frac{\sup_{z\in\partial\Omega} ||z-x||^{m-j}}{h^{m-j}} , \tag{38b}$$

from which the global estimate

$$||D^j (V_m D^m v)(x)|| \leq c(x) \frac{h^{m-j} |v|_m}{(m-j)!} \tag{38c}$$

directly follows.

The proof reveals the surprising fact that (37) essentially follows by *only one application of Schwarz's inequality in* $L^2(\Omega)$ from the integral representation (implied by Theorem 1)

$$(V_m D^m v)(x) = \frac{S^{-1}}{(m-1)!} \int_\Omega D^m v(y) \cdot K(x,y) dy,$$

where the (tensor-valued!) kernel $K(x,y)$ is defined as

$$K(x,y) := \{\int_0^{\tau_x(y)} (1-t)^{-n-1} dt\} (x-y)^m,$$

$\tau_x(y)$ denoting the altitude of the upper boundary surface of the $(n+1)$-dimensional cone in $\mathbb{R}^n \times \mathbf{R}$ with base $\Omega \times \{0\}$ and vertex $\{x\} \times \{1\}$. This explains in which precise sense (37) may be regarded as an *optimal* result. As illustrated in Section 3.1.2., results that would be optimal in the absolute sense are manifestly out of practical reach, since obtaining them would amount to solving extremely complicated boundary value problems. However, though $K(x, y)$ is not an m-th Fréchet derivative (of a so-called *Rodrigues function*), it can often be used to derive interesting *Peano-like integral representations for remainders* together with upper bounds of practical value.

3.3 Key Estimates in $H^m(\Omega)$ for Mean-square Approximation

Theorem 3. *Let* Ω *be a bounded open convex set in* $\mathbb{R}^n$ *with Lebesgue measure* S *and Euclidean diameter* h. *If* $v \in H^m(\Omega)$ *(with integer* ≥ 1), *then we have, for every multi-index* $\alpha := (\alpha_1,\ldots,\alpha_n)$ *such that* $|\alpha| := \alpha_1 + \ldots + \alpha_n < m$, *the following bound:*

$$|\partial^\alpha (V_m D^m v)|_0 \leq \frac{S^{-1/2}}{(m-|\alpha|)!} \{(m-|\alpha|) \sup_{x \in \Omega} \int_\Omega \|x-a\|^{2(m-|\alpha|)} da$$

$$\int_0^1 (1-t)^{m-|\alpha|-1} \min(t^{-n},(1-t)^{-n}) dt\}^{1/2} |\partial^\alpha v|_{m-|\alpha|} \,.$$

A reasonable simplification leads to the result of more practical value:

$$|\partial^\alpha (V_m D^m v)|_0 \leq d \, \frac{h^{m-j} |\partial^\alpha v|_{m-j}}{(m-j)!} \,, \tag{40a}$$

where $j := |\alpha| < m$ *and*

$$d := \{\frac{(m-j)n}{2(m-j)+n} \int_0^1 (1-t)^{m-j-1} \min(t^{-n},(1-t)^{-n}) dt\}^{1/2}, \tag{40b}$$

from which the global estimate

$$|V_m D^m v|_j \leq d \, \frac{h^{m-j} |v|_m}{(m-j)!} \tag{40c}$$

directly follows. However, for $j := |\alpha| < m-n/2$, *somewhat sharper bounds are known, for example,*

$$|V_m D^m v|_j \leq \left\{\frac{(m-j)n}{4(m-j)^2-n^2}\right\}^{1/2} \frac{h^{m-j}|v|_m}{(m-j)!} \qquad (41)$$

as it trivially results from Theorem 2.

It should be noted that, as regards the expected sharpness of these bounds, the situation is apparently less satisfactory than it was throughout Section 3.2.2., in so far as more than just one application of basic integral inequalities seems to be here indispensable.

REFERENCES

1. Arcangéli, R., et Gout, J. L., "Sur l'évaluation de l'erreur d'interpolation de Lagrange dans un ouvert de $\mathbb{R}^n$", R.A.I.R.O. Analyse numérique 10, 1976, pp. 5-27.

2. Babuška, I., and Aziz, A. K., "On the angle condition in the finite element method", SIAM J. Numer. Anal. 13, 1976, pp. 214-226.

3. Barnhill, R. E., and Gregory, J. A., "Interpolation remainder theory from Taylor expansions on triangles", Numer. Math. 25, 1976, pp. 401-408.

4. Bramble, J. H., and Hilbert, S. R., "Estimation of linear functionals on Sobolev spaces with application to Fourier transforms and spline interpolation", SIAM J. Numer. Anal. 7, 1970, pp. 112-124.

5. Bramble, J. H., and Hilbert, S. R., "Bounds for a class of linear functionals with applications to Hermite interpolation", Numer. Math. 16, 1971, pp. 362-369.

6. Ciarlet, P. G., "The finite element method for elliptic problems", Amsterdam: North-Holland Publishing Company, 1978.

7. Ciarlet, P. G., and Raviart, P. A., "General Lagrange and Hermite interpolation in $\mathbb{R}^n$ with applications to finite element methods", Arch. Rational Mech. Anal. 46, 1972, pp. 177-199.

8. Ciarlet, P. G., and Wagschal, C., "Multipoint Taylor formulas and applications to the finite element method", Numer. Math. 17, 1971, pp. 84-100.

9. Descloux, J., Personal Communication.

10. Dupont, T., and Scott, R., "Constructive polynomial approximation in Sobolev spaces", Recent Advances in Numerical Analysis (C. de Boor and G. Golub, eds.), pp. 31-44, New York: Academic Press, 1978.

11. Dupont, T., and Scott, R., "Polynomial approximation of functions in Sobolev spaces", Math. Comp. 34, 1980, pp. 441-463.

12. Gout, J. L., "Estimation de l'erreur d'interpolation d'Hermite dans $\mathbf{R}^n$", Numer. Math. 28, 1977, pp. 407-429.

13. Jamet, P., "Estimations d'erreur pour des éléments finis droits presque dégénérés", R.A.I.R.O. Analyse numérique 10, 1976, pp. 43-61.

14. Meinguet, J., "Realistic estimates for generic constants in multivariate pointwise approximation", Topics in Numerical Analysis II (J. J. H. Miller ed.),pp. 89-107, London: Academic Press, 1975.

15. Meinguet, J., "Structure et estimations de coefficients d'erreurs", R.A.I.R.O. Analyse numérique 11, 1977, pp. 355-368.

16. Meinguet, J., "A practical method for estimating approximation errors in Sobolev spaces", Multivariate Approximation (D. C. Handscomb ed.), pp. 169-187, London: Academic Press, 1978.

17. Meinguet, J., "A convolution approach to multivariate representation formulas", ISNM 51: Multivariate Approximation Theory (W. Schempp and K. Zeller eds.), pp. 198-210, Basel: Birkhäuser Verlag, 1979.

18. Meinguet, J., "From Dirac distributions to multivariate representation formulas", Approximation Theory and Applications (Z. Ziegler ed.), pp. 225-248, New York: Academic Press, 1981.

19. Meinguet, J., "Sharp "a priori" error bounds for polynomial approximation in Sobolev spaces", ISNM 61: Multivariate Approximation Theory II(W. Schempp and K. Zeller eds.), pp. 255-274, Basel: Birkhäuser Verlag, 1982.

20. Meinguet, J., and Descloux, J., "An operator-theoretical approach to error estimation", Numer. Math. 27, 1977, pp. 307-326.

21. Morrey, C. B, "Multiple integrals in the calculus of variations", New York: Springer-Verlag, 1966.

22. Natterer, F., "Berechenbare Fehlerschranken für die Methode der Finiten Elemente", ISNM 28: Finite Elemente und Differenzverfahren, pp. 109-121, Basel: Birkhäuser Verlag, 1975.

23. Sard, A., "Linear approximation", Providence: American Mathematical Society, 1963.

24. Sigillito, V. G., "Explicit "a priori" inequalities with applications to boundary value problems", London: Pitman Publishing, 1977.

25. Sobolev, S. L., "Applications of functional analysis in mathematical physics", Providence: American Mathematical Society, 1963.

26. Strang, G., and Fix, G. J., "An analysis of the finite element method", Englewood Cliffs: Prentice-Hall, 1973.

27. Synge, J. L., "The hypercircle in mathematical physics", New York: Cambridge University Press, 1957.

SURFACE SPLINE INTERPOLATION: BASIC THEORY AND COMPUTATIONAL ASPECTS

Jean Meinguet

Université Catholique de Louvain

ABSTRACT

One of the most efficient methods to date for global interpolation of scattered data has come to be called "surface spline interpolation". It turns out that the underlying mathematical theory has for natural setting some functional semi-Hilbert space whose reproducing kernel is known in closed form and can be computed economically. Solving the interpolation problem thus amounts to minimizing some Sobolev seminorm under interpolatory constraints or eventually, to solving a positive definite linear system. Our purpose here is to give a motivated and self-contained presentation of this interesting material.

1. INTRODUCTION

The problem of fitting appropriate surfaces to arbitrarily spaced data arises in many applications in science and technology. A variety of numerical methods have been devised for representing and constructing such surfaces. For a survey, see e.g. the "classical" references [1, 13] and the comprehensive technical report [4] (condensed in the recent paper [5]), which is essentially devoted to a critical comparison (in terms of timing, storage, accuracy, visual pleasantness, ease of implementation, etc.) of some 29 methods (either *local* or *global* in nature) for *interpolation of scattered data*.

It turns out that the particular method of *surface spline interpolation* (also known as *thin plate spline interpolation*)

S. P. Singh et al. (eds.), Approximation Theory and Spline Functions, 127–142.

we will here discuss thoroughly is not only one of the most efficient methods to date for *global* interpolation (see [4], p. 82), but also one of the few methods to derive from a truly elegant mathematical structure which is perfectly understood by now (see e.g. [3] and, for a deliberately more constructive approach, [7, 8, 9] and also [10, 11]). This paper is devoted to a motivated, *self-contained* presentation of that most interesting method. As regards the underlying mathematical theory (see Section 2), it can be said that a prominent role is played by the deep Theorem 1, which is given here (for the first time) a reasonably *simple, constructive* proof. Owing to this basic theorem, the interpolants to be considered must belong to a *functional semi-Hilbert space of continuous functions*, the *reproducing kernel* of which is *explicitly known* and involves no functions more complicated than logarithms (see Section 3). It follows that solving the interpolation problem amounts strictly to solving a Cramer system of linear equations with a Gram matrix (whose elements, just like the reproducing kernel itself, can be easily computed).

The solution of a linear algebraic system with a *positive definite symmetric* matrix of coefficients (which is precisely the case here) is a quite simple problem (at least in principle). Indeed, the classical *Cholesky algorithm* is *very economical* (in comparison with Gaussian elimination, with suitable pivotal strategy, for more general matrices) and, moreover, enjoys remarkable *numerical stability*. Though this most important fact has been known to numerical analysts for a long time, a detailed *rounding-error analysis of the Cholesky process in standard floating-point arithmetic* is essentially lacking in the current technical literature, the very recent paper [12] excepted. In Section 4, we will summarize (without proof) the main results of the two self-contained round-off analyses presented in [12], to wit: novel *a posteriori* error bounds (of practical use) and refined *a priori* results (of more theoretical significance, see also [14]).

For simplicity, all functions and vector spaces considered in this paper are real.

2. THE BASIC INTERPOLATION PROBLEM

Roughly speaking, the interpolation problem to be addressed here can be formulated as follows: *Given a finite set* $A := (a_i)_{i \in I}$ *of distinct points of* $\mathbb{R}^n$ *and associate (real) values* $(\alpha_i)_{i \in I}$, *construct a (continuous) function* $v: \mathbb{R}^n \to \mathbb{R}$ *so that* $v(a_i) = \alpha_i$, $i \in I$.

In order to force *well-posedness* (i.e., existence, uniqueness and stability of solutions of this problem), it proves quite natural to require in addition, from any candidate for solution, the minimization of some appropriate expression such as, typically, the quadratic seminorm $|\cdot|$ that is generated by the *rotation invariant* semi-inner product

$$(v, w) := \sum_{i_1,\dots,i_m=1}^{n} \int_{\mathbb{R}^n} \partial_{i_1\dots i_m} v(x)\, \partial_{i_1\dots i_m} w(x)\,dx \qquad (1)$$

where $\partial_{i_1\dots i_m} := \partial^m/\partial x_{i_1}\dots\partial x_{i_m}$; it should be noted that the kernel of this (Sobolev-like) seminorm $|\cdot|$ is simply the vector space P of dimension

$$M = \binom{m+n-1}{n} \qquad (2)$$

of all polynomials in n variables of (total) degree $\leq m - 1$. Needless to say, the introduction of such a characteristic property of an "optimal" solution of the interpolation problem is strongly motivated by the nice *intrinsic association between spline interpolation and orthogonal projection* that is known to hold in the univariate case (i.e., for $n = 1$). Moreover, it should be realized that $|v|^2$ may be physically interpreted (at least if $m = n = 2$ and under some simplifying assumptions) as the *bending energy of a thin plate of infinite extent*, v denoting then the deflection normal to the rest position (supposed of course to be plane). In view of these considerations, interpolants of minimum seminorm $|\cdot|$ can be appropriately termed *surface splines* (as proposed in [6], where this ingenious interpolating device was introduced).

By virtue of the well known *orthogonal projection theorem*, such an optimal interpolation problem is certainly well-posed if the (non-empty) set of interpolants can be regarded as a *closed* linear variety of a *Hilbert space* (of *norm* $|\cdot|$). It turns out that this assumption is fulfilled for the abstract setting provided by the *space of Beppo Levi type* (of order m over $\mathbf{R}^n$), which is simply the vector space

$$X := \{v \in \mathcal{D}' : \partial_{i_1\dots i_m} v \in L_2 \text{ for } i_1,\dots,i_m \in [1, n]\} \qquad (3)$$

equipped with the seminorm $|\cdot|$, $\mathcal{D}'$ denoting as usual the space of all the (Schwartz) distributions in $\mathbf{R}^n$ and all partial derivatives being interpreted in the distributional sense. Indeed we have the following (purely mathematical) result, which is partly reminiscent of the classical *Sobolev imbedding theorem* and which proves essential here.

Theorem 1. *Suppose*

$$m > n/2. \tag{4}$$

Then the semi-inner product space X, *defined by* (3) *with* (1), *is a functional semi-Hilbert subspace of the space* C^0 *of continuous functions in* $\mathbb{R}^n$ *(equipped with the topology of compact convergence).*

This means in other words that, always under assumption (4):

(a) every $v \in X$ is a bona fide *function*, defined and (even) *continuous everywhere* in $\mathbb{R}^n$ (so that C^0 may be substituted for D' in definition (3)).

(b) for any *direct sum decomposition*

$$X = P \oplus X_0 \tag{5}$$

that is *topological in* C^0 (i.e., such that the associated linear *projector* P of X onto P along X_0 is *continuous in the sense of the* C^0 *topology*), the (M-codimensional) subspace X_0 *is a Hilbert space for* $|\cdot|$.

(c) the natural injection of any such X_0 into the Fréchet space C^0 is continuous (i.e., *convergence in* X_0 *implies convergence in* C^0).

This most important theorem was given a first constructive proof in [7], from convenient *representation formulas* in X obtained by resorting to such basic mathematical tools as *convolutions* and *Fourier transforms* of distributions. The somewhat simpler *constructive proof* presented hereafter is based on the alternative representation formula we introduced in [10], to wit:

$$v = \sum_{i_1,\ldots,i_m=1}^{n} \partial_{i_1\ldots i_m}(\alpha e) * \partial_{i_1\ldots i_m} v \tag{6}$$
$$+ \Delta^m[(1 - \alpha)e] * v, \ \forall v \in X.$$

Here:

(a) e, the so-called *fundamental solution* of the *iterated (distributional) Laplacian* Δ^m *in* $\mathbb{R}^n$ (which simply means that $\Delta^m e$ is the n-dimensional Dirac distribution), is the locally integrable function defined in the complement of the origin in $\mathbb{R}^n$ by the formulas:

$$e(x) := \begin{cases} c|x|^{2m-n}\ell n|x|, & \text{if } 2m \geq n \text{ and } n \text{ is even}, \\ d|x|^{2m-n} & \text{otherwise}, \end{cases} \tag{7a}$$

with

$$c := \frac{(-1)^{n/2+1}}{2^{2m-1}\pi^{n/2}(m - 1)!(m - n/2)!}, \tag{7b}$$

$$d := \frac{(-1)^m\Gamma(n/2 - m)}{2^{2m}\pi^{n/2}(m - 1)!}, \tag{7c}$$

$|x|$ denoting as usual the radial coordinate (or Euclidean norm) of $x \in \mathbb{R}^n$.

(b) α denotes an arbitrarily selected *cutoff function*, that is, a function in D (i.e., the space of infinitely differentiable functions with compact support in $\mathbb{R}^n$) that is unity in some neighborhood of the origin.

(c) any expression of the form $v * w$ is to be interpreted as the *convolution of the distributions* v and w, which is the distribution uniquely defined (if existent at all!) by the equation

$$\langle v * w, \phi\rangle := \langle v_\xi, \langle w_\eta, \phi(\xi + \eta)\rangle\rangle, \ \forall\phi \in D,$$

where $\langle.,.\rangle$ denotes the *duality pairing* between dual topological vector spaces (such as, typically, D' and D); such a convolution is *commutative, associative* and *commutes with differentiation* (at least conditionally: for example, whenever v is compactly supported), and has for *unit* the Dirac distribution (or measure) δ.

In view of these facts (borrowed from the classical theory of distributions), it is rather easy to verify that:

(a) *each term of the sum in the representation formula* (6) *belongs to the Banach space* C_0 (i.e., the space of all uniformly continuous functions in $\mathbb{R}^n$ that vanish at infinity, equipped with the maximum norm) *and depends continuously on the m-th partial derivative of* v *it contains*, say $\partial_{i_1 \ldots i_m} v \in L_2$; indeed, such a term is the convolution of this L_2 function with $\partial_{i_1 \ldots i_m}(\alpha e) \equiv \alpha \partial_{i_1 \ldots i_m} e (\mathrm{mod}\ D)$, which is also a L_2 function whenever $m > n/2$ (as already mentioned in [10], the latter function is in fact $O(|x|^{m-n}|\ell n|x||)$, as $x \to 0$).

(b) *the last term* on the right *in* (6), say w, *is a* C^∞ *function in* $\mathbb{R}^n$; as a matter of fact, since $\Delta^m[(1-\alpha)e] \in D$, w can be regarded simply as a *regularization* of $v \in D'$.

(c) every m-th partial derivative of w belongs to C_0 and depends continuously on the same partial derivative of v (indeed, $\partial_{i_1 \ldots i_m} w$ is the convolution of $\Delta^m[(1-\alpha)e]$, to be regarded here as a L_2 function having compact support, with the L_2 function $\partial_{i_1 \ldots i_m} v$); it follows that *the remainder term in the Taylor formula of the m-th order for* w (about any particular point of $\mathbb{R}^n$), which only differs from w by a polynomial of degree $\leq m-1$, *is a* C^0 *function depending continuously on all the m-th partial derivatives of* v.

(d) *any Cauchy sequence in* (any admissible) X_0 *for* $|\cdot|$ *is a Cauchy sequence in the Fréchet space* C^0, *whose limit necessarily belongs to* X_0; indeed, all the m-th partial derivatives of such a C^0 limit must be in L_2, qua limits in D' of Cauchy sequences of L_2 functions. This essentially completes the proof of Theorem 1.

In view of this theorem, the optimal (or surface spline) interpolation problem to be analyzed and solved eventually, namely *Problem* (P), can be formulated precisely as follows: *Find* $u \in V$, *where* V *denotes the linear variety of all* X-*interpolants to the data* $(a_i, \alpha_i) \in \mathbb{R}^n \times \mathbb{R}$, $i \in I$, i.e.,

$$V: = \{v \in X: v(a_i) = \alpha_i, \forall i \in I\}, \tag{8}$$

such that

$$|u| = \inf_{v \in V} |v|, \tag{9}$$

it being always understood that $m > n/2$ *and that* $A: = (a_i)_{i \in I}$ *is a finite set of distinct points containing a P-unisolvent subset*, i.e., a set $B: = (a_j)_{j \in J}$ of M points of A such that there exists a unique $p \in P$ satisfying the interpolating conditions

$$p(a_j) = \alpha_j, \forall j \in J, \tag{10}$$

for arbitrarily prescribed (real) scalars α_j, $j \in J$ (the latter assumption is clearly necessary for u to be unique). We have then the following result.

Theorem 2. *If* $m > n/2$, *then Problem* (P) *is well-posed in the sense that its solution exists, is unique, and depends continuously on the data* (a_i, α_i), $i \in I$.

As a matter of fact, this can be regarded as a trivial consequence of Section 3, where indeed more precise results (important by themselves!) are established.

3. AN EFFICIENT METHOD FOR SOLUTION

Let p_i, $i \in J$, denote the unique solution in P of the interpolation problem (10) for $\alpha_j: = \delta_{ij}$, $\forall j \in J$, where δ_{ij} is the Kronecker symbol. Then, for every $v \in X$ and always *provided that* $m > n/2$,

$$Pv: = \sum_{i \in J} v(a_i)p_i \tag{11}$$

is the (uniquely defined) *P-interpolant of* $v \in X$ *on* B, expressed in the Lagrange form. The mapping $P: X \to X$ is clearly a linear projector of X onto *P* with kernel

$$X_0: = \{v \in X: v(a_j) = 0, \forall j \in J\}, \quad (12)$$

which is continuous in the sense of the C^o topology (so that the associated direct sum decomposition (5) of X is topological in C^o).

Equipped with the inner product (1), X_0 as defined by (12) is a *functional Hilbert space*, i.e., a Hilbert space whose elements are bona fide functions and which is such that, for each $x \in \mathbb{R}^n$, the *evaluation functional* $\delta_{(x)}: v \mapsto v(x)$ on X_0 is linear and bounded; this readily follows from Theorem 1, which states moreover that the elements of X are continuous functions in $\mathbb{R}^n$ (so that $\delta_{(x)}$ can be regarded simply as the Dirac measure at the point x). Hence the basic *representation formula in the space of Beppo Levi type* X *for* $m > n/2$:

$$v(x) = (Pv)(x) + (v, k_x), \forall v \in X, \forall x \in \mathbb{R}^n, \quad (13)$$

where $k_x \in X_0$ denotes the (necessarily unique) *Fréchet-Riesz representer* of $\delta_{(x)}$. It turns out that, as explained hereafter, a closed form expression for k_x can be readily obtained, which involves no functions more complicated than logarithms and is easily coded (which is most welcome in the matter of applications!)

In view of the elementary identity

$$(\phi - P\phi)(x) = \langle \delta_{(x)} - \sum_{j \in J} p_j(x)\delta_{(a_j)}, \phi \rangle, \forall \phi \in D, \forall x \in \mathbb{R}^n,$$

and since (by definition of the differentiation for distributions)

$$(\phi, k_x) = \langle (-1)^m \Delta^m k_x, \phi \rangle, \forall \phi \in D, \forall x \in \mathbb{R}^n,$$

it follows from (13) (which indeed may be restricted to *D*) that, for each $x \in \mathbb{R}^n$, k_x must be a solution (actually the *unique* solution) of the problem: Find $k_x \in X_0$ such that

$$(-1)^m \Delta^m k_x = \delta_{(x)} - \sum_{j \in J} p_j(x) \delta_{(a_j)} \quad \text{in } \mathcal{D}'. \tag{14}$$

On the basis of the superposition principle of linear operators, we get directly a particular solution of the (distributional) partial differential equation (14), viz.,

$$h_x(y) := (-1)^m [e(x - y) - \sum_{j \in J} p_j(x) e(a_j - y)], \ \forall y \in \mathbb{R}^n, \tag{15}$$

where e is (since $m > n/2$) the everywhere continuous function defined by (7a, b, c). *The crux of the matter is that*

$$h_x \in X \quad \text{when} \quad m > n/2. \tag{16}$$

Although this important result can be proved directly (see Theorem 2 in [7]), or also by exploiting basic properties of the Fourier transformation of distributions (see [3] and [7]), we deem it justified for the sake of transparency to outline here a *new, simpler proof* of (16).

Let $\Omega \subset \mathbb{R}^n$ denote a bounded open ball, centered at the origin and containing the P-unisolvent set $B = (a_j)_{j \in J}$ and the arbitrarily prescribed point x. Then:

(a) the restriction to Ω of every m-th *distributional* partial derivative of h_x is square integrable whenever $m > n/2$; in view of (15), this amounts essentially to verifying that every m-th *ordinary* partial derivative (in $\Omega \setminus \{0\}$) of $e(y)$ is in $L_2(\Omega)$, which indeed is the case since (as already recalled above)

$$\partial_{i_1 \ldots i_m} e(y) = O(|y|^{m-n} |\ln|y||), \quad \text{as} \quad y \to 0.$$

(b) the restriction to $\mathbb{R}^n \setminus \Omega$ of every m-th partial derivative of the $C^\infty(\mathbb{R}^n \setminus \Omega)$ function h_x is (always) square integrable; indeed, for any y fixed in $\mathbb{R}^n \setminus \Omega$, we have the *Peano-like integral representation formula:*

$$(-1)^m \partial_{i_1 \ldots i_m} h_x(y) = \int_0^1 \frac{(1 - t)^{m-1}}{(m - 1)!} \{D^m \partial_{i_1 \ldots i_m} e(tx - y) \cdot x^m$$

$$- \sum_{j \in J} p_j(x) D^m \partial_{i_1 \ldots i_m} e(ta_j - y) \cdot (a_j)^m \} dt,$$

as it readily follows from (15), interpreted as the remainder at the variable point $x \in \Omega$ for the P-interpolant on B of the $C^\infty(\Omega)$ function $(-1)^m e(\cdot - y)$, by making use of the Taylor formula of the m-th order for $\partial_{i_1 \ldots i_m} e(\cdot - y) \in C^\infty(\Omega)$ about the origin (D^m denotes the m-th Fréchet derivative); since at any rate

$$D^{2m} e(y) = O(|y|^{-n} \ell n |y|), \quad \text{as} \quad y \to \infty,$$

the proof of (16) is virtually complete.

Hence it finally follows that, for each $x \in \mathbb{R}^n$, $k_x \in X_0$ is the (continuous) function in $\mathbb{R}^n$:

$$\begin{aligned} k_x(y) := (h_x - Ph_x)(y) = (-1)^m [e(x - y) - \sum_{i \in J} p_i(x) e(a_i - y) \\ - \sum_{j \in J} p_j(y) e(x - a_j) + \sum_{i \in J} \sum_{j \in J} p_i(x) p_j(y) e(a_i - a_j)]. \end{aligned} \tag{17}$$

The set $\{k_x : x \in \mathbb{R}^n\}$ is the so-called *reproducing kernel* (or *kernel function*) of the functional Hilbert space X_0; it can be regarded equivalently as the (continuous) function $(x, y) \mapsto k(x, y) := k_x(y)$ in $\mathbb{R}^n \times \mathbb{R}^n$.

As a matter of fact, this nice result can still be extended significantly (as already noted in [9]):

Theorem 3. *Let* ℓ *denote the largest integer such that*

$$0 \leq \ell < m - n/2. \tag{18}$$

Then every distributional derivative of order $\leq \ell$ *of* $v \in X$, *say* $\partial^\nu v$ *for* $\nu := (\nu_1, \ldots, \nu_n) \in \mathbb{N}^n$ *and* $|\nu| := \nu_1 + \ldots + \nu_n \leq \ell$, *can be expressed in the form*

$$(\partial^\nu v)(x) = (\partial^\nu (Pv))(x) + (v, \partial_x^\nu k_x), \; \forall x \in \mathbb{R}^n, \tag{19}$$

where Pv *is defined by* (11) *and the notation* $\partial^\nu_x k_x$ *means that the operator* ∂^ν *is applied to the reproducing kernel* (17) *considered as a function of* x. *All the derivatives* $\partial^\nu(v - Pv)$ *of order* $\leq \ell$ *are accordingly bounded and uniformly continuous functions that vanish at infinity.*

As emphasized elsewhere (see e.g. [7, 8, 9]), reproducing kernels may prove extremely useful in approximation theory and in numerical analysis.

Owing to the orthogonal projection theorem, we have for the *solution* u *of Problem* (P) the simple *geometric characterization:*

$$u = V_0^{\perp} \cap V; \tag{20}$$

here V, the translation defined by (8) of

$$V_0 := \{v \in X: v(a_i) = 0, \forall i \in I\}, \tag{21}$$

is known to be the linear variety of all X-*interpolants on* A *to the prescribed values* $(\alpha_i)_{i \in I}$, whereas $V_0^{\perp}$, the orthogonal supplementary of V_0 in X, is the vector subspace of all *optimal* (i.e., of minimum seminorm $|\cdot|$) X-*interpolants on* A *to arbitrary values.* The following theorem (borrowed from [8]) extends to the elements of the latter subspace, that is to the *surface splines on* A, the *fundamental identity* used as a cornerstone for an intrinsic theory of univariate splines.

Theorem 4. *Any* $w \in V_0^{\perp}$ *has a unique representation of the form*

$$w(y) = (-1)^m \sum_{i \in I} \gamma_i e(a_i - y) + q(y), \forall y \in \mathbb{R}^n, \tag{22a}$$

where $q \in P$ *and the coefficients* γ_i *are (real) scalars satisfying the relations*

$$\sum_{i \in I} \gamma_i p(a_i) = 0, \forall p \in P. \tag{22b}$$

The corresponding fundamental identity is

$$|v - w|^2 = |v|^2 - |w|^2 - 2 \sum_{i \in I} \gamma_i [v(a_i) - w(a_i)] \qquad (23a)$$

or equivalently,

$$(v, w) = \sum_{i \in I} \gamma_i v(a_i), \ \forall v \in X, \ \forall w \in V_o^{\perp} . \qquad (23b)$$

It should be noticed that for $n = 1$ the definition (22a, b) of *surface splines* (relative to the set A of given interpolation points) reduces strictly to the classical definition of *natural splines* (relative to the set A of given *knots* in $\mathbb{R}$). The reader interested in further *intrinsic properties* of surface splines (such as the *minimum norm property*, the *best approximation property*, the *Schoenberg theorem* and the related *hypercircle inequality*) is referred to [9].

As regards the *efficient construction of the solution of Problem* (P), the representation formula:

$$(u - Pu)(y) = \sum_{i \in I \setminus J} \gamma_i k_{a_i}(y), \ \forall u \in V_o^{\perp}, \ \forall y \in \mathbb{R}^n, \qquad (24)$$

which stems simply from the fact that $P \cap V_o = \{0\}$ (remember that A contains a P-unisolvent subset!) implies by (13) that

$$X_o = V_o \oplus \operatorname{span}\{(k_{a_i})_{i \in I \setminus J}\},$$

should be definitely preferred to (22a, b). Indeed, by expressing now that $u \in V$, we get for determining the (real) coefficients γ_i in (24) a Cramer system of linear equations, viz.,

$$\sum_{j \in I \setminus J} k(a_i, a_j)\gamma_j = \alpha_i - \sum_{j \in J} \alpha_j p_j(a_i), \ \forall i \in I \setminus J, \qquad (25)$$

whose coefficient matrix is (real) *symmetric* and *positive definite* (as Gram matrix of a sequence of linearly independent elements $k_{a_j} \in X_o$). Solving (25) is therefore a most simple problem (at least in principle!): standard algorithms, based on the idea of *Cholesky factorization*, are known to be *numerically stable* (see Section 4) and *very economical* as regards the number of arithmetic operations; moreover, as explained in detail in [8], the whole solution process of Problem (P) can be described in a *recursive form*, the interpolation data being then exploited in sequence.

On the other hand, the apparently more natural approach that is based on the representation formula (22a, b) requires solving the linear algebraic system (with respect to the coefficients γ_j and γ'_ν):

$$\sum_{j \in I} (-1)^m e(a_i - a_j)\gamma_j + \sum_{|\nu| \leq m-1} (a_i)^\nu \gamma'_\nu = \alpha_i, \ \forall i \in I, \quad (26a)$$

$$\sum_{j \in I} (a_j)^\nu \gamma_j = 0, \quad \text{for all} \quad \nu \in \mathbb{N}^n \quad \text{with} \quad |\nu| \leq m - 1, \quad (26b)$$

whose coefficient matrix is (real) *symmetric* but certainly *not positive definite* (its diagonal elements are indeed 0). It is well known that in such cases Gaussian elimination without a suitable form of pivoting can be numerically unstable or can fail. In recent years, a number of schemes which share some or all of the advantages of Cholesky's method have been proposed for solving *symmetric indefinite systems*; for an interesting *comparison of some computer implementations*, see [2]; this leads to the conclusion that the algorithms proposed by Aasen and Bunch-Kaufman appear to be substantially superior to their potential competitors.

4. ON THE PRACTICAL CHOLESKY PROCESS

We will summarize here (without proof) the main results of the two self-contained *rounding-error analyses of the Cholesky process in standard floating-point arithmetic* which are presented in detail in [12]. They hold valid under the following (realistic) assumption about the (relative) rounding error ε made in any *elementary operation* $*$ (i.e., arithmetic operations +, -, x, /, together with the square root function) using *standard floating-point arithmetic:*

$$f\ell(x * y) = (x * y)(1 + \varepsilon), \quad \text{with} \quad |\varepsilon| \leq h,$$

where x and y are *digital* (i.e., machine numbers), $f\ell(x * y)$ denotes the machine result and h is the *machine precision* or Wilkinson's *macheps* (i.e., the machine constant equal to the smallest positive machine number for which $1 + h \neq 1$).

Theorem 5. *(Novel a posteriori error bounds). Let* $\bar{x}$ *denote the approximate solution, computed by the Cholesky algorithm in standard floating-point arithmetic, of the (real) symmetric digital linear system* $Ax = b$ *of order* n. *Then* $\bar{x}$ *can be interpreted as the exact solution of the perturbed system*

$$(A + \delta A)\overline{x} = b,$$

where the perturbation

$$\delta A: = (b - A\overline{x})\overline{x}^T/(\overline{x}^T\overline{x})$$

satisfies

$$|\delta A| < \frac{(2n + 3)h}{1 - (n + 1)h} |\overline{R}^T| \; |\overline{R}| \quad \text{if} \quad (n + 1)h \leq 1/2,$$

$\overline{R}$ *denoting the computed (upper triangular) Cholesky factor of* A.

Theorem 6. *(Refined a priori results)*

If A *is a (real) positive definite digital matrix of order* n, *then provided*

$$(1 + f_{n-1})^{n-1} < 1 + 1/\text{cond}(A) \tag{27a}$$

where

$$f_{n-1}: = \frac{h}{1 - 5h} [(n - 1)^{1/2} + 2], \tag{27b}$$

the Cholesky (upper triangular) factor R *can be computed without breakdown and the computed* $\overline{R}$ *is such that*

$$||\overline{R}^T\overline{R} - A|| < \frac{h}{1 - 5h} \{\frac{2}{3}(n - 1)^{3/2} + 2n + \frac{1}{2}(n - 1)^{1/2} - \frac{1}{6}\}$$

$$(1 + f_{n-1})^{n-1}||A||.$$

Here $\text{cond}(A): = ||A|| \, ||A^{-1}||$ *and* $||\cdot||$ *denotes the matrix spectral norm.*

It should be noted that (27a) gives a simple *criterion for the "numerical positive definiteness"* of positive definite digital matrices.

REFERENCES

1. Barnhill, R. E., "Representation and approximation of surfaces", Mathematical Software III (J. R. Rice ed.), pp. 69-120, New York, Academic Press, 1977.

2. Barwell, V. and George, A., "A comparison of algorithms for solving symmetric indefinite systems of linear equations", ACM Transactions on Mathematical Software 2, 1976, pp. 242-251.

3. Duchon, J., "Interpolation des fonctions de deux variables suivant le principe de la flexion des plaques minces", R.A.I.R.O. Analyse numérique 10, 1976, pp. 5-12.

4. Franke, R., "A critical comparison of some methods for interpolation of scattered data", Report NPS-53-79-003, Naval Postgraduate School, Monterey, 1979.

5. Franke, R., "Scattered data interpolation: tests of some methods", Math. Comp. 38, 1982, pp. 181-200.

6. Harder, R. L. and Desmarais, R. N., "Interpolation using surface splines", J. Aircraft 9, 1972, pp. 189-191.

7. Meinguet, J., "An intrinsic approach to multivariate spline interpolation at arbitrary points", Polynomial and Spline Approximation (B. N. Sahney ed.), pp. 163-190, Dordrecht, D. Reidel Publishing Company, 1979.

8. Meinguet, J., "Multivariate interpolation at arbitrary points made simple", Journal of Applied Mathematics and Physics (ZAMP) 30, 1979, pp. 292-304.

9. Meinguet, J., "Basic mathematical aspects of surface spline interpolation", ISNM 45, Numerische Integration (G. Hämmerlin ed.), pp. 211-220, Basel, Birkhäuser Verlag, 1979.

10. Meinguet, J., "A convolution approach to multivariate representation formulas", ISNM 51, Multivariate Approximation Theory (W. Schempp and K. Zeller eds), pp. 198-210, Basel, Birkhäuser Verlag, 1979.

11. Meinguet, J., "From Dirac distributions to multivariate representation formulas", Approximation Theory and Applications (Z. Ziegler ed.), pp. 225-248, New York, Academic Press, 1981.

12. Meinguet, J., "Refined error analyses of Cholesky factorization", SIAM J. Numer. Anal. 20, 1983, pp. 1243-1250.

13. Schumaker, L. L., "Fitting surfaces to scattered data", Approximation Theory II (G. G. Lorentz, C. K. Chui and L. L. Schumaker eds), pp. 203-268, New York, Academic Press, 1976.

14. Wilkinson, J. H., "A priori error analysis of algebraic processes", Proceedings of International Congress of Mathematicians (Moscow, 1966), pp. 629-640, Moscow, Mir Publishers, 1968.

INTERPOLATION OF SCATTERED DATA: DISTANCE MATRICES AND CONDITIONALLY POSITIVE DEFINITE FUNCTIONS

Charles A. Micchelli

§1. INTRODUCTION

For practical problems of data fitting in two dimensions two methods seem to be most popular: Thin Plate Splines (TPS) Duchon (1976) and Hardy's Multiquadric Surfaces (MQS) Hardy (1971), (1982), see also Franke (1982). The theory for TPS has been developed in a series of papers (see Duchon (1976), Meinguet (1979)). However, beyond its numerical performance little seems to be known about MQS. For instance, in his lecture notes for a recent meeting, Franke (1983) raised (based on extensive numerical experience) the following conjecture.

Conjecture. (Franke) *Given any distinct points* $x^1, \ldots, x^n$ *in the plane*

$$(-1)^{n-1}\det \sqrt{1 + \|x^i - x^j\|^2} > 0. \tag{1}$$

Here $\|x\|^2 = x_1^2 + x_2^2$, $x = (x_1, x_2)$ is the Euclidean norm of x. This conjecture says, in particular, that there is a unique surface $f(x) = c_1 \sqrt{1 + \|x - x^1\|^2} + \ldots c_n \sqrt{1 + \|x - x^n\|^2}$ which interpolates (data) $y_1, \ldots, y_n$ at $x^1, \ldots, x^n$. (Apparently, it was not even known that interpolation by MQS was always possible (Wahba 1982).)

As an extension of MQS, Barnhill and Stead (1983) explored

S. P. Singh et al. (eds.), Approximation Theory and Spline Functions, 143–145.

surfaces based on the kernel $K_\mu(x, y) = (1 + \|x - y\|^2)^{-\mu}$, where μ is a real number, in contour plotting for three-dimensional interpolation. Thus they used

$$f(x) = c_1 k_\mu(x, x^1) + \ldots + c_n k_\mu(x, x^n) \tag{2}$$

to interpolate scattered data

$$f(x^i) = y_i, \; i = 1, \ldots, n \tag{3}$$

and asked for those values of μ for which (3) has a unique solution. A similar question was raised by N. Dyn in a recent conversation for the kernel $k(x, y) = \log(1 + \|x - y\|^2)$, (see Dyn, Levin, Ruppa (1983).

The purpose of this paper is to address these questions. We place them into a unified context so that we can draw upon ideas from the theory of positive definite functions (Stewart (1976)) and distance matrices (Blumenthal (1970)). A full account will appear elsewhere.

In this announcement, we just mention the following result. Let A be the class of $n \times n$ real symmetric matrices such that

$$\sum_{i=1}^{n} \sum_{j=1}^{n} c_i c_j A_{ij} \le 0$$

whenever $\sum_{j=1}^{n} c_j = 0$ (conditionally negative definite) and

$$A_{ij} > \frac{1}{2}(A_{ii} + A_{jj}), \quad i, j = 1, \ldots, n, \; i \ne J.$$

Theorem. Suppose $F'(t)$ _is completely monotonic but not constant on_ $(0, \infty)$ and $F(0) > 0$. _Then for any_ $A \in A$,

$$(-1)^{n-1} \det F(A_{ij}) > 0.$$

The choice $F(t) = (1 + t)^{1/2}$ and $A_{ij} = \|x^i - x^j\|^2$, $x^1, \ldots, x^n \in R^p$ proves Franke's conjecture for _any_ R^p.

References

1. Barnhill, R., and Stead, S., (1983), "Multistage trivariate surfaces", preprint.

2. Blumenthal, (1970), "Theory and applications of distance geometry", Chelse Publishing.

3. Duchon, J., (1976), "Splines minimizing rotation-invariant semi-norms in Sobolev spaces", Constructive theory of functions of several variables, Oberwolfach, W. Schempp and K. Zeller, Springer, Berlin-Heidelberg.

4. Dyn, N., Levin, D., and Rippa, S., (1983), "Numerical procedures for global surface smoothing of noisy scattered data", preprint.

5. Franke, R., (1982), "Scattered data interpolation: Tests of some methods", Mathematics of Computation, 38, pp. 381-400.

6. Franke, R., (1983), "Lecture notes on: Global basis functions for scattered data".

7. Hardy, R. L., (1971), Multiquadric equations of topography and other irregular surfaces", J. Geophys. Res., C.

8. Hardy, R. L., (1982), "Surface fitting with biharmonic and harmonic models", Proceedings of the NASA workshop on surface fitting, Center for Approximation Theory, Texas A & M University, College Station, Texas.

9. Meinguet, J., 1979, "An intrinsic approach to multivariate spline interpolation at arbitrary points: polynomial and spline approximation", B. W. Sahney ed., pp. 163-1890, D. Reidel Publishing Company Dordrecht.

10. Stewart, J., (1976), Positive definite functions and generalizations, an historical survey", Rocky Mountain Journal of Mathematics", 6, pp. 409-434.

11. Wahba, G., (1982), "Private communication".

SEMI-NORMS IN POLYNOMIAL APPROXIMATION

G. M. Phillips* and P. J. Taylor

Universities of St. Andrews and Stirling

1. INTRODUCTION

S. N. Bernstein [1] proved the following result.

Theorem 1. If $f \in C^{n+1}[a, b]$, then the semi-norm

$$E_n(f) = \inf_{q \in P_n} \|f - q\|_\infty, \tag{1}$$

where $\|\cdot\|_\infty$ denotes the maximum norm on $[a, b]$, satisfies

$$E_n(f) = \frac{1}{(n+1)!} |f^{(n+1)}(\xi)| \cdot E_n(x^{n+1}), \quad a < \xi < b. \tag{2}$$

Bernstein obtained Theorem 1 via the intermediate result which we now state.

Theorem 2. If $|f^{(n+1)}(x)| \le g^{(n+1)}(x)$, $a \le x \le b$, then $E_n(f) \le E_n(g)$.

Phillips [5] gave a more direct proof of Theorem 1 based on the fact that the minimax polynomial of degree n for f on $[a, b]$ interpolates f at $n + 1$ points of $[a, b]$. This interpolating property, which also holds for all best L_p approximations, was used by Phillips [6] to generalize Theorem 1 from minimax approximations to best L_p approximations.

S. P. Singh et al. (eds.), Approximation Theory and Spline Functions, 147–150.

2. GENERALIZATION OF BERNSTEIN'S THEOREM

Holland, Phillips and Taylor [2] generalized Theorem 1 to a certain class of semi-norms. Prompted by Bernstein's Theorem 2, they proposed:

Definition 1. A semi-norm E_n on $C^{n+1}[a, b]$ is said to satisfy Property B of order n if

$$|f^{(n+1)}(x)| \leq g^{(n+1)}(x), \quad a \leq x \leq b,$$

implies that $E_n(f) \leq E_n(g)$.

Of course, Theorem 2 shows that the semi-norm (1) satisfies Property B. It is easily verified (see [2]) that if a semi-norm E_n satisfies Property B of order n, then

$$E_n(q) = 0, \quad \forall q \in P_n;$$

$$E_n(f + q) = E_n(f), \quad \forall f \in C^{n+1}[a, b], \quad q \in P_n.$$

The generalization of Theorem 1 given by Holland, Phillips and Taylor [2] is:

Theorem 3. If E_n is a semi-norm which satisfies Property B of order n, then for each $f \in C^{n+1}[a, b]$, there exists some $\xi \in (a, b)$ such that

$$E_n(f) = \frac{1}{(n+1)!} |f^{(n+1)}(\xi)| \cdot E_n(x^{n+1}).$$

The proof follows closely that of Theorem 1 as given in Meinardus [4].

Kimchi and Richter-Dyn [3] showed that Property B is satisfied by all semi-norms E_n of the form (1) where $||\cdot||_\infty$ is replaced by any monotone norm. (Recall that $||\cdot||$ is monotone if $|f(x)| \leq |g(x)|$, $a \leq x \leq b$, implies that $||f|| \leq ||g||$. This includes, for example, all of the p-norms.) If we combine Kimchi and Richter-Dyn's result and Theorem 3 we obtain immediately a generalization of Theorem 1:

Theorem 4. If $f \in C^{n+1}[a, b]$, then the semi-norm

$$E_n(f) = \inf_{q \in P_n} \|f - q\| ,$$

where $\|\cdot\|$ is any monotone norm on $[a, b]$, satisfies

$$E_n(f) = \frac{1}{(n+1)!} |f^{(n+1)}(\xi)| \cdot E_n(x^{n+1}).$$

3. SOME RELATED RESULTS

In this section we will take $[a, b]$ as $[-1, 1]$. (This is for convenience and does not involve loss of generality.) Let us write

$$E_n(f) = \max_{-1 \le x \le 1} |f(x) - q(x)|, \tag{3}$$

where q interpolates f at certain fixed points $x_0, x_1, \ldots, x_n$ of $[-1, 1]$. Then an argument similar to that used by Phillips [5] to prove Theorem 1 shows that

$$E_n(f) = \frac{1}{(n+1)!} |f^{(n+1)}(\xi)| \cdot E_n(x^{n+1}), \quad -1 < \xi < 1. \tag{4}$$

Clearly, the smallest value attained (uniquely) by $E_n(x^{n+1})$ in (4), taken over all possible sets of interpolating points $x_0, x_1, \ldots, x_n$, is $1/2^n$. This corresponds to the interpolating points being chosen as the zeros of the Chebyshev polynomial T_{n+1}. Thus we have

$$\max_{-1 \le x \le 1} |f(x) - q(x)| = \frac{1}{(n+1)!} \cdot \frac{1}{2^n} \cdot |f^{(n+1)}(\xi)|, \tag{5}$$

where q interpolates f at the zeros of T_{n+1}. Note that (5) gives precisely the same result for the maximum error of interpolation at the Chebyshev zeros as (2) gives for the maximum error of minimax approximation. Of course, the value of ξ may differ between (2) and (5). However, (2) and (5) show that if $f^{(n+1)}$ does not vary very much, then the accuracy of interpolation at the Chebyshev zeros is close to that of minimax approximation. This observation is compatible with practical experience: interpolation at the Chebyshev zeros is commonly used as a more easily computed substitute for the minimax polynomial.

There is another commonly used substitute for the minimax polynomial. This is the polynomial, say q^*, such that

$$f(t_i) - q^*(t_i) = (-1)^i \cdot e, \quad 0 \le i \le n+1,$$

where e is some constant and the t_i are the extreme points of T_{n+1}. Thus $f - q^*$ equioscillates on the point set consisting of the extreme points of T_{n+1}. Phillips and Taylor [7] have shown that (5) also holds with this q^* in place of q.

REFERENCES

1. Bernstein, S. N., "Lecons sur les propriétés extrémales et la meilleure approximation des fonctions analytiques d'une variable réelle", Gauthier-Villars, Paris, 1926.

2. Holland, A. S. B., Phillips, G. M. and Taylor, P. J., "Generalization of a theorem of Bernstein", Proceedings of Conference on Approximation Theory in honour of G. G. Lorentz, edited by E. W. Cheney, Academic Press, New York, 1980.

3. Kimchi, E. and Richter-Dyn, N., "Restricted range approximation of k-convex functions in monotone norms", SIAM J. Numer. Anal., 15, p. 1030, 1978.

4. Meinardus, G., "Approximation of functions: theory and numerical methods", Springer-Verlag, New York, 1967.

5. Phillips, G. M., "Estimate of the maximum error in best polynomial approximations", Comp. J. 11, p. 110, 1968.

6. Phillips, G. M., "Error estimates for best polynomial approximations", contrib. to Approximation Theory, edited by A. Talbot, Academic Press, London and New York, 1970.

7. Phillips, G. M., and Taylor, P. J., "Polynomial approximation using equioscillation on the extreme points of Chebyshev polynomials", J. Approx. Theory 36, p. 257, 1982.

*Paper presented by this author.

ON SPACES OF PIECEWISE POLYNOMIALS IN TWO VARIABLES

Larry L. Schumaker

Center for Approximation Theory,
Texas A & M University,
College Station, TX 77843, U.S.A.

1. INTRODUCTION

The purpose of this paper is to survey the progress which has been made in the last several years in developing a theory for spaces of piecewise polynomials in two variables. The ultimate goal for this area would be to have a complete analog of the univariate theory, but as we shall see, much remains to be done.

We begin by introducing the spaces of interest. Suppose Ω is a closed subset of $\mathbb{R}^2$, and that $\Delta = \{\Omega_i\}_i^n$ is a collection of open subsets such that

$$\Omega_i \cap \Omega_j = \phi, \quad \text{all} \quad i,j = 1, \ldots, n, \ i \neq j \tag{1.1}$$

$$\Omega = \bigcup_{i=1}^{n} \overline{\Omega}_i . \tag{1.2}$$

We call Δ a partition of Ω. Typical partitions are shown in Figure 1. Given a positive integer d, we define the space of polynomials of degree d (in two variables) by

$$P_d = \{p(x, y) = \sum_{i=0}^{d} \sum_{j=0}^{d-i} a_{ij} x^i y^j , \ a_{ij} \in \mathbb{R}\}. \tag{1.3}$$

Throughout this paper, we shall be interested in the following space of smooth piecewise polynomials defined on the partition Δ of Ω:

S. P. Singh et al. (eds.), Approximation Theory and Spline Functions, 151–197.

Definition 1.1

Given a nonnegative integer r and d, let

$$S_d^r(\Delta) = \{s \in C^r(\Omega) : s\big|_{\Omega_i} \in P_d, \quad i = 1, \ldots, n\}. \tag{1.4}$$

Clearly, $S_d^r(\Delta)$ is a linear space which reduces to P_d if $r \geq d$. Hence, we shall concentrate on the case $0 \leq r < d$.

The space $S_d^r(\Delta)$ defined in (1.4) is the natural extension to two variables of the well-known univariate splines (associated with a partition of an interval). Thus, we may refer to $S_d^r(\Delta)$ as a <u>space of splines of degree</u> d <u>and smoothness</u> r. In order to construct a theory of multivariate splines analogous to the well-known univariate theory (cf. [33]), we must deal with the following problems:

- find the dimension of $S_d^r(\Delta)$
- construct a basis for $S_d^r(\Delta)$
- construct local-support basis elements
- construct a partition of unity
- define quasi-interpolants
- examine the approximation power of $S_d^r(\Delta)$
- use $S_d^r(\Delta)$ for interpolation, data fitting, and in finite-element applications

While some progress has been made in each of these areas, much of this program remains incomplete. In this paper, we shall restrict our attention to the algebraic problems of identifying the dimension of $S_d^r(\Delta)$ and constructing bases for it.

The paper is organized as follows: In Section 2, we present some general results on dimension. In Sections 3 - 6, we specialize these results to various types of rectangular and triangular partitions where we can also give explicit bases. In Section 7, we deal with so-called cross-cut partitions. We conclude the paper with remarks and references.

We close this section with one additional definition. For many applications, it is important to deal with subspaces of $S_d^r(\Delta)$ which satisfy certain boundary conditions.

Definition 1.2

Given an integer ρ with $0 \leq \rho \leq r$, let

$$S_d^{r,\rho}(\Delta) = \{s \in C^\rho(\mathbf{R}^2) \cap S_d^r(\Delta) : s \equiv 0 \text{ on } \mathbb{R} \setminus \Omega\}. \quad (1.5)$$

Such spaces are of particular interest in finite-element applications. For example, with $\rho = 0$, we are dealing with the subspace of splines which vanish on the boundary, while with $\rho = 1$, the subspace consists of those splines which vanish along with their normal derivatives along the boundary of Ω.

2. SOME RESULTS ON DIMENSION

Throughout the remainder of this paper, we shall restrict our attention to partitions with the property

all edges of the partition inside of Ω are straight lines. (2.1)

We refer to a partition with property (2.1) as a rectilinear partition. Examples of typical rectilinear and non-rectilinear partitions are shown in Figure 1. Note that the boundary of a rectilinear partition of a set Ω is not required to consist of straight lines.

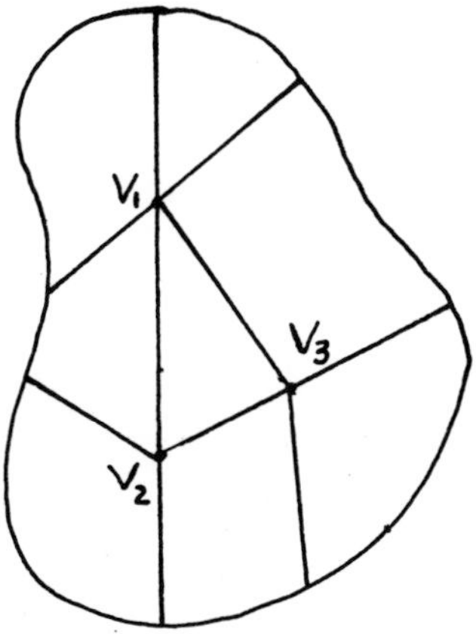

a. Rectilinear

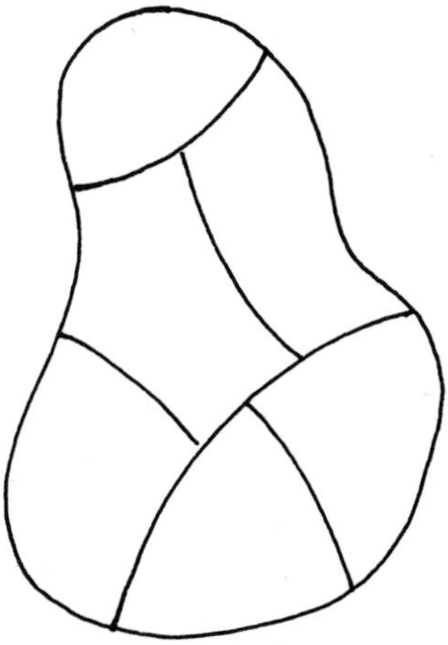

b. Non-rectilinear

Figure 1. Typical partitions.

The advantage of working with rectilinear partitions is that it is easy to see how two polynomials must join together across a line in order to belong to $C^r(\mathbb{R}^2)$. In particular, if Ω_0 and Ω_1 are separated by a straight line with equation $y + \alpha x = 0$, then polynomials p_0 and p_1 defined on Ω_0 and Ω_1, respectively, have r continuous derivatives across the boundary line if and only if

$$p_1(x,y) - p_0(x,y) = \sum_{j=1}^{d-r} \sum_{k=1}^{j} c_{jk}\, \phi_{jk}(x,y)_+, \tag{2.2}$$

where

$$\phi_{j,k}(x,y)_+ = \begin{cases} \phi_{j,k}(x,y), & (x,y) \in \Omega, \\ 0, & (x,y) \in \Omega_0, \end{cases} \tag{2.3}$$

and

$$\phi_{j,k}(x,y) = x^{j-k}(y+\alpha x)^{r+k}, \tag{2.4}$$

$j=1, \ldots, d-r$, $k = 1, \ldots, j$. This observation plays an important role in the proofs of several theorems given below. For a discussion of how two polynomials must join across a general algebraic curve, see [8].

Before stating our main result on dimension, we introduce some additional notation. Given a rectilinear partition Δ of a set Ω (cf. Figure 1), let

v = number of vertices of the partition inside of Ω. (2.5)

e = number of edges of the partition inside of Ω. (2.6)

Given a particular ordering of the vertices, then associated with each interior vertex, we also define

e_i = number of edges attached to the ith vertex with different slopes. (2.7)

$\tilde{e}_i$ = number of edges attached to the ith vertex with different slopes, and not yet counted. (2.8)

To illustrate these concepts, consider the following example.

Example 2.1

Let Δ be the partition shown in Figure 1.a. For this partition, $v = 3$, $e = 10$, and

$$e_1 = 3 \qquad \tilde{e}_1 = 3$$
$$e_2 = 3 \qquad \tilde{e}_2 = 3$$
$$e_3 = 3 \qquad \tilde{e}_3 = 2.$$

. It is clear from the definition that in general

$$e_i \leq \tilde{e}_i, \quad i = 1, \ldots, v. \tag{2.9}$$

This example shows that it is possible for $\tilde{e}_i < e_i$ to hold for some i.

We are now ready for the main result of this section.

Theorem 2.2

If Δ is a rectilinear partition of a set Ω, then

$$\alpha + \beta e - \gamma v + \sum_{i=1}^{v} \sigma_i \leq \dim S_d^r(\Delta) \tag{2.10}$$
$$\leq \alpha + \beta e - \gamma v + \sum_{i=1}^{v} \tilde{\sigma}_i,$$

where

$$\alpha = (d+1)(d+2)/2 \tag{2.11}$$

$$\beta = (d-r)(d-r+1)/2 \tag{2.12}$$

$$\gamma = \alpha - (r+1)(r+2)/2 \tag{2.13}$$

and

$$\sigma_i = \sum_{j=1}^{d-r} (r+j+1-j \cdot e_i)_+ \tag{2.14}$$

$$\tilde{\sigma}_i = \sum_{j=1}^{d-r} (r+j+1-j \cdot \tilde{e}_i)_+, \tag{2.15}$$

$i = 1, \ldots, v$.

Discussion

The lower bound in (2.10) was established in [32] by induction on v. For a "standard cell" (the case $V = 1$ with no "flaps", cf. Figure 2), the argument is based directly on the observation (2.2) and gives the exact dimension (i.e., the upper and lower bounds coincide). An exact formula for the general case of $v = 1$ follows easily. The induction involves a dissection process in which the partitioned set Ω is regarded as the concatenation of a cell together with a partitioned set $\tilde{\Omega}$ with one less interior vertex (cf. e.g., Figure 3).

The upper bound in (2.10) was established in [34]. The idea of the proof is to construct a set of linear functionals $\{\lambda_i\}_1^N$ defined on $S_d^r(\Delta)$ such that

$$\lambda_i s = 0, \quad i = 1, \ldots, N \quad \text{implies} \quad s \equiv 0. \tag{2.16}$$

Then it follows that $\dim S_d^r(\Delta) \leq N$. The desired functionals are defined in [34] as point functionals (possibly involving derivatives).

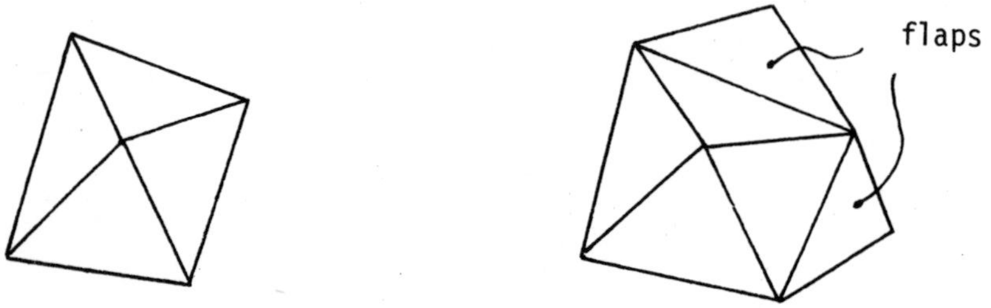

(a) standard cell (b) cell with 2 flaps

Figure 2. Cells

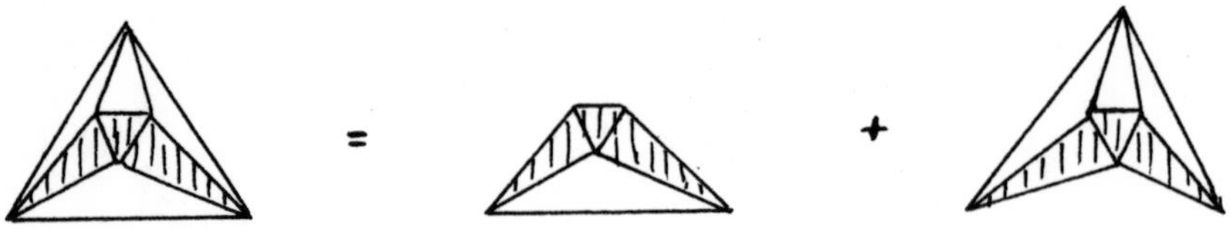

$v = 3$ $v = 1$ $v = 2$

Figure 3. The dissection process

Clearly, the bounds given in Theorem 2.2 are numerically computable for any given rectilinear partition. The quantities α, β, γ defined in (2.11) - (2.13) are computable directly from the degree d of the splines and the smoothness r. The quantities e and v depend on the general geometric "shape" of the partition, while each e_i and $\tilde{e}_i$ depends explicitly on the angles at which the edges meet at the ith vertex for $i = 1, \ldots, v$. In addition, we note that the quantities $\tilde{e}_i$ (and thus the upper bound) may depend on the ordering of the vertices (for an example, see [34]).

If Δ is a partition such that

$$e_i = \tilde{e}_i, \quad i = 1, \ldots, v, \tag{2.17}$$

then the upper and lower bounds in (2.10) coincide and Theorem 2.2 gives the exact dimension of $S_d^r(\Delta)$. Several situations where this happens will be discussed in later sections. On the other hand, it is not difficult to give examples where the upper and lower bounds do not coincide. The first such example was given in [27]:

Example 2.3

Let Δ be the partition shown in Figure 3. Then $6 \le \dim S_2^1(\Delta) \le 7$.

Discussion

In this case, $d = 2$, $r = 1$ and $\alpha = 6$, $\beta = 1$, $\gamma = 3$. For this partition, $v = 3$ and $e = 9$. Regardless of how we order the three vertices, $e_1 = e_2 = e_3 = 4$ and $\tilde{e}_1 = 4$, $\tilde{e}_2 = 3$, $\tilde{e}_3 = 2$. This yields $\sigma_1 = \sigma_2 = \sigma_3 = \tilde{\sigma}_1 = \tilde{\sigma}_2 = 0$ and $\tilde{\sigma}_3 = 1$.

In Example 2.3, we have seen that the upper and lower bounds do not always coincide. This same example serves to illustrate the fact that, in general, $\dim S_d^r(\Delta)$ may depend explicitly on the geometry. Indeed, in addition to the six polynomials $\{1, x, x^2, y, xy, y^2\}$, the space $S_2^1(\Delta)$ of Example 2.3 also contains a non-polynomial element precisely when the partition Δ describes a perfectly symmetric figure (cf. [27]). This is particularly disturbing from a numerical standpoint, since if we start with the 7-dimensional space $S_2^1(\Delta)$ with Δ being the symmetric partition of Figure 3, then any arbitrarily small perturbation of Δ (for example, moving one vertex) results in a space of dimension 6.

The possible dependence of $\dim S_d^r(\Delta)$ on the geometry of the partition provides a natural impetus for the study of special partitions. Several special partitions of particular interest will be treated in the following sections. For some additional results on general triangulations, see Section 6.

3. RECTANGULAR PARTITIONS

Given nonnegative integers k, $\tilde{k}$ and real numbers $a < b$ and $\tilde{a} < \tilde{b}$, let

$$\Delta_k = \{a = x_0 < x_1 < \ldots < x_{k+1} = b\} \tag{3.1}$$

$$\tilde{\Delta}_{\tilde{k}} = \{\tilde{a} = \tilde{x}_0 < \tilde{x}_1 < \ldots < \tilde{x}_{\tilde{k}+1} = \tilde{b}\} \tag{3.2}$$

Associated with these univariate partitions, let

$$\Omega_{ij} = \{(x, y) : x_i < x < x_{i+1} \text{ and } y_j < y < y_{j+1}\}, \tag{3.3}$$

$i = 0, 1, \ldots, k$ and $j = 0, 1, \ldots, \tilde{k}$. Then $\Delta_{k\tilde{k}} = \{\Omega_{ij}\}_{i=0,j=0}^{k,\tilde{k}}$ defines a partition of the rectangle

$$\Omega = \{(x, y) : a \leq x \leq b, \ \tilde{a} \leq y \leq \tilde{b}\}. \tag{3.4}$$

Throughout this section, we are interested in the space $S_d^r(\Delta_{k\tilde{k}})$ associated with the rectangular partition $\Delta_{k\tilde{k}}$ defined above. Before proceeding, we note that although $\Delta_{k\tilde{k}}$ is a tensor product partition, the space $S_d^r(\Delta_{k\tilde{k}})$ is <u>not</u> a space of tensor-product splines (cf. [33, p. 484]). Indeed, here the pieces are polynomials of <u>total degree</u> d rather than tensor-product polynomials (see Remark 3). We begin by quoting a result on dimension.

Theorem 3.1

For all $0 \leq r$,

$$\dim S_d^r(\Delta_{k\tilde{k}}) = \frac{k\tilde{k}}{2}(d - 2r)(d - 2r - 1)_+ + \frac{(k + \tilde{k})}{2}(d - r + 1)(d - r)_+ + \frac{(d + 1)(d + 2)}{2}. \tag{3.5}$$

Discussion

For the partition Δkk, we have $V = k\tilde{k}$ and $e = 2k\tilde{k} + (k + \tilde{k})$. Moreover, with the natural typewriter ordering of the vertices, $e_i = \tilde{e}_i = 2$, $i = 1, \ldots, v$. This is condition (2.17), and it follows that the upper and lower bounds in Theorem 2.2 coincide to give the desired dimension. This result also follows from a theorem in [8] concerning simple cross-cut partitions (cf. Theorem 7.1 below).

Our next result gives a basis of one-sided splines for the space dealt with in Theorem 3.1.

Theorem 3.2

The functions

$$\{x^m y^n : 0 \le m + n \le d\} \tag{3.6}$$

$$\{x^m y^n (x - x_i)_+^{r+1} : 0 \le m + n \le d - r - 1\}_{i=1}^{k} \tag{3.7}$$

$$\{x^m y^n (y - y_j)_+^{r+1} : 0 < m + n < d - r - 1\}_{j=1}^{\tilde{k}} \tag{3.8}$$

$$\{x^m y^n (x - x_i)_+^{r+1} (y - y_j)_+^{r+1} : 0 \le m + n \quad + n \le d - 2r - 2\}_{i=1,j=1}^{k,\tilde{k}} \tag{3.9}$$

form a basis for $S_d^r(\Delta_{k\tilde{k}})$.

Discussion

This result was established in Chui and Wang [8] as an application of results on simple cross-cut partitions (cf. Theorem 7.1 below). It can also be proved directly. Indeed, it is easily checked that the functions (3.6) - (3.9) are linearly independent and that the total number equals the dimension given in (3.5). The splines in (3.7) are identically zero to the left of the line $x = x_i$ while those in (3.8) are zero below the line $y = y_j$. Thus, they may be regarded as analogs of the one-sided univariate splines. The splines in (3.9) are non-zero only in the cone $\{(x, y) : x \ge x_i, y \ge y_j\}$ We may call them <u>cone splines</u>. They are not present in the case where $2r \ge d - 2$.

In order to construct a basis of local support splines for $S_d^r(\Delta_{k\tilde{k}})$, we may use tensor products of appropriate univariate B-splines. First, we give a lemma on univariate splines satisfying boundary conditions. Let

$$S_q^{r,\alpha,\beta}(\Delta_k) = \{s \in C^r[a, b] : s\big|_{(x_i,\ x_{i+1})} \tag{3.10}$$

is a univariate polynomial of degree q
for $i = 0, 1, \ldots, k$ and such that
$D^i s(a) = 0, \quad i = 0, 1, \ldots, \alpha$ and
$D^j s(b) = 0, \quad j = 0, 1, \ldots, \beta\}$.

Lemma 3.3

If $0 \leq d \leq r$, then

$$\eta_q := \dim S_d^{r,\alpha,\beta}(\Delta_k) = (q - \alpha - \beta - 1)_+ \tag{3.11}$$

while if $0 \leq r < d$, then

$$\begin{aligned} \eta_q &:= \dim S_d^{r,\alpha,\beta}(\Delta_k) \\ &= [(k + 1)(q - r) + r - \alpha - \beta - 1]_+. \end{aligned} \tag{3.12}$$

Moreover, if $r \geq \alpha + \beta + 1$, then there exist functions $\{\phi_i\}_1^\infty$ such that for all $q \geq \alpha + \beta + 2$, $\{\phi_i\}_1^{\eta_q}$ is a basis for S with

$$\phi_i \in P_{\alpha+\beta+1+i}, \quad i = 1, \ldots, r - \alpha - \beta - 1 \tag{3.13}$$

$$\phi_{\eta_{q-1}+1}, \ldots, \phi_{\eta_q} \text{ are B-splines of degree } q. \tag{3.14}$$

Finally, if $r < \alpha + \beta + 1$, then the same statement holds for $q \geq r + \ell$, where ℓ is the unique integer such that

$$(\ell - 1)(k + 1) \leq \alpha + \beta + 1 - r < \ell(k + 1) \tag{3.15}$$

Discussion

This lemma was established in [16]. Its usefulness is embodied in the fact that we have bases of B-splines of lowest possible degree. In particular, when $r \geq \alpha + \beta + 1$, the number of ϕ_i's of <u>exact</u> degree q is given by

$$\nabla\eta_q = \eta_q - \eta_{q-1} = \begin{cases} 0, & q = 1, \ldots, \alpha + \beta + 1 \\ 1, & q = \alpha + \beta + 2, \ldots, r \\ k+1, & q = r + 1, \ldots . \end{cases} \tag{3.16}$$

In the case $r < \alpha + \beta + 1$, the number of ϕ_i's of exact degree q is given by

$$\nabla\eta_q = \begin{cases} 0 & , \quad q = 1, 2, \ldots, r + \ell - 1 \\ \ell(k+1) + r - \alpha - \beta - 1, & \quad q = r + \ell \\ k + 1 & , \quad q = r + \ell + 1, \ldots . \end{cases} \tag{3.17}$$

We can now give an alternative basis for $S_d^r(\Delta_{k\tilde{k}})$ with a considerable number of locally supported elements.

Theorem 3.4

A basis for $S_d^r(\Delta_{k\tilde{k}})$ is given by

$$\{\phi_i(x)\tilde{\phi}_j(y)\}_{i=\eta_{q-1}+1,\ j=1,\ q=0}^{\eta_q,\ \tilde{\eta}_{d-q},\ d} \tag{3.18}$$

where the η_q's and ϕ_i's are associated with the univariate spline space $S_d^r(\Delta_k)$ as in Lemma 3.3 while the $\tilde{\eta}_q$'s and $\tilde{\phi}_j$'s are associated with the space $S_d^r(\Delta_{\tilde{k}})$.

Discussion

This result was established in [16]. Note that it gives the following alternate expressions for dimension:

$$\dim S_d^r(\Delta_{k\tilde{k}}) = \sum_{q=0}^{d} \tilde{\eta}_{d-q} \nabla\eta_q = \sum_{q=0}^{d} \eta_{d-q} \nabla\tilde{\eta}_q .$$

The following example illustrates the difference between the bases given in Theorems 3.2 and 3.4

Example 3.5

Let $k = \tilde{k} = 1$. Then $\dim S_3^1(\Delta_{11}) = 16$. The one-sided basis of Theorem 3.2 consists of the functions $\{1, x, x^2, x^3, y, xy, x^2y, y^2, y^2x, y^3\}$, $(x - x_1)_+^2$, $y(x - x_1)_+^2$, $(x - x_1)_+^3$, $(y - y_1)_+^2$, $x(y - y_1)_+^2$, and $(y - y_1)_+^3$. The basis given in Theorem 3.4 consists of the functions $\{1, y, \tilde{N}_2^3(y),$ $\tilde{N}_3^3(y)$, $\tilde{N}_2^4(y)$, $\tilde{N}_5^4(y)\}$, $\{x, xy, x\tilde{N}_2^3(y), x\tilde{N}_3^3(y)\}$, $\{N_2^3(x),$ $N_3^3(x)$, $yN_2^3(x)$, $yN_3^3(x)\}$ and $\{N_2^4(x), N_5^4(x)\}$.

Discussion

Here, the N_i^q are the usual B-splines associated with the extended knot sequence $a = y_1 = \ldots = y_q$, $y_{q+1} = \ldots = y_{2q-2}$, $y_{2q-1} = \ldots = y_{3q-2} = b$. The $\tilde{N}_i^q$'s are defined similarly. The necessary B-splines are depicted in Figure 4.

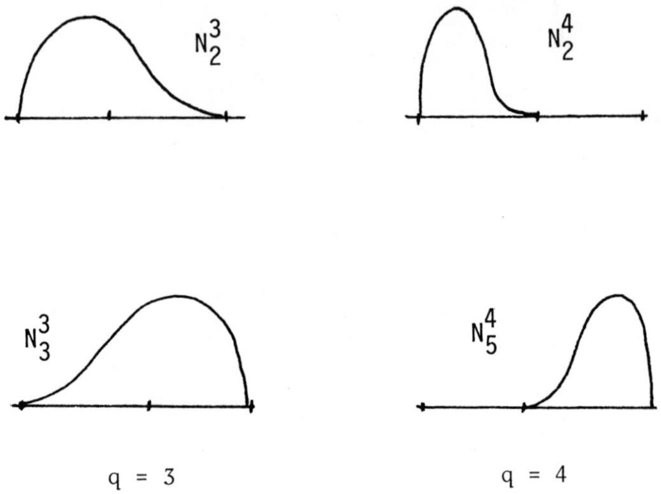

Figure 4. The B-splines in Example 3.5

In the remainder of this section, we examine spaces of splines on the rectangular partition $\Delta_{k\tilde{k}}$ which also satisfy boundary conditions. Let

$$S_d^{r,\alpha,\beta,\tilde{\alpha},\tilde{\beta}}(\Delta_{k\tilde{k}}) = \{s \in S_d^r(\Delta_{k\tilde{k}}) : \tag{3.19}$$

$$D_x^i s(a, \tilde{x}) = 0, \quad i = 0, \ldots, \alpha \quad \text{and}$$

$$D_x^j s(b, \tilde{x}) = 0, \quad j = 0, \ldots, \beta \quad \text{for all}$$

$$\tilde{a} \le \tilde{x} \le \tilde{b} \quad \text{while} \quad D_y^i s(y, \tilde{a}) = 0, \quad i = 0, \ldots, \tilde{\alpha}$$

$$\text{and} \quad D_y^j s(y, \tilde{b}) = 0, \quad j = 0, \ldots, \tilde{\beta} \quad \text{for all}$$

$$a \le x \le b\}.$$

This is the space of splines which vanishes along with their first α normal derivatives along the edge $x = a$ with similar boundary conditions on the other three edges of the rectangle Ω.

Theorem 3.6

For all $0 \le \alpha, \tilde{\alpha}, \beta, \tilde{\beta}$,

$$\dim S_d^{r,\alpha,\beta,\tilde{\alpha},\tilde{\beta}}(\Delta_{k\tilde{k}}) = \sum_{q=1}^{d} \tilde{\eta}_{d-q} \nabla\eta_q, \tag{3.20}$$

where η_q and $\tilde{\eta}_q$ are defined as in Lemma 3.3. Moreover, a basis for this space is given by the functions

$$\{\phi_i(x)\tilde{\phi}_j(y)\}_{i=\eta_{q-1}+1,\ j=1\ ,\ q=1,}^{\eta_q\qquad ,\ \tilde{\eta}_{d-q},\ d} \tag{3.21}$$

where the ϕ_i's and $\tilde{\phi}_j$'s are also defined in Lemma 3.3.

Discussion

For a proof of this result, see [16]. A lower bound for the dimension is obtained by showing that the functions in (3.21) are linearly independent and belong to S. An upper bound is obtained by choosing linear functionals as in (2.16).

Example 3.7

Let $d = 3$ and $r = \alpha = \tilde{\alpha} = \beta = \tilde{\beta} = 0$. Then

$$\dim S_3^{0,0,0,0,0}(\Delta_{11}) = 5.$$

Discussion

This space consists of piecewise cubics which vanish on the boundary of Ω and are globally continuous. A basis for this space is given by $N_2^2(x)\tilde{N}_2^2(y)$, $N_2^2(x)\tilde{N}_2^3(y)$, $N_2^2(x)\tilde{N}_3^3(y)$, $N_2^3(x)\tilde{N}_2^2(y)$, $N_3^3(x)\tilde{N}_2^2(y)$. Here, the N_i^2 are the linear B-splines associated with the knot sequence $y_1 = y_2 = a$, $y_3 = x_1$, $y_4 = y_5 = b$. The $\tilde{N}_i^q$'s are defined similarly.

Theorem 3.6 deals with splines satisfying very general boundary conditions. If we set $\alpha = \tilde{\alpha} = \beta = \tilde{\beta} = \rho$, then writing $S_d^{r,\rho}(\Delta)$ for the space (3.19), we have:

Corollary 3.8

If $d \geq 2r + 2$, then

$$\dim S_d^{r,\rho}(\Delta_{k\tilde{k}}) = (k+1)(\tilde{k}+1)(d-2r)(d-2r-1) + (k+\tilde{k}+2)(d-2r-1)(r-2\rho-1) + (r-2\rho-1)^2. \tag{3.22}$$

In particular,

$$\dim S_d^{0,0}(\Delta_{k\tilde{k}}) = k\tilde{k}\,\frac{d(d-1)}{2} + (k+\tilde{k})\frac{(d-1)(d-2)}{2} + \frac{d^2-5d+6}{2}. \tag{3.23}$$

4. TYPE 1 TRIANGULATIONS

In this section, we deal with a partition of a rectangle $\Omega = [a, b] \otimes [\tilde{a}, \tilde{b}]$ into triangles which is of particular interest for several reasons (cf. Remark 5). To define this partition, we take a rectangular partition as in Section 3 and triangulate it by drawing in the diagonals sloping up and to the right in each subrectangle. We call the resulting partition a _type-1 triangulation_ and denote it by $\Delta^1_{k\tilde{k}}$. If the x's and $\tilde{x}$'s defining this partition (cf. (3.1) - (3.2)) are equally spaced, we say $\Delta^1_{k\tilde{k}}$ is _uniform type 1_. We shall write $\Delta 1$ for a uniform type-1 partition covering the entire plane. Figure 5 shows typical uniform and non-uniform type-1 partitions.

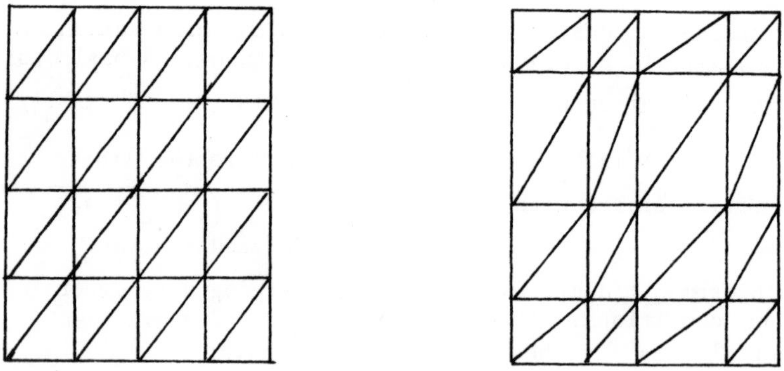

(a) Uniform (b) Nonuniform

Figure 5. Type-1 triangulations.

In most of the remainder of this section, we deal with the

uniform case. We begin with a result on dimension.

Theorem 4.1

Let $\Delta^1_{k\tilde{k}}$ be a uniform type-1 triangulation. Then for all $0 < r < d$,

$$\dim S^r_d(\Delta^1_{k\tilde{k}}) = k\tilde{k}(d^2 - 3rd + 2r^2 + \sigma) + (k + \tilde{k})(d^2 - 2rd + d - r + r^2) + (2d^2 + 4d - 2rd - r + r^2 + 2)/2, \quad (4.1)$$

where

$$\sigma = \sum_{j=1}^{d-r} (r + 1 - 2j)_+. \quad (4.2)$$

Discussion

To establish this result, we may apply Theorem 2.2. Here, we have $e = 3k\tilde{k} + 2(k + \tilde{k}) + 1$, $v = k\tilde{k}$, and ordering the vertices in the usual lexicographical way, $\sigma_i = \tilde{\sigma}_i = \sigma$ for each vertex. This result was first established in [32] for $d \le 4$ (for the general case, see [34]). It was established independently in [12] by using a result on cross-cut partitions (cf. Theorem 7.2 below).

We now give a one-sided basis for this space.

Theorem 4.2

A basis for $S^r_d(\Delta^1_{k\tilde{k}})$ is given by the union of the elements

$$\{x^m y^n : 0 \le m + n \le d\} \quad (4.3)$$

$$\{x^m y^n (x - x_i)^{r+1}_+ : 0 \le m + n \le d - r - 1\}_{i=1}^{k} \quad (4.4)$$

$$\{x^m y^n (y - y_j)^{r+1}_+ : 0 \le m + n \le d - r - 1\}_{j=1}^{\tilde{k}} \quad (4.5)$$

$$\{x^m y^n (y - x - x_i)^{r+1}_+ : 0 \le m + n \le d - r - 1\}_{i=0}^{k} \quad (4.6)$$

$$\{x^m y^n (y - x - y_j)^{r+1}_+ : 0 \le m + n \le d - r - 1\}_{j=1}^{\tilde{k}} \quad (4.7)$$

$$\{s_{\nu ij}\}_{\nu=1,}^{\delta} \,{}_{i=1,}^{k} \,{}_{j=1,}^{\tilde{k}} \tag{4.8}$$

where

$$\delta = (d - 2r + [\frac{r+1}{2}])(d - r - [\frac{r+1}{2}])_+ . \tag{4.9}$$

Here, $[\frac{r+1}{2}]$ is the largest integer less than or equal to $(r+1)/2$, and the splines $s_{\nu ij}$ in (4.8) are supported in the cone $c_{ij} = \{(x, y) : x \geq x_i, y \geq y_j\}$.

Discussion

This result follows from Theorem 4.2 of [12] where a basis of one-sided splines is constructed for general cross-cut partitions. It is not immediately obvious that the total number of functions given in (4.3) - (4.8) is equal to the expression for the dimension of S_d^r given in (4.1). To check this, we note that the number of basis functions given here is

$$\frac{(d+1)(d+2)}{2} + [2(k + \tilde{k}) + 1]\frac{(d-r)(d-r+1)}{2}$$

$$+ k\tilde{k}(d - 2r + \theta)(d - r - \theta),$$

where $\theta = [\frac{r+1}{2}]$. Rearranging this sum gives

$$k\tilde{k}(d^2 - 3rd + 2r^2 + r\theta - \theta^2)$$

$$+ (k + \tilde{k})(d^2 - 2rd + d - r + r^2)$$

$$+ (2d^2 + 4d - 2rd - r + r^2 + 2)/2.$$

Now, by considering the cases of r even and r odd, it is not hard to show that $r\theta - \theta^2 = \sigma$, where σ is defined in (4.2).

In analogy with univariate spline theory, it is natural to try to take linear combinations of the one-sided splines presented in Theorem 4.2 to construct locally supported basis elements in $S_d^r(\Delta_{k\tilde{k}}^1)$. These would be analogs of the classical B-splines. The first question using this approach is which one-sided splines have to be combined (i.e., what is the smallest <u>support set</u> for the B-spline). The following lemma helps answer this question (for an analogous result in the univariate case, see Lemma 4.7 in [33]).

Lemma 4.3

A necessary condition for a spline s of degree d and global smoothness $C^r(\mathbb{R}^2)$ to vanish identically outside of a polygon with a corner vertex with θ edges attached to it is that

$$d > \frac{\theta r + 1}{\theta - 1} . \qquad (4.10)$$

Discussion

To see that condition (4.10) is necessary, we need only consider a slice along a straight line cutting through the polygon of support - see Figure 6. Along such a line, the bivariate spline reduces to a univariate spline of degree d with θ knots at which $r + 1$ continuity conditions must hold. This implies $(\theta-1)(d+1)$ coefficients and $\theta(r+1)$ conditions; i.e., $(\theta-1)(d+1) > \theta(r+1)$ is required. This inequality reduces immediately to (4.10). The necessity of condition (4.10) seems to have been known by several researchers who did not bother to formalize the result. For a formal statement, see [1, 12].

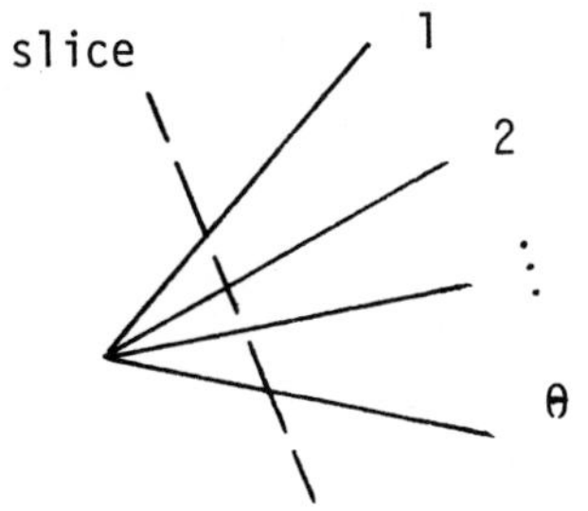

Figure 6. A corner of a support polygon

If we apply Lemma 4.3 to type-1 partitions where a typical corner has $\theta = 3$, we find that a necessary condition for the construction of locally supported elements is that

$$d > \frac{3r + 1}{2} . \qquad (4.11)$$

The following example deals with the case $r = 1$.

Example 4.4

Let Δ^1 be a uniform type-1 partition. Then there exist cubic C^1 splines B^1 and B^2 whose support sets are as shown in Figure 7. These are the smallest support sets for splines in this space.

Discussion

The spline B^1 was first constructed by Fredrickson [20] where the explicit polynomial pieces are given. Chui and Wang [9, 13] derived B^1 independently; they describe it in terms of an appropriate set of nodal values and give a correction to Fredrickson's list of polynomial pieces. B^1 is positive throughout its support set. The spline B^2 is simply B^1 rotated by 180 degrees.

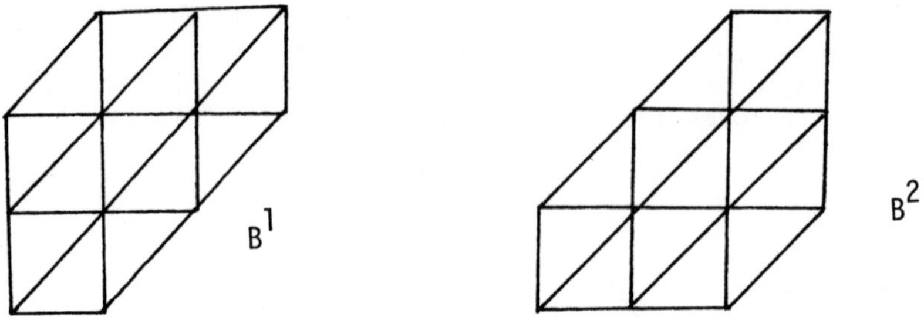

Figure 7. The supports of two B-splines in $S_3^1(\Delta^1_{k\tilde{k}})$

In working on a uniform partition, once we have one B-spline, we can get others by simply taking translates. In the univariate case, this process results in a basis, but as the next theorem shows, this is not the case in the bivariate setting.

Theorem 4.5

Let $\Delta^1_{k\tilde{k}}$ be the uniform type-1 partition of the rectangle $\Omega = [0, k+1] \otimes [0, \tilde{k}+1]$. Let B^1 and B^2 be the B-splines of Example 4.4, and let

$$B = \{B^1_{ij} : \text{support } (B^1_{ij}) \cap \Omega \neq \phi\} \tag{4.12}$$

$$\cup \{B^2_{ij} : \text{support } (B^2_{ij}) \cap \Omega \neq \phi\},$$

where $B^p_{ij}(x, y) = B^p(x - i, y - j)/_\Omega$ for $p = 1, 2$ and all i, j. Then the set B contains $2k\tilde{k} + 6(k + \tilde{k}) + 16$ elements while

$$\dim S^1_3(\Delta^1_{k\tilde{k}}) = 2k\tilde{k} + 6(k + \tilde{k}) + 13. \tag{4.13}$$

A basis for $S^1_3(\Delta^1_{k\tilde{k}})$ can be obtained from B by dropping three (appropriate) elements from B.

Discussion

The dimensionality statement (4.13) follows from Theorem 4.1 with $d = 3$ and $r = 1$. The count on B is easy. Chui and Wang [13] have given three explicit linear conditions satisfied by the elements of B, and give rules on which B-splines can be dropped to obtain a basis.

We now quote two results on subspaces of $S^1_3(\Delta^1_{k\tilde{k}})$ satisfying boundary conditions. Our first result concerns the subspace of splines which vanish along with their normal derivatives on the boundary.

Theorem 4.6

Let $\Delta^1_{k\tilde{k}}$ be a uniform type-1 partition of a rectangle Ω, and let

$$\tilde{B} = \{\phi \in B : \text{support } (\phi) \subseteq \Omega\}, \tag{4.14}$$

where B is the set of B-splines in (4.12). Then

$$\dim S^{1,1}_3(\Delta^1_{k\tilde{k}}) = 2(k - 1)_+(\tilde{k} - 1)_+, \tag{4.15}$$

and a basis is given by the set $\tilde{B}$.

Discussion

This result is established in Theorem 6.1 of [17]. The dimensionality result is obtained by looking at the space of splines $s \in S^1_3(\Delta^1_{k\tilde{k}})$ which are annihilated by a set of linear functionals forcing s to vanish along with its normal derivative along the boundary.

Our next result deals with the larger subspace of $S^1_3(\Delta^1_{k\tilde{k}})$ which vanish on the boundary. Here, the situation is more complicated. First, we need some additional B-splines. After some algebra, it can be shown (cf. [17]) that there exist B-splines in $S_3^{1,0}(\Delta^1_{k\tilde{k}})$ with support sets as shown in Figure 8. The double lines in these figures mark the edges where the spline is only C^0 instead of C^1. The obvious translates and rotations of these splines are also in $S_3^{1,0}(\Delta^1_{k\tilde{k}})$. For example, there are two elements of types B^a and B^b, one of each in the upper-left and lower-right corners, respectively.

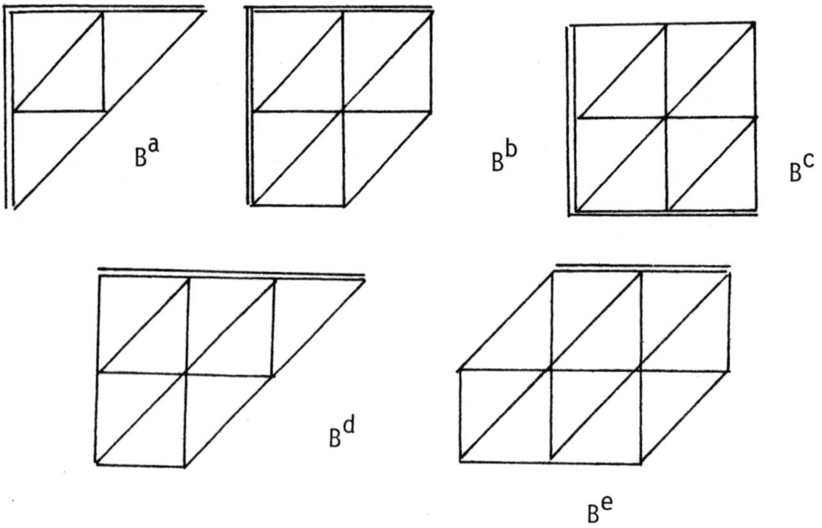

Figure 8. The supports of some B-splines in $S_3^{1,0}(\Delta^1_{k\tilde{k}})$

Theorem 4.7

If $\Delta^1_{k\tilde{k}}$ is a uniform type-1 partition, then

$$\dim S_3^{1,0}(\Delta^1_{k\tilde{k}}) = 2(k\tilde{k} + k + \tilde{k}) + 1. \tag{4.16}$$

Let

$$B^0 = \tilde{B} \cup \{B \in S_3^{1,0}(\Delta^1_{k\tilde{k}}) : B \text{ is a translate} \tag{4.17}$$

and/or rotation of one of the B-splines B^a, B^b, B^c, B^d, B^e shown in Figure 8}.

Then B^0 is a set of $2(k\tilde{k} + k + \tilde{k})$ linearly independent elements in $S_3^{1,0}(\Delta^1_{k\tilde{k}})$. In order to get a basis for this space, one additional linearly independent element B^* is needed. It is impossible to select B^* with local support; more precisely, B^* must satisfy

$$D_x S(0, \cdot) \not\equiv 0 \quad \text{and} \quad D_x S(k + 1, 0) \not\equiv 0 \tag{4.18}$$

$$\underline{\text{or}} \quad D_y S(\cdot, 0) \not\equiv 0 \quad \text{and} \quad D_y S(\cdot, \tilde{k} + 1) \not\equiv 0.$$

Discussion

This result was established in [17]. If we set loc $S_3^{1,0}(\Delta^1_{k\tilde{k}})$ = span B^0 and call a spline $s \in S_3^{1,0}(\Delta^1_{k\tilde{k}})$ <u>global</u> provided that (4.18) holds, then (cf. [17]) if s is not global, then $s \in$ loc $S_3^{1,0}(\Delta^1_{k\tilde{k}})$. The space of splines $S_3^{1,0}$ is a simple example of a reasonable space which has <u>no local basis</u>.

To illustrate further the kind of anomalies which can occur in bivariate spline theory, we now consider quartic C^2 splines (this is the next natural space to consider on type-1 partitions since $r = 2$ in (4.10) implies $d > 7/2$, and so the smallest d for which local support splines exist is $d = 4$). On a uniform type-1 partition, it was shown in [13] that there exists a quartic C^2 spline B with support as shown in Figure 9. B is positive throughout its support, and there is no quadratic C^2 spline with smaller support. It is a box spline - see Remark 7 and Theorem 4.9 below.

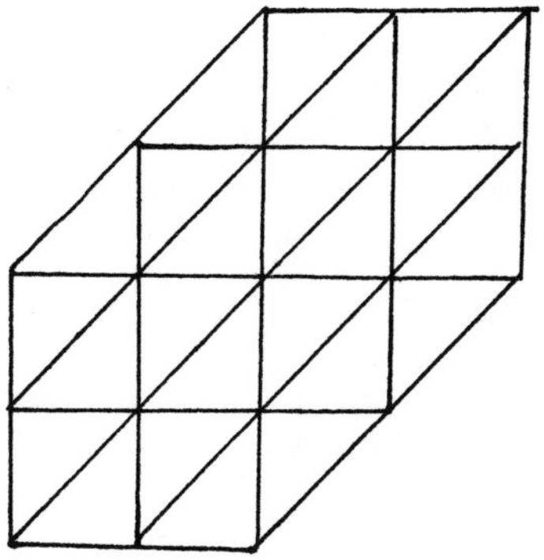

Figure 9. The support of a B-spline in $S_4^2(\Delta^1)$.

Theorem 4.8

If $\Delta^1_{k\tilde{k}}$ is a uniform type-1 partition of $\Omega = [0, k+1] \otimes [0, \tilde{k}+1]$, then

$$\dim S^2_4(\Delta^1_{k\tilde{k}}) = k\tilde{k} + 6(k + \tilde{k}) + 18. \tag{4.19}$$

Let

$$B = \{B_{ij} : \text{support } B_{ij} \cap \Omega \neq 0\}, \tag{4.20}$$

where $B_{ij} = \phi(x - i, y - j)/_\Omega$ for all i, j, and B is the element in Figure 9. Then B contains $k\tilde{k} + 4(k + \tilde{k}) + 14$ linearly independent elements. A basis for $S^2_4(\Delta^1_{k\tilde{k}})$ can be obtained from B by adding the $2(k + \tilde{k}) + 4$ one-sided splines

$$\{(y - j)^4_+\}^k_{i=0} \cup \{(x - j)^4_+\}^{\tilde{k}}_{j=0} \tag{4.21}$$

and

$$\{(x - y - i)^4_+\}^k_{i=-\tilde{k}-1}. \tag{4.22}$$

Discussion

This result is due to Chui and Wang [13]. They also show that the one-sided splines in (4.21) - (4.22) cannot be replaced by locally supported splines. More precisely, if $s \in S^2_4(\Delta^1_{k\tilde{k}})$ is not global, then $s \in \text{loc } S^2_4(\Delta^1_{k\tilde{k}}) := \text{span } B$.

In view of Lemma 4.3 (cf. (4.10), the most interesting spaces of splines on uniform type-1 partitions are the spaces

$$S^1_3,\quad S^2_4,\quad S^3_6,\quad S^4_7,\quad S^5_9,\quad \ldots\ .$$

The first two spaces have been treated above. The following result deals with the rest of the sequence.

Theorem 4.9

Let Δ^1 be a uniform type-1 partition. Then, for all $\mu = 1, 2, \ldots,$ there exist splines $B^\mu \in S^{2\mu}_{3\mu+1}(\Delta')$ with support as shown in Figure 9, but with $\mu + 1$ segments on each boundary edge. These B-splines have minimal support and are positive throughout it. Moreover, for $\mu = 1, 2, \ldots,$ there exist splines C^μ, $D^\mu \in S^{2\mu-1}_{3\mu}(\Delta^1)$ with support on the sets shown in Figure 10. These B-splines are positive on their support sets, which are minimal.

Discussion

This theorem is due to de Boor and Hollig [3] where these splines are constructed by taking convolutions. Their paper also contains results on partitions of unity, local independence, and approximation power. The splines B^μ are box splines; they correspond to the spaces S^2_4, S^4_7, ... The splines C^μ and D^μ relate to the spaces S^1_3, S^3_6, ..., but are not box splines.

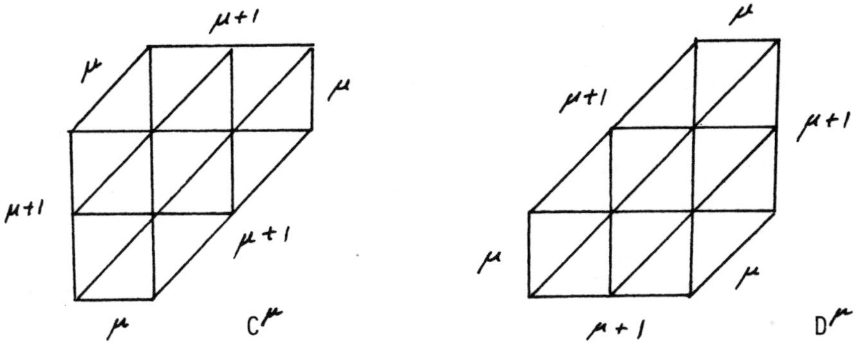

Figure 10.

So far in this section, we have dealt exclusively with uniform type-1 partitions. The situation is much more complicated for the non-uniform case. Even the dimensionality problem is difficult. We have

Theorem 4.10

Let $\Delta^1_{k\tilde{k}}$ be a (possibly non-uniform) type-1 partition of a rectangle Ω. Then for all $1 < d$,

$$\dim S^1_d(\Delta^1_{k\tilde{k}}) = k\tilde{k}(d^2 - 3d + 2) + (k + \tilde{k})(d^2 - d) + (d^2 + d + 1). \tag{4.23}$$

Discussion

This result was established in [34] by using Theorem 2.2. Indeed, if the vertices are put in lexicographical order, then $\tilde{e}_i \geq 3$ and hence, $\sigma_i = \tilde{\sigma}_i = 0$ for all vertices. In [34], we gave an example with $r = 2$ and $d = 3$ where the upper and lower bounds of Theorem 2.2 do not agree. Recently, Bamberger and Chui (unpublished) have determined $\dim S_d^r$ for $r \leq 4$ and for $d > 2r$, $r = 5, \ldots$. They also give an improved lower bound in the general case although the exact dimension remains an open question.

Recently, Chui and Wang [14, 15] looked at the question of constructing locally supported B-splines on non-uniform type-1 partitions. That this is a difficult task is shown by the following result from [14, 15].

Theorem 4.11

There exists a spline $s \in S_3^{1,1}(\Delta^1)$ on the non-uniform type-1 partition shown in Figure 11 if and only if

$$\frac{(x_{i+1} - x_i)^2}{(x_{i+2} - x_{i+1})(x_i - x_{i-1})} = \frac{(y_{j+1} - y_j)^2}{(y_{j+2} - y_{j+1})(y_j - y_{j-1})}, \tag{4.24}$$

all $i, j = 2$.

Discussion

A spline with support as in Figure 11 is uniquely determined up to a constant multiple. Moreover, there cannot exist any cubic C^1-splines with smaller support [14, 15]. It is possible to construct locally supported splines without condition (4.24), but they must have larger support sets.

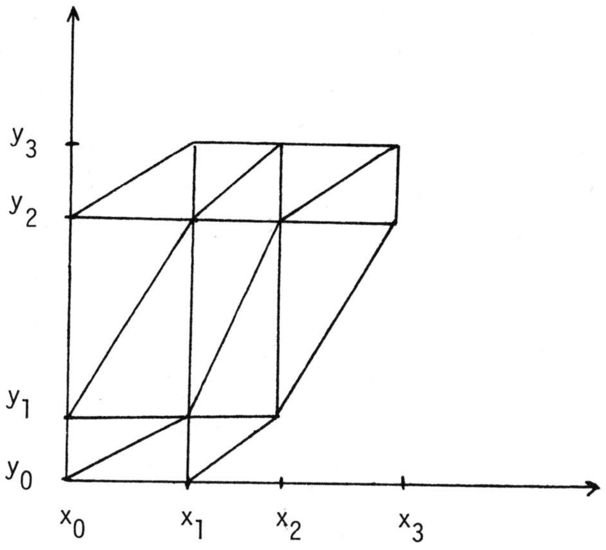

Figure 11. The support of Cubic C^1 B-splines.

5. TYPE-2 TRIANGULATIONS

In addition to the type-1 triangulation discussed in Section 4, there is another natural triangulation associated with a rectangular partition of a rectangle Ω. This second type of triangulation is obtained from the rectangular partition $\Delta_{k\tilde{k}}$ defined in (3.1) - (3.4) by drawing in diagonals in both directions. A type-2 partition extending infinitely in both directions (covering the whole plane) will be denoted by Δ^2. We denote it by Δ^2_{kk}. Typical uniform and non-uniform type-2 triangulations are shown in Figure 12. Most of this section will be devoted to the uniform case.

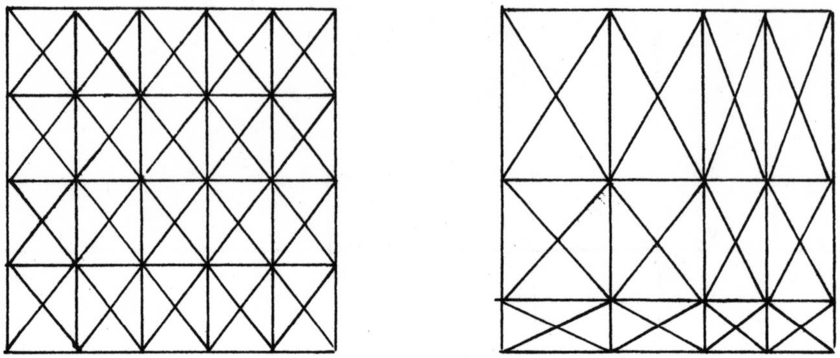

Uniform Non-uniform

Figure 12. Type-2 triangulations

We begin with a result on dimension.

Theorem 5.1

Let $\Delta^2_{k\tilde{k}}$ be a uniform type-2 triangulation of a rectangle Ω. Then

$$\begin{aligned}\dim S^r_d(\Delta^2_{k\tilde{k}}) &= k\tilde{k}(2d^2 - 6rd + 4r^2 + \sigma_g + \sigma_c) \\ &+ (k + \tilde{k})(2d^2 - 5rd + d - r + 3r^2 + \sigma_c) \\ &+ (4d^2 + 4d - 8rd - r + 5r^2 + 2\sigma_c + 2)/2,\end{aligned} \tag{5.1}$$

where

$$\sigma_g = \sum_{j=1}^{d-r} (r + 1 - 3j)_+ \tag{5.2}$$

$$\sigma_c = \sum_{j=1}^{d-r} (r + 1 - j)_+ . \tag{5.3}$$

Discussion

This result follows from Theorem 2.2 if we note that $e = 6k\tilde{k} + 5(k + \tilde{k}) + 4$ and $v = 2k\tilde{k} + (k + \tilde{k}) + 1$. In particular, we note that $k\tilde{k}$ of the vertices are on the rectangular grid (where $\tilde{e}_i = e_i = 4$) while $k\tilde{k} + (k + \tilde{k}) + 1$

of them are in the centers of the subrectangles (where $e_i = \tilde{e}_i = 2$). This proof was given in [34] (see also [32] for $d \le 4$). An independent proof based on general results for cross-cut partitions (cf. Theorem 7.2 below) was given in [12].

Using results on cross-cut partitions, we can now give a basis for $S_d^r(\Delta^2_{k\tilde{k}})$ consisting of one-sided splines.

Theorem 5.2

A basis for $S_d^r(\Delta^2_{k\tilde{k}})$ is given by the functions (4.3) - (4.7) coupled with:

$$\{x^m y^n (y + x - x_i)_+^{r+1} \;:\; 0 \le m + n \le d - r - 1\}_{i=1}^{k+1} \tag{5.4}$$

$$\{x^m y^n (y + x - y_j)_+^{r+1} \;:\; 0 \le m + n \le d - r - 1\}_{j=1}^{\tilde{k}} \tag{5.5}$$

$$\{s_{\nu ij}\}_{\nu=1,\ i=1,\ j=1}^{\tilde{\delta},\ \ k,\ \ \tilde{k}} \tag{5.6}$$

$$\{\tilde{s}_{\nu ij}\}_{\nu=1,\ i=1,\ j=1}^{\varepsilon,\ \ k+1,\ \ \tilde{k}+1}, \tag{5.7}$$

where

$$\tilde{\delta} = (d - r - [\tfrac{r+1}{3}])_+ \cdot (3d - 5r + 3[\tfrac{r+1}{3}] + 1)/2 \tag{5.8}$$

$$\varepsilon = (d - 2r - 1)_+ (d - 2r)/2. \tag{5.9}$$

The $s_{\nu ij}$ are splines which vanish outside of cones with vertices at (x_i, y_j), while the $\tilde{s}_{\nu ij}$ are splines which vanish outside of cones with vertices at $(\frac{x_i + x_{i+1}}{2}, \frac{y_j + y_{j+1}}{2})$.

Discussion

This result follows from Theorem 4.2 of [12] which deals with general cross-cut partitions - see Theorem 7.2 below. The total number of basis functions is

$$N = \frac{(d+1)(d+2)}{2} + \frac{(d-r)(d-r+1)}{2}[3k + 3\tilde{k} + 2] \tag{5.10}$$
$$+ k\tilde{k}(\delta + \tilde{\delta} + \varepsilon) + (k + \tilde{k} + 1)\varepsilon,$$

where δ is given in (4.9). It is not obvious that N is equal to the expression for the dimension given in (5.1), but a careful analysis of cases shows that they are equal.

We turn now to the question of constructing locally supported basis elements. Assuming that we are able to construct an element whose support has corners at vertices on the original grid partition (where the number of edges with different slopes which meet there is four), Lemma 4.3 tells us that we need $d > \frac{4r+1}{3}$. We now examine the first interesting case of $r = 1$ which suggests that we consider $d = 2$; i.e., quadratic splines. The following example shows that there indeed do exist locally supported B-splines in $S_2^1(\Delta_{k\tilde{k}}^2)$.

Example 5.3

If Δ^2 is a uniform type-2 partition, then there exists a unique quadratic C^1 spline ϕ with support on the set shown in Figure 13. The spline ϕ is positive throughout the interior of its support set.

Discussion

The spline ϕ was first constructed by Zwart [41]. It was later discovered independently by Powell [29], and Chui and Wang [11], where explicit formulae for the polynomial pieces are given as well as a description in terms of nodal values. We note that there is no quadratic C^1 spline with smaller support on Δ^2.

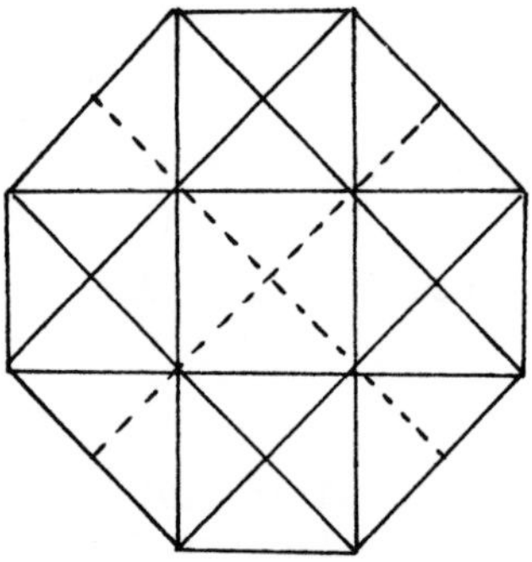

Figure 13. The support of a quadratic C^1 B-spline

We can now give a theorem about the possibility of using translates of the B-spline of Example 5.3 to construct a basis for $S_2^1(\Delta_{k\tilde{k}}^2)$.

Theorem 5.4

Let $\Delta_{k\tilde{k}}^2$ be the uniform type-2 partition of the rectangle $\Omega = [0, k+1] \otimes [0, \tilde{k}+1]$. Then

$$\dim S_2^1(\Delta_{k\tilde{k}}^2) = k\tilde{k} + 3(k + \tilde{k}) + 8 \tag{5.11}$$

Let

$$B = \{B_{ij} : \text{support } (B_{ij}) \cap \Omega \neq \phi\}, \tag{5.12}$$

where $B_{ij}(x, y) = \phi(x - i, y - j)$ for all i, j. Then the set B contains $k\tilde{k} + 3(k + \tilde{k}) + 9$ elements. These elements are linearly dependent; in particular,

$$\sum_i \sum_j (-1)^{i+j} B_{ij}(x, y) \equiv 0, \quad \text{all } (x, y) \in \Omega. \tag{5.13}$$

They form a partition of unity; i.e.,

$$\sum_i \sum_j B_{ij}(x, y) \equiv 1, \quad \text{all } (x, y) \in \Omega.$$

To obtain a basis for $S_2^1(\Delta_{k\tilde{k}}^2)$, we may drop any one element from B.

Discussion

The dimensionality statement follows by setting $r = 1$ and $d = 2$ in Theorem 5.1. The count on B is trivial. The dependency condition (5.13) and the fact that dropping one element from B leads to an independent set are established in Chui and Wang [11].

We now quote two results on subspaces of $S_2^1(\Delta_{k\tilde{k}}^2)$ satisfying boundary conditions. Our first result concerns the subspace of splines which vanish along with their normal derivatives along the boundary of Ω.

Theorem 5.5

Let $\Delta_{k\tilde{k}}^2$ be a uniform type-2 partition, and let

$$B^1 = \{s \in B : \text{support } (s) \subseteq \Omega\}, \tag{5.14}$$

where B is the set of B-splines in (5.12). Then

$$\dim S_2^{1,1}(\Delta_{k\tilde{k}}^2) = (k - 1)_+(\tilde{k} - 1)_+, \tag{5.15}$$

and a basis is given by B^1.

Discussion

For the proof of this result, see [18].

We turn now to the larger subspace $S_2^{1,0}(\Delta_{k\tilde{k}}^2)$ of $S_2^1(\Delta_{k\tilde{k}}^2)$ consisting of splines which vanish on the boundary of Ω. Here, the situation is more complicated, and we need to introduce some additional locally supported B-splines. It is shown in [18] that there exist quadratic C^1 B-splines with support on the sets shown in Figure 14. These B-splines are globally C^1 except across the edges marked with double lines where they are only C^0, and thus (along with suitable translations and rotations) are suitable for use along the boundary of Ω.

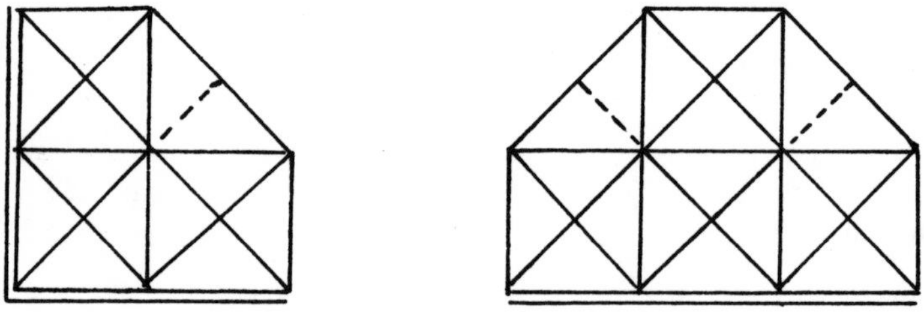

B_c B_s

Figure 14. Supports of some B-splines in $S_2^{1,0}(\Delta_{k\tilde{k}}^2)$

Theorem 5.6

If $\Delta_{k\tilde{k}}^2$ is a uniform type-2 partition, then

$$\dim S_2^{1,0}(\Delta_{k\tilde{k}}^2) = k\tilde{k} + k + \tilde{k}. \tag{5.16}$$

Let

$$B^0 = B^1 \cup \{s \in S_2^{1,0}(\Delta_{k\tilde{k}}^2) : B \text{ is a translation and/or rotation of the splines } B_c \text{ or } B_s \text{ shown in Figure 14}\}. \tag{5.17}$$

Then B^0 contains $k\tilde{k} + (k + \tilde{k}) + 1$ elements, and dropping any one element from B^1, we obtain a basis for $S_2^{1,0}(\Delta_{k\tilde{k}}^2)$.

Discussion

For a proof of this result, see [18].

We now discuss some results for spaces of smoother splines on uniform type-2 partitions. As noted above, in order for locally supported elements to exist on this type of partition, we need $d > \frac{4r+1}{3}$. Thus, the spaces of interest are

$$S_2^1,\ S_4^2,\ S_5^3,\ S_6^4,\ S_8^5,\ S_9^6,\ S_{10}^7,\ \ldots .$$

The following result deals with a subsequence of this sequence (every fourth space).

Theorem 5.7

Let Δ^2 be a uniform type-2 partition. Then, for each $\mu = 1, 2, \ldots,$ there exists a spline $B^\mu \in S_{4\mu-2}^{3\mu-2}(\Delta^2)$ whose support is the set shown in Figure 13 but with μ segments on each boundary edge. B^μ is uniquely determined up to a constant multiple and is positive throughout its support set. No other elements of these spline spaces have smaller support sets.

Discussion

The case $\mu = 1$ is precisely the case discussed in Example 5.3 and Theorems 5.4 - 5.6. The result for $\mu > 1$ is due to de Boor and Hollig [4] who note that the B^μ can be constructed as box splines (cf. Remark 7) by a convolution process. This theorem deals with the spaces S_2^1, S_6^4, S_{10}^7, $\ldots$. A variety of other results (such as local independence, a partition of unity, and approximation power) for these B-splines can be found in [4].

The following theorem deals with some spline spaces where the degree is larger than necessary for the existence of locally supported splines.

Theorem 5.8

Let Δ^2 be a uniform type-2 partition. Then for each $\mu = 1, 2, \ldots,$ there exist splines A^μ and $\tilde{A}^\mu$ in $S_{4\mu-3}^{3\mu-3}(\Delta^2)$ with supports as shown in Figure 15. These splines are positive throughout their support set, but are not generally splines of smallest possible support.

Discussion

This result is due to de Boor and Hollig [4]. These B-splines are constructed by convolution. This paper also includes results on partitions of unity, local independence, and the approximation power of linear combinations of these B-splines. The spaces covered by this theorem are S_1^0, S_5^1, S_9^4, ...

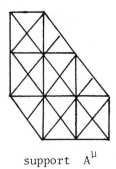

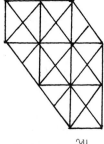

support A^μ Support $\tilde{A}^\mu$

Figure 15. Supports of some B-splines in $S_{4\mu-3}^{3\mu-3}(\Delta^2)$

We give one more result for uniform type-2 partitions.

Theorem 5.9

If Δ^2 is a uniform type-2 partition, then there exist splines $C, D \in S_3^1(\Delta^2)$ with support sets on 4 and 9 rectangles, respectively. These splines are positive throughout their support.

Discussion.

Explicit formulae for C and D are given in Chui and Wang [10], along with a partition of unity and some approximation results.

All of the results presented so far in this section have been for uniform type-2 triangulations. We now present two recent theorems concerning the non-uniform case.

Theorem 5.10

Let $\Delta^2_{k\tilde{k}}$ be an arbitrary type-2 triangulation. Then for $r = 0, 1, 2,$ and all $d > r$,

$$\dim S^r_d(\Delta^2_{k\tilde{k}}) = k\tilde{k}(2d^2 - 6rd + 4r^2 + \sigma_c) \qquad (5.18)$$
$$+ (k + \tilde{k})(2d^2 - 5rd + d - r + 3r^2 + \sigma_c)$$
$$+ (4d^2 + 4d - 8rd - r + 5r^2 + 2 + 2\sigma_c)/2,$$

where σ_c is defined in (5.3).

Discussion

This theorem was proved in [34]. It follows directly from Theorem 2.2 A similar result holds with $r = 3$; the only difference is that the dimension is increased by $\rho := $ number of grid vertices where the upward sloping vertices are equal. The dimensionality of these spaces for $r > 3$ remains an open question. It is easy to construct an example with $r = 4$, $d = 5$, where the upper and lower bounds of Theorem 2.2 differ.

We close this section with statements concerning the possibility of constructing locally supported splines on non-uniform type-2 partitions and using them to get a basis.

Theorem 5.11

Given any arbitrary type-2 partition, there exists a unique quadratic C^1 spline B with support on an octahedron as in Figure 16.

Discussion

This B-spline is constructed explicitly in Chui and Wang [14, 15]. It is uniquely determined up to a constant, and it is impossible to construct a quadratic C^1 spline with a smaller support. In addition, they show that given a general type-2 partition of a rectangle Ω, it is possible to define B-splines B_{ij} with support as octahedrons (extending outside of Ω if necessary) such that

$$\sum\sum(-1)^{i+j}(x_{i+1} - x_i)(y_{j+1} - y_j)B_{ij}(x, y) \equiv 0, \qquad (5.19)$$
$$\text{all } (x, y) \in \Omega.$$

This is the analog of (5.13) in the uniform case, and shows the dependency of the B-splines. Finally, they also show that the B_{ij} form a partition of unity:

$$\sum\sum B_{ij}(x, y) \equiv 1, \quad \text{all} \quad (x, y) \in \Omega. \tag{5.20}$$

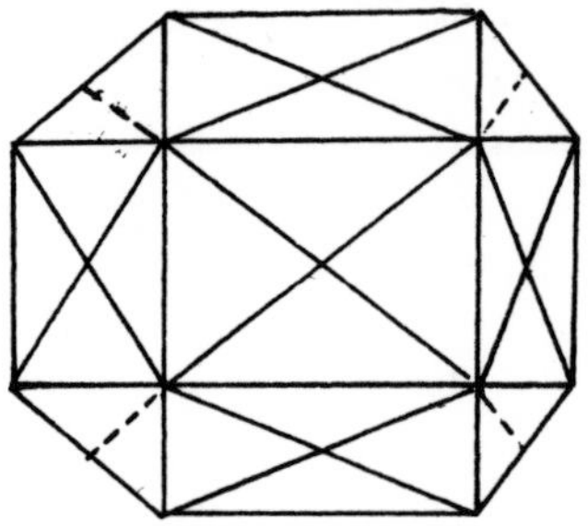

Figure 16. The support of a quadratic C^1 B-spline

Theorem 5.12

Let $\Delta^2_{k\tilde{k}}$ be an arbitrary type-2 triangulation and let $B = \{B_{ij} : \text{support } (B_{ij}) \cap \Omega \neq \phi\}$, where the B_{ij} are quadratic C^1 B-splines given in Theorem 5.11. Then

$$\dim S^1_2(\Delta^2_{k\tilde{k}}) = k\tilde{k} + 3(k + \tilde{k}) + 8, \tag{5.21}$$

and a basis can be obtained by dropping any one B-spline from B.

Discussion

This result is due to Chui and Wang [15].

6. SOME RESULTS FOR GENERAL TRIANGULATIONS

For this section, we present several results for general triangulations, but for restricted choices of d. Our first result concerns C^0 splines. Throughout this section, we shall assume that Δ is a <u>triangulation</u> of a set Ω, and shall use the notation (cf. Section 2)

$$\begin{aligned} e &= \text{number of interior edges in } \Delta \\ v &= \text{number of interior vertices in } \Delta \\ E &= \text{total number of edges of } \Delta \\ V &= \text{total number of vertices of } \Delta \end{aligned} \tag{6.1}$$

Theorem 6.1

For all $d > 0$,

$$\dim S_d^0(\Delta) = \frac{(d+1)(d+2)}{2} + \frac{d(d+1)}{2} e - \frac{(d^2+3d)}{2} v \tag{6.2}$$

Discussion

This result follows directly from Theorem 2.2 since $\sigma_i = \tilde{\sigma}_i = 0$, all i.

We now show how to construct a local basis for $S_d^0(\Delta)$. There are three basic kinds of B-splines we can use (cf. Figure 17):

- splines which have support on a single triangle.
- splines which have support on a diamond.
- splines which have support on a star.

For details on constructing these splines, see [17, 18]. There are $(d-1)(d-2)/2$ linearly independent splines with support on each triangle, $d - 1$ linearly independent

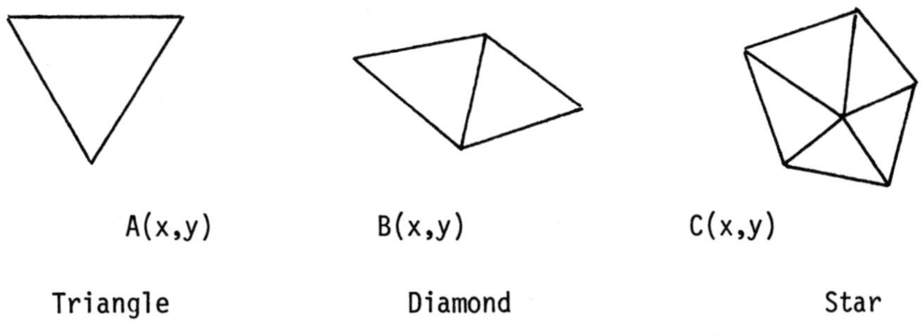

Figure 17. Supports of splines in $S_d^0(\Delta)$

splines with support on each diamond (but not on any subset of it), and exactly one spline with support on each star (but not on any subset of it).

Theorem 6.2

Let

$$\{A_{ij\nu}\}_{i=0,\ j=0,\ \nu=1}^{d-3,\ i,\ \ T} \tag{6.3}$$

$$\{B_{i\nu}\}_{i=1,\ \nu=1}^{d-1,\ E} \tag{6.4}$$

$$\{C_i\}_{i=1}^{V} \tag{6.5}$$

be the collection of locally supported splines in $S_d^0(\Delta)$ associated with a triangulation Δ, where the A-splines have supports on triangles, the B-splines have supports on diamonds, and the C-splines have supports on stars. Then this set of splines form a basis for $S_d^0(\Delta)$.

Discussion

It is not hard to construct a dual set of linear functionals by putting $(d-1)(d-2)/2$ point functionals in each triangle, $d - 1$ point functionals on each edge, and 1 point functional at each vertex. The total number of elements in (6.3) - (6.5) is $\frac{(d-1)(d-2)}{2}T + (d-1)E + V$. This does not appear to be the same as the expression (6.2) for $\dim S_d^0(\Delta)$, but they are, in fact, identical as can be seen by making use of the identities

$$T = e - v + 1 \tag{6.6}$$

$$V = 3 - 2v + 3 \tag{6.7}$$

$$E = 2e - 3v + 3 \tag{6.8}$$

These, in turn, are immediate consequences of the following well-known relations for any triangulation:

$$3T = E + e \tag{6.9}$$

$$E - e = V - v \tag{6.10}$$

$$T = V + v - 2\ . \tag{6.11}$$

Concerning the subspace $S_d^{0,0}(\Delta)$ of splines satisfying zero boundary conditions, we have

Theorem 6.3

For any triangulation,

$$\dim S_d^{0,0}(\Delta) = \frac{(d-1)(d-2)}{2} T + (d-1)e + v . \qquad (6.12)$$

A basis is given by

$$\{A_{ijv}\}_{i=0,\ j=0,\ v=1}^{d-3,\ i,\ T} \cup \{B_{iv}\}_{i=1,\ v=1}^{d-1,\ e} \cup \{C_i\}_{i=1}^{v} . \qquad (6.13)$$

Discussion

The splines listed in (6.13) are precisely those basis elements from Theorem 6.2 which vanish on the boundary.

There is a similar result for C^1 splines on arbitrary partitions, provided that we consider splines of sufficiently high degree.

Theorem 6.4

For any triangulation Δ and all $d \geq 5$,

$$\dim S_d^1(\Delta) = \frac{(d-4)(d-5)}{2} T + (2d-7)E + 3V^* + 4(V - V^*) \qquad (6.14)$$

V^* : = number of interior vertices where at least three edges with different slopes are attached. (6.15)

Moreover, there exists a basis for $S_d^1(\Delta)$ consisting of splines

$$\{A_{ij\nu}\}_{i=0,\ j=0,\ \nu=1}^{d-6,\ i,\ T} \qquad (6.16)$$

$$\{B_{i\nu}\}_{i=1,\ \nu=1}^{2d-7,\ E} \qquad (6.17)$$

$$\{C_{i\nu}\}_{i=1,\ \nu=1}^{3,\ V^*} \cup \{\tilde{C}_{i\nu}\}_{i=1,\ \nu=1}^{4,\ V-V^*} , \qquad (6.18)$$

where the A-splines have support on single triangles, the B-splines have support on diamonds, and the C-splines have support on stars (cf. Figure 17).

Discussion

This result is due to Morgan and Scott [26]. A dual basis

for the splines (6.16) - (6.18) can be constructed by placing $(d-5)(d-4)/2$ point evaluations in each triangle, $2d - 7$ on each edge, and either 3 or 4 at each vertex. We put 3 at each nonsingular interior vertex (i.e., those satisfying (6.15)), and 4 at each singular interior vertex and at each boundary vertex.

Theorem 6.4 gives an explicit expression for the dimension of C^1 spline spaces on arbitrary triangulations provided $d \geq 5$. Example 2.3 (with $d = 2$) showed that such explicit expressions are impossible in general (as the dimension may depend on the exact geometry). To my knowledge, the cases $d = 3, 4$ still remain open - i.e., it is now known whether the dimension of $S_d^1(\Delta)$ depends on the geometry or not. We conclude this section with a result on a subspace with boundary conditions.

Theorem 6.5

For all $d \geq 5$,

$$\dim S_d^{1,1}(\Delta) = \frac{(d-4)(d-5)}{2} T + (2d-7)e + 3V^* + 4(v - V^*). \tag{6.19}$$

A basis for this space can be obtained by taking only those basis splines in (6.16) - (6.18) whose supports lie inside Ω.

7. CROSS-CUT PARTITIONS

In this section, we briefly survey recent results on still another kind of partition where the questions of dimensionality and construction of one-sided bases for associated spline spaces can be answered. Let Ω be the closure of any open simply connected subset of $\mathbb{R}^2$. Then, given L straight lines cutting through Ω, it is apparent that these lines partition Ω into pieces satisfying (1.1) - (1.2). The resulting partition Δ is called a cross-cut partition (cf. [6, 9, 12]. The special case where Δ is such that at most two lines intersect at any given point inside Ω is called a simple cross-cut partition.

Figure 18 shows a typical cross-cut partition. The rectangular type-1 and type-2 partitions discussed in Sections 3 - 5 are examples of cross-cut partitions. On the other hand, most triangulations (cf. Figures 2 and 3) are not cross-cut partitions.

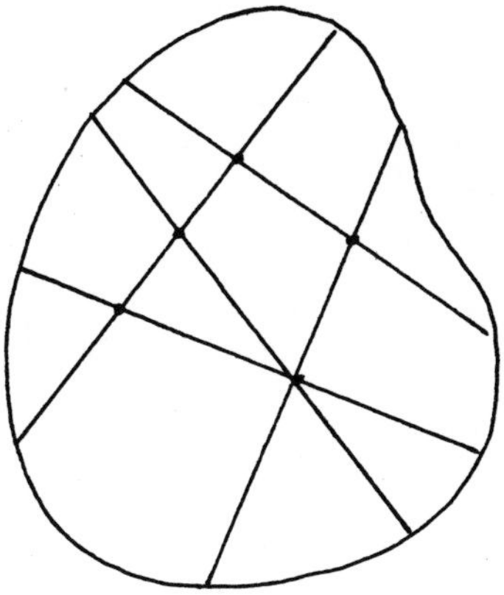

Figure 18. A cross-cut partition

Theorem 7.1

Suppose Δ is a simple cross-cut partition of Ω defined by L lines forming e edges and v vertices. Then, for all $0 \leq r < d < 2r + 1$,

$$\dim S_d^r(\Delta) = \alpha + \beta(e - 2v), \tag{7.1}$$

and for all $2r + 1 \leq d$,

$$\dim S_d^r(\Delta) = \alpha + \beta e - \frac{(d^2 + 3d - 2r^2 - 4r)}{2} v \tag{7.2}$$

where $\alpha = (d+1)(d+2)/2$ and $\beta = (d-r)(d-r+1)/2$. Moreover, a basis for S can be constructed in the form

$$\{x^m y^n\} \quad 0 \leq m + n \leq d \tag{7.3}$$

$$\{x^m y^n (\ell_\nu^{r+1})_+, \ \nu = 1, \ \ldots, \ L\} \quad 0 \leq m + n \leq d - r - 1 \tag{7.4}$$

$$\{x^m y^n (\ell_\nu^{r+1})_+ (\ell_\mu^{r+1})_+, \ (\nu,\mu) \in \Lambda\} \tag{7.5}$$
$$0 \leq m + n \leq d - 2r - 2$$

where $\Lambda = \{(\nu, \mu)$ such that lines ν and μ intersect$\}$ and where ℓ_ν is the equation of the νth line and the $+$ has the usual one-sided meaning.

Discussion

This result is due to Chui and Wang [6, 8]. When $d < 2r + 1$, none of the functions (7.5) need be included. The functions in (7.4) and (7.5) are one-sided splines. The support sets for typical elements in (7.4) and (7.5) are shown in Figure 19. This result holds even if Ω is non-convex.

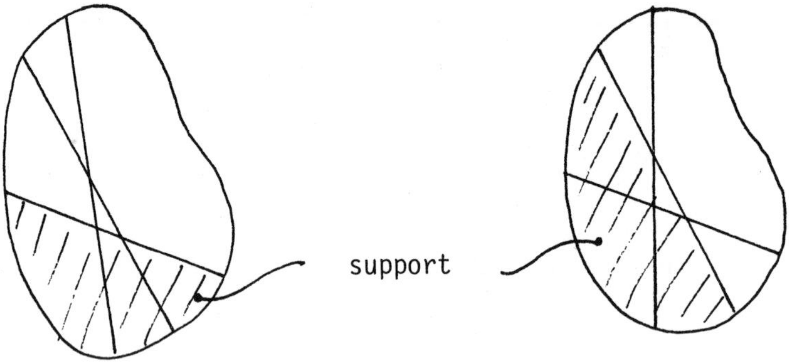

type (7.4) type (7.5)

Figure 19. Support of one-sided splines.

The situation for the general cross-cut case is slightly more complicated.

Theorem 7.2

If Δ is a general cross-cut partition, then

$$\dim S_d^r(\Delta) = \alpha + \beta L + \sum_1^V d(e_i), \tag{7.6}$$

where in general,

$$d(\eta) = \left(d - r - \left[\frac{r+1}{\eta-1}\right]\right)_+ \cdot \left(d(\eta - 1) - r(\eta+1) + (\eta - 3) + (\eta - 1)\left[\frac{r+1}{\eta-1}\right]\right)/2 \tag{7.7}$$

and e_i is (as before) the number of edges with different slopes attached to the ith vertex.

Discussion

This result is due to Chui and Wang [12]. Here,

$\alpha = (d+1)(d+2)/2$ and $\beta = (d-r)(d-r+1)/2$, as usual. For simple cross-cut partitions, $e_i = 2$, all i, and since $d(2) = (d-2r-1)_+ (d-2r)$, Theorem 7.1 follows as a special case once we take account of the fact that for simple cross-cut partitions, $L = e - 2v$ (cf. [8]).

A basis for $S_d^r(\Delta)$ in the general cross-cut case has been constructed by Chui and Wang. Since the notation is quite complicated, we refer the reader to [12] for details.

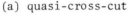

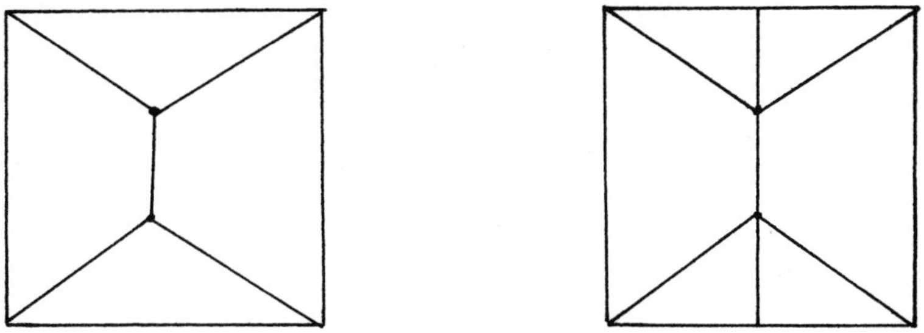

(a) quasi-cross-cut (b) non-quasi-cross-cut

Figure 20. Two more general partitions

8. REMARKS

1. Practitioners of the finite element method (see [25, 28]) for solving boundary-value problems for partial differential equations have been using spaces of piecewise polynomials defined on partitions (mostly triangulations) of a set $\Omega \subseteq \mathbb{R}^2$ for a long time. The finite element approach usually involves constructing locally supported "elements" (cf. [39, 40]) and then approximating by their linear span. Generally, finite elements have been constructed with support on a single triangle or on a "star"; cf. Figure 17.

2. Generally, the space spanned by a collection of finite elements is a proper subset of the full space of smooth piecewise polynomials on a partition Δ. One of the first practitioners of the finite element theory to suggest looking at the full space was Strang [35]. In [35], he gives a conjecture for $\dim S_d^r(\Delta)$ on triangulations which in fact is equal to the lower bound in Theorem 2.2, but with the factor $\sigma_1, + \dots + \sigma_V$ missing.

3. The space of tensor-product polynomials in two variables is defined by $P_{(d_1,d_2)} = \{p(x, y) = \sum_{i=0}^{d_1} \sum_{j=0}^{d_2} c_{ij} x^i y^j\}$. This is different from the space of polynomials of total degree d defined in (1.3) and used throughout this paper. For results on tensor-product splines, see [33].

4. Another approach to studying spaces of smooth piecewise polynomials is to construct locally supported elements as projections (or shadows) of objects (usually simplices or parallelopipeds) in higher dimensions. Elements produced in this way are called multidimensional B-splines. For a survey of the explosive development of this subject over the past few years, see Dahmen and Micchelli [19]. In general, this approach leads to partitions which are quite complicated, but for some special cases, one gets elements corresponding to uniform type-1 and type-2 partitions as discussed in Sections 4 and 5. Since one starts with certain basic splines and looks at their span, this theory generally deals with subspaces of the spline spaces considered in this paper.

5. Type-1 and type-2 partitions seem to be sufficiently natural to have been studied independently by a variety of researchers (cf. e.g., [3, 4, 9, 18, 20, 21, 27, 29, 30, 32, 41]). Type-1 partitions are also known as "uni-diagonal meshes" and as "three direction meshes" in the uniform case. Similarly, type-2 partitions have also been referred to as "criss-cross triangulations" and as "four direction meshes" in the uniform case.

6. In addition to the various terminologies in use, the notation in the papers referenced here varies from paper to paper. For example, in [32], m was used for the order of the polynomials (one more than the degree). Some papers (e.g., [32]) dealing with rectangular and type-1/type-2 partitions use $k - 1$ instead of k lines to define the mesh. Thus, care is required in comparing results.

7. Multivariate B-splines which are generated as shadows of parallelopipeds are called box splines. For some results on box splines, see [2 - 5] and the survey [19]. The quartic C^2 spline shown in Figure 9 is an example of a box spline.

8. In several theorems, we have given bases for $S_d^r(\Delta)$ with local support and forming a partition of unity (i.e., adding up to identically 1 throughout Ω). The importance

of having a partition of unity is that often it can be used to define useful approximation operators called quasi-interpolants. Typically, if $1 \equiv \sum\sum B_{ij}(x, y)$, then we can define $Qf(x, y) = \sum\sum(\lambda_{ij}f)B_{ij}(x, y)$, where $\lambda_{ij}f$ are linear functionals operating on f. Often, $\lambda_{ij}f$ can be taken as simply the value of f at an appropriate point. For examples, see [1 - 4, 9, 11].

9. Quasi-interpolation operators (cf. Remark 8) were used in [6, 8] to study the question of when spaces of splines $S_d^r(\Delta)$ are dense in $C(\Omega)$ as some measure of the fineness of Δ goes to zero. For other results on the approximation power of $S_d^r(\Delta)$, see [1 - 4, 19], and the references therein. There has been considerable work on the approximation power of the spaces spanned by multivariate B-splines - see the survey [19] and references therein.

10. So far, there has not been much work on the problem of interpolation using the spaces $S_d^r(\Delta)$. For some special results, see [22 - 24, 29]. Somewhat more has been done for linear combinations of multi-variate B-splines; see e.g., [5] and the references therein.

11. In Sections 3 - 5, we have given several results for subspaces of $S_d^r(\Delta)$ satisfying boundary conditions. In some cases, the dimensionality of these subspaces can be determined by choosing a minimal set of linear functionals to force the boundary conditions and subtracting the number of these functionals from the dimension of $S_d^r(\Delta)$ itself. This does not always work, however; see the examples in [17, 18].

12. In Section 7, we surveyed results on cross-cut partitions. Chui and Wang [12] have also determined the dimension of $S_d^r(\Delta)$ in the case where Δ is similar to, but more general than a cross-cut partition. In particular, they have treated partitions Δ where each edge in the interior of Ω lies on a ray which terminates on the boundary of Ω. Such partitions are called quasi-cross-cut partitions. For an example, see Figure 20. It is easy to construct examples of rectilinear partitions which are not quasi-cross-cut partitions. So far, no basis has been constructed for spaces defined on general quasi-cross-cut partitions.

13. As we have seen in this survey, there remain many open questions in the theory of multivariate splines, and work is still in progress on several of them. For example, Chui has recently obtained some results on spaces satisfying periodic boundary conditions (a case of interest mentioned in [27, 35]). I have partial results on $S_d^r(\Delta)$ for arbitrary triangulations provided d is sufficiently large compared with r (cf. Theorem 6.4 for the case $r = 1$). Several researchers are working on the cases of non-uniform type-1 and type-2 partitions. So far, there is essentially nothing available for more than 2 variables.

References

1. de Boor, C. and DeVore, R. "Approximation by smooth multivariate splines", Trans. Amer. Math. Soc. 276 (1983), pp. 775-788.

2. de Boor, C. and Hollig, K. "B-splines from Parallelpipeds", J. d'Anal. 42 (1982-83), pp. 99-115.

3. de Boor, C. and Hollig, K. "Bivariate splines of minimal support on a regular mesh", J. Comp. and Appl. Math. 9 (1983), pp. 13-28.

4. de Boor, C. and Hollig, K. "Bivariate box splines on a four direction mesh", manuscript.

5. de Boor, C., Hollig, K., and Riemenschneider, S. "Bivariate cardinal interpolation by splines on a three dimension mesh", Ill. J. Math.

6. Chui, C. K. and Wang, R. H. "Bases of bivariate spline spaces with cross-cut grid partitions", J. Math. Research and Exposition 2 (1982), pp. 1-4.

7. Chui, C. K. and Wang, R. H. "A generalization of univariate splines with equally spaced knots to multivariate splines", J. Math. Research and Exposition 2 (1982), pp. 99-104.

8. Chui, C. K. and Wang, R. H. "On smooth multivariate spline functions", Math. Comp. 41 (1983), pp. 131-142.

9. Chui, C. K. and Wang, R. H. "Bivariate cubic B-splines relative to cross-cut triangulations", Chinese Anals. To appear.

10. Chui, C. K. and Wang, R. H. "Multivariate B-splines on triangulated rectangles", J. Math. Anal. Appl. 92 (1983), pp. 533-551.

11. Chui, C. K. and Wang, R. H. "On a bivariate B-spline basis", Sci. Sinica. To appear.

12. Chui, C. K. and Wang, R. H. "Multivariate spline spaces", J. Math. Anal. Appl. 94 (1983), pp. 197-221.

13. Chui, C. K. and Wang, R. H. "Spaces of bivariate cubic and quartic splines on type-1 triangulations", J. Math. Anal. Appl. (1983). To appear.

14. Chui, C. K. and Wang, R. H. "Bivariate B-splines on triangulated rectangles", in Approximation Theory IV, C. K. Chui, L. L. Schumaker, and J. D. Ward, eds., Academic Press, N.Y. (1983), pp. 413-418.

15. Chui, C. K. and Wang, R. H. "Concerning C^1 B-splines on triangulations of non-uniform rectangular partitions", J. Approx. Th. and Applic., to appear.

16. Chui, C. K. and Schumaker, L. L. "On spaces of piecewise polynomials with boundary conditions", I. Rectangles, in Multivariate Approximation Theory II, ed. by W. Schempp and K. Zeller, Birkhauser, Basel (1982), pp. 69-80.

17. Chui, C. K., Schumaker, L. L., and Wang, R. H. "On spaces of piecewise polynomials with boundary conditions", II. Type-1 triangulations, in Second Edmonton Conference on Approximation Theory, CMS Vol. 3, American Math. Soc., Providence (1983), pp. 51-66.

18. Chui, C. K., Schumaker, L. L., and Wang, R. H. "On spaces of piecewise polynomials with boundary conditions", III. Type-2 triangulations, in Second Edmonton Conference on Approximation Theory, CMS Vol. 3, American Math. Soc., Providence (1983), pp. 67-80.

19. Dahmen, W. and Michelli, C. A. "Recent progress in Multivariate Splines", in Approximation Theory IV, C. K. Chui, L. L. Schumaker, and J. D. Ward, eds., Academic Press, N.Y. (1983) pp. 27-121.

20. Fredrickson, P. "Triangular spline interpolation", Report #670, Whitehead Univ. (1970).

21. Fredrickson, P. "Generalized triangular splines", Report #7-71, Lakehead Univ. (1971).

22. Heindl, G. "Uber verallgemeinerte Stammfunktionen und LC-Functionen in R^n", dissertation, Tech. Univ. Munich (1968).

23. Heindl, G. "Spline-Functionen mehrerer Veranderlicher.I.", Bayerische Akad. 6 (1970), pp. 49-63.

24. Heindl, G. "Interpolation and approximation by piecewise quadratic C^1 functions of two variables", in Multivariate Approximation Theory, W. Schempp and K. Zeller, eds., Birkhauser, Basel (1970), pp. 146-161.

25. Mitchell, A. R. and Wait, R. The Finite Element Method in Partial Differential Equations, Wiley, N.Y. (1977).

26. Morgan, J. and Scott, R. "A nodal basis for C^1 piecewise polynomials of degree $n \geq 5$", Math. Comp. 29 (1975), pp. 736-740.

27. Morgan, J. and Scott, R. "The dimension of piecewise polynomials", manuscript (1977), unpublished.

28. Oden, J. T. and Reddy, J. N. An Introduction to the Mathematical Theory of Finite Elements, Wiley, N.Y. (1976).

29. Powell, M. J. D. "Piecewise quadratic surface fitting for contour plotting", in Software for Numerical Analysis, D. J. Evans, ed., Academic Press, N.Y. (1974), pp. 253-271.

30. Sablonniere, P. "De l'existence de spline a support borne sur une triangulation equilaterale de plan", ANO-30 O.E.R. d'I.E.F.A. Informatique, Univ. Lille (1981).

31. Schumaker, L. L. "Fitting surfaces to scattered data", in Approximation Theory II, Lorentz, Chui and Schumaker, eds., Academic Press, N.Y. (1976), pp. 203-268.

32. Schumaker, L. L. "On the dimension of spaces of piecewise polynomials in two variables", in Multivariate Approximation Theory, W. Schempp and K. Zeller, eds., Birkhauser, Basel (1979), pp. 396-412.

33. Schumaker, L. L. Spline Functions: Basic Theory, Wiley-Interscience, N.Y. (1981).

34. Schumaker, L. L. "Bounds on the dimension of spaces of multivariate piecewise polynomials", Rocky Mt. J. 14 (1983) pp. 251-264.

35. Strang, G. "The dimension of piecewise polynomials, and one-sided approximation", in Lecture Notes 365, Springer-Verlag, N.Y., (1974), pp. 144-152.

36. Wang, R. H. "The structural characterization and interpolation for multivariate splines", Acta Math. Sinica 18 (1975), pp. 91-106.

37. Wang, R. H. "On the analysis of multivariate splines in the case of arbitrary partition", Sci. Sinica (Math. I) (1979), pp. 215-226.

38. Wang, R. H. "On the analysis of multivariate splines in the case of arbitrary partition II", Num. Math. of China 2 (1980), pp. 78-81.

39. Zenisek, A. "Interpolation polynomials on the triangle", Numer. Math. 15 (1970), pp. 283-296.

40. Zienkiewicz, O. C. "The finite element method: from intuition to generality", Appl. Mech. Rev. 23 (1970), pp. 249-256.

41. Zwart, P. "Multi-variate splines with non-degenerate partitions", SIAM J. Numer. Anal. 10 (1973), pp. 665-673.

Supported in part by NASA Grant 4764-2.

BIRKHOFF INTERPOLATION ON THE ROOTS OF UNITY

A. Sharma

1. INTRODUCTION

The recent monograph of G. G. Lorentz et al [4] on Birkhoff Interpolation devotes one chapter to lacunary interpolation on the roots of unity. However, the general problem of Birkhoff interpolation on the roots of unity has not received enough attention. Even for a three row incidence matrix E we do not know any simple criterion for settling its regularity on the cube roots of unity.

In the special case when E has Hermite sequences of length p and q in the first and third row and only two non-zero entries in the middle row, then for real nodes, only sufficient conditions for regularity of E are known ([1], [4]).

The object of this note is to give a brief summary of some recent results on this problem in a joint paper [2]. Since the order of the rows is immaterial when the nodes are on the unit circle, we shall say that a matrix is almost Hermitian if all rows except one are Hermitian. It has been shown [2] that a n-row almost Hermitian matrix with only two non-zero entries in one row is regular on the cube roots of unity. In §2 we state the notation and the main result for three row almost Hermitian matrices. In §3 we give the proof only for $p = q$ and $n = 3$. For the general case $n > 3$ we refer the reader to [2]. In §4 we state some cases of non-regularity and some open problems.

S. P. Singh et al. (eds.), Approximation Theory and Spline Functions, 199–205.

2. PRELIMINARIES

Let $0 \leq k_k < k_2 < \ldots < k_r$ be r given integers and let $\{p_i\}_1^{n-1}$ be $n - 1$ non-zero integers. We shall denote by $E_n(\{p_i\}_1^{n-1}; k_1, \ldots, k_r)$ the incidence matrix which has $n - 1$ Hermitian sequences of length $p_1, \ldots, p_{n-1}$ and one non-Hermitian row with non-zero entries in columns $k_1, \ldots, k_r$. For the sake of brevity we shall denote this matrix by $E_n(\{p_i\}; K(r))$ where $K(r) = \{k_i\}_1^r$ and sometimes by E_n. When $p_1 = \ldots = p_{n-1} = p$, we shall denote the matrix by $E_n(p; k_1 \ldots, k_r)$ or $E_n(p; K(r))$.

We shall say that E_n is regular on the n-th roots of unity if the interpolation problem defined by E_n is uniquely solvable on the n-th roots of unity. We shall suppose that E_n satisfies strong Polya condition. Without loss of generality we take $n - 1$ Hermitian sequences to correspond to $\omega, \omega^2, \ldots, \omega^{n-1}$ where $\omega^n = 1$. Set

$$Q(z) = \prod_{j=1}^{n-1} (z - \omega^j)^{p_j}, \quad P(z) = Q(z) \sum_{\nu=0}^{r-1} a_\nu z^\nu. \tag{2.1}$$

If $Q(z) = \prod_{j=1}^{N} b_j(z - 1)^j$, $N = p_1 + p_2 + 0 \ldots + p_{n-1}$, then from the requirement $P^{(k_j)}(1) = 0$, $j = 1, 2, \ldots, r$, we obtain the following

Theorem A. *If E_n satisfies strong Polya condition, then E_n is regular on the n-th roots of unity if and only if*

$$\Delta_r(E_n) \neq 0 \tag{2.2}$$

where $\Delta_r(E_n)$ is a determinant of order r given by

$$\Delta_r(E_n) = \begin{vmatrix} b_{k_1} & b_{k_1-1} & \cdots & b_{k_1-r+1} \\ b_{k_2} & b_{k_2-1} & \cdots & b_{k_2-r+1} \\ \cdots & \cdots & \cdots & \cdots \\ b_{k_r} & b_{k_r-1} & \cdots & b_{k_r-r+1} \end{vmatrix} . \tag{1.3}$$

In the next section we shall prove the following when $r = 2$ and $n = 3$:

Theorem 1. *If $1 \le p < q$ are given integers and if $E_3(p, q; k_1, k_2)$ satisfies strong Polya condition then it is regular on the cube root of unity.*

Since regularity of any interpolation problem is independent of translation and rotation we may, by linear translation bring the cube roots of unity $1, \omega, \omega^2$ into $1, i\sqrt{3}$ and -1. Then $Q(z) = (z - 1)^p(z + 1)^q$ and condition (2.2) becomes

$$S_{k_1} \neq S_{k_2}, \quad 0 \le k_1 \le p + q - 1, \quad k_1 < k_2 \le p + q \tag{2.4}$$

where

$$S_k = S_k(p, q) = \frac{1}{k} \frac{Q^{(k)}(i\sqrt{3})}{Q^{(k-1)}(i\sqrt{3})} .$$

The numbers $S_k = S_k(p, q)$ can be calculated recursively from the following relations which are easy to prove:

$$S_k = -\frac{(p + q + 2 - 2k)i\sqrt{3} + q - p}{4k} - \frac{p + q + 2 - k}{4kS_{k-1}} ,$$

$$k = 2, 3, \ldots, p + q \tag{2.5}$$

$$S_1 = \frac{(p + q)i\sqrt{3} + q - p}{-4}, \quad S_{p+q+1} = 0.$$

A table of these numbers was prepared by Bob Norfolk (Kent State University) and W. Aiello (University of Alberta). (See [2]). This table led us to prove

Theorem 2. If we set $S_k = a_k + i\sqrt{3}\, b_k$ where S_k is given by (2.5), then the numbers b_k are negative. Moreover, we have

$$|b_k| > |b_{k+1}|, \; k = 0, 1, \ldots, p + q. \tag{2.6}$$

Theorem 1 follows from Theorem 2 since (2.6) implies (2.4).

3. PROOF OF THEOREM 2 (Case $p = q$)

In this case $Q(z) = (z^2 - 1)^p$ and the zeros of $Q^{(k-1)}(z)$ are symmetric about the origin. We shall first take the case when k is even.

(a) ($k = 2\ell$). Let the non-negative zeros of $Q^{(k-1)}(z)$ be denoted by

$$0 = \xi_0 < \xi_1 < \ldots < \xi_{p-\ell} \le 1$$

where the last zero $\xi_{p-\ell}$ may have multiplicity $p - k + 1$ if $p > k - 1$. If $p \le k - 1$, then $\xi_{p-\ell}$ is a simple zero.

Let us now denote the zeros of $Q^{(k)}(z)$ by $\pm\eta_1$, $\pm\eta_2, \ldots, \pm\eta_{p-\ell}$ with their absolute values arranged in increasing order. Then we know that

$$0 = \xi_0 < \eta_1 < \xi_1 < \eta_2 < \ldots < \eta_{p-\ell} < \xi_{p-\ell} \le 1. \tag{3.1}$$

If $\xi_{p-\ell}$ has multiplicity $p - k + 1$, then $\eta_{p-\ell}$ will be of multiplicity $p - k$. Thus

$$Q^{(k-1)}(z) = Cz \prod_{j=1}^{p-\ell} (z^2 - \xi_j^2) \quad \text{and} \quad Q^{(k)}(z) = C' \prod_{j=1}^{p-\ell} (z^2 - \eta_j^2).$$

Since $S_k = Q^{(k)}(i\sqrt{3})/kQ^{(k-1)}(i\sqrt{3})$, we have

$$S_k = a_k + i\sqrt{3}\, b_k = -\frac{i\sqrt{3}}{k}\left[\frac{1}{3} + 2\sum_{j=1}^{p-\ell} \frac{1}{\xi_j^2 + 3}\right]. \tag{3.2}$$

Similarly, we have

$$S_{k+1} = a_{k+1} + i\sqrt{3}\, b_{k+1} = -\frac{2i\sqrt{3}}{k+1} \sum_{j=1}^{p-\ell} \frac{1}{\eta_j^2 + 3} . \tag{3.3}$$

Thus from (3.2) and (3.3), we see that $a_k = a_{k+1} = 0$ and

$$\begin{aligned} b_k &= -\frac{1}{k}\,\frac{1}{3} + 2 \sum_{j=1}^{p-\ell} \frac{1}{\xi_j^2 + 3} , \\ b_{k+1} &= -\frac{2}{k+1}\, 2 \sum_{j=1}^{p-\ell} \frac{1}{\eta_j^2 + 3} . \end{aligned} \tag{3.4}$$

From (3.1) we have

$$0 < \eta_1^2 < \xi_1^2 < \ldots < \eta_{p-\ell}^2 < \xi_{p-\ell}^2 \le 1.$$

Here we suppose for the sake of simplicity that $\xi_{p-\ell}$ is a simple zero. For if not, then $\eta_{p-\ell}$ will have multiplicity one less and the argument will still go through. Thus from (3.5), we have

$$(\xi_j^2 + 3)^{-1} > (\eta_{j+1}^2 + 3)^{-1}, \; j = 1, 2, \ldots, p - \ell - 1.$$

It follows from (3.4) and (3.5) that in order to prove (2.7) it suffices to show that

$$\frac{1}{3} + \frac{2}{\xi_{p-\ell}^2 + 3} > \frac{2}{\eta_1^2 + 3}$$

which is easily proved, since the left hand side above is greater than $\frac{1}{3} + \frac{2}{4} > \frac{2}{3} > 2(\eta_1^2 + 3)^{-1}$.

(b) <u>(k = 2ℓ + 1)</u>. In this case $Q^{(k-1)}(z)$ does not vanish at the origin and so its zeros are $\{\pm\xi_j\}_1^{p-\ell}$ where

$$0 < \xi_1 < \xi_1 < \ldots < \eta_{p-\ell} \le 1.$$

(Here again, as in the case above, we shall suppose $\xi_{p-\ell}$ to be a simple zero.) The zeros of $Q^{(k)}(z)$ will be 0, $\pm\eta_1, \ldots, \pm\eta_{p-\ell-1}$ where $0 < \eta_1 < \eta_2 < \ldots < \eta_{p-\ell-1}$. Moreover, we have the following inequalities:

$$0 = \eta_0 < \xi_1 < \eta_1 < \xi_2 < \ldots < \eta_{p-\ell-1} < \xi_{p-\ell} \le 1,$$

and $\eta_{p-\ell-1}$ will be a simple zero by our supposition. Since $\xi_j^2 < \eta_j^2$ $(j = 1, 2, \ldots, p - \ell - 1)$ and since

$$\frac{2}{\xi_{p-\ell}^2 + 3} > \frac{2}{4} > \frac{1}{3}$$

we see that

$$2 \sum_{j=1}^{p-\ell} \frac{1}{\xi_j^2 + 3} > \frac{1}{3} + 2 \sum_{j=1}^{p-\ell-1} \frac{1}{\eta_j^2 + 3},$$

whence we get (2.7).

4. SOME RESULTS AND EXAMPLES

On using the notion of Polya frequency sequences [3], we can prove (See [2] for proof):

Theorem 3. *If the almost Hermitian matrix* $E_3(p; \{k_i\}_1^r, r \le 4)$ *satisfies strong Polya condition, then it is regular on the cube roots of unity.*

The conditions $r \le 4$ is necessary as is seen by taking $r = 5$, $k_j = j$, $j = 1, 2, 3, 4, 5$. In this case the non-trivial polynomial $A_6(x) = (x - 1)^6 - (\omega - 1)^6$ with $\omega^3 = 1$ satisfies

$$A_6(\omega) = A_6(\omega^2) = A_6^{(k_j)}(1) = 0, \; j = 1, 2, 3, 4, 5.$$

It seems surprising that $E_3(1, 2; K(10))$ is not regular on the cube roots of unity when

$$K(10) = \{j\}_1^5 \cup \{j\}_7^{11}$$

while

$$E_3(1, 2; K(\ell)) \text{ is regular on the cube roots of unity when } K(\ell) = \{j\}_1^{\ell}.$$

There is a close relation between interpolation on the roots of unity and trigonometric interpolation [5]. It would be interesting to find similar conditions of regularity for trigonometric interpolation on equidistant nodes. For a fuller discussion of these problems we refer to [2].

REFERENCES

1. DeVore, R. A., Meir, A., and Sharma, A., "Strongly and weakly non-poised H-B interpolation problems", Canad. J. Math. 25, 1973, pp. 1040-1050.

2. Fabrykowski, J., Sharma, A., and Zassenhaus, H., "Some Birkhoff interpolation problems on the roots of unity", (to appear J. Linear Algebra and its Applications), p. 1-35.

3. Karlin, S., "Total positivity", Stanford University Press, 1968.

4. Lorentz, G. G., Jetter, K., and Riemenschneider, S. D., "Birkhoff interpolation", Ed. G. C. Rota, Addison-Wesley, 1983.

5. Sharma, A., Smith, P., and Tzimbalario, J., "Polynomial interpolation on roots of unity with applications", Approximation and Function Spaces, Proc. International Conference, Gdansk, Poland, 1979, pp. 667-685.

APPLICATIONS OF TRANSFORMATION THEORY: A LEGACY FROM ZOLOTAREV (1847-1878)*

John Todd

California Institute of Technology

"And out of olde bokes, in good feyth,
Cometh all this newe science that men lere"

CHAUCER, THE PARLIAMENT OF FOWLS

INTRODUCTION

What we are concerned with is, roughly, the generalization to the elliptic case of the familiar multiple angle formulas of elementary trigonometry such as

$$\cos 2\theta = 2\cos^2\theta - 1; \quad \tan 2\theta = \frac{2\tan\theta}{1-\tan^2\theta};$$

$$\sin 2\theta = 2\sin\theta\cos\theta = 2\sin\theta\sqrt{(1-\sin^2\theta)}$$

(which are respectively polynomial, rational, algebraic). More generally we have

$$\cos n\theta = 2^{n-1}[\cos^n\theta - \tfrac{1}{4}n\cos^{n-2}\theta + \dots]$$

which we can also express as a Chebyshev polynomial:

$$T_n(x) = \cos(n \arccos x) = 2^{n-1}[x^n - \tfrac{1}{4}nx^{n-2} + \dots]$$

$$= 2^{n-1}\tilde{T}_n(x).$$

S. P. Singh et al. (eds.), Approximation Theory and Spline Functions, 207–245.

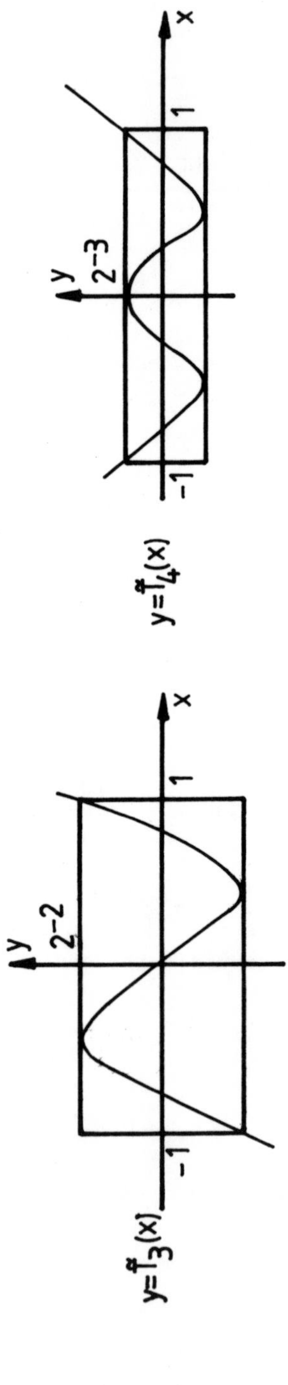

Figure 1

Zolotarev stated and solved in 1868, 1877, 1878, four problems in approximation theory, or the constructive theory of functions. These problems turned up again in practical contexts in different areas and different countries and were solved independently during the last 50 years. We state these problems in §2. We describe in 3A, B, C some applications of (Z3). Finally, we give some indication of the solutions: in §4, we first discuss an elementary problem which indicates the method for dealing with (Z1) and in §5, we discuss the general method of solution. The solutions all depend on the theory of transformation of elliptic functions, a subject beyond the scope of the usual texts, [cf. 22.421, W8W] and the usual sylllabi. Greenhill [1892, p. x, Introduction] notes the reintroduction of elliptic functions "... excluding the theta functions and the theory of transformation" in the regulations for Schedule II, Part I of the Mathematical Tripos at Cambridge, beginning in May 1893.

Although Chebyshev was well aware of the inspiration afforded by applications, as indicated by the following quotation [Chebyshev, 1899, I, p. 239] there seems to be no reference to the potentialities of the work of Zolotarev.

"Le rapprochement de la théorie et de la pratique donnent les résultats les plus féconds. La pratique n'est pas la seule à tirer profit de ces rapports: réciproquement les sciences elles-mêmes se développent sous l'influence de la pratique. C'est elle qui leur découvre de nouveaux sujets d'étude et des points de vue nouveaux sur les sujets connus depuis longtemps."

There is a short biography of Zolotarev by Ozigova [1966]. Actually he is perhaps more celebrated for his work in algebra and number theory than in approximation theory.

For an account of Chebyshev's visit to England in 1852 and other relevant matters, see the Inaugural Lecture of A. Talbot [1971].

It is worth noting that Zolotarev wrote in the Minutes of the Meeting of the Council of the St. Petersburg University for the second half of the academic year 1869/70: "In mathematics it is incomparably harder to find a problem and state in correctly than to solve it; as soon as a problem is stated correctly its solution is found in one way or another."

See Kuznetsov [1971].

§2. ZOLOTAREV'S FOUR PROBLEMS

To see the place of the first problem we go back to the Chebyshev polynomials. It is well known that the best approximation to zero in $[-1,1]$ by a monic polynomial in the Chebyshev norm is $\tilde{T}_n(x)$. In fact

$$\text{(T1)} \quad \min_{(a)} \max_{-1\le x\le 1} |x^n + a_1 x^{n-1} + \ldots + a_n|$$

is 2^{1-n} *and is achieved by*

$$\tilde{T}_n(x) = \Pi(x - x_r)$$

where

$$x_r = \cos((2r + 1)\pi/2n), \; r = 0, 1, \ldots, n - 1.$$

There are several related problems which we state:

(T2)(Markov) *Determine*

$$\min_{(a)} \max_{-1\le x\le 1} |a_0 x^n + a_1 x^{n-1} + \ldots + a_n|$$

where $a_r = 1$ *for some* r, $1 \le r \le n$.

(T3)(Chebyshev) *Determine*

$$\min_{(a)} \max_{-1\le x\le 1} |a_0 x^n + a_1 x^{n-1} + \ldots + a_n|$$

where for some ξ *outside* $[-1,1]$, $a_0\xi^n + a_1\xi^{n-1} + \ldots + a_n = \eta$, η *given*.

We note that the T_n's, in compensation for their smallness inside $[-1,1]$, are largest outside:

(T4)(Chebyshev) *If* $p_n(x)$ *is a polynomial of degree* n *such that* $\max_{-1\le x\le 1} |p_n(x)| = 1$ *then for* ξ *outside* $[-1,1]$ *we have*

$$|p_n(\xi)| \le |T_n(\xi)|.$$

In (T1) Chebyshev fixed the *first* coefficent. Zolotarev asked the same question only requiring that the first *two* coefficients be fixed.

(Z1) Determine

$$\min_{(a)} \max_{-1\le x\le 1} |x^n - n\sigma x^{n-1} + a_2 x^{n-2} + \ldots + a_n|$$

where σ is a parameter.

This being solved, it is natural to ask the same question only fixing the first *three* coefficients. This was solved by Achiezer in 1928. The final stage was results about the case when r coefficients were fixed: these were obtained by Meiman in 1960. For details see the reviews and translations of his papers.

The second problem of Zolotarev is related to (Z1) just as (T3) is to (T1).

(Z2) Determine

$$\min_{(a_2,\ldots,a_n)} \max_{-1\le x\le 1} |x^n + a_1 x^{n-1} + \ldots + a_n|$$

where a_1 is determined so that $\xi^n + a_1\xi^{n-1} + \ldots + a_n = \eta$, where ξ, ($\xi < -1$ or $\xi > 1$) and η are given.

The other two problems of Zolotarev are concerned with *rational* approximation. [Compare these with (T4).] The relations between the problems have been discussed in detail by Achiezer.

(Z3) Find the rational function $y = \phi(x)/\psi(x)$, where the degrees of the polynomials ϕ, ψ do not exceed n, which satisfies

$$|y(x)| \le 1, \quad -1 \le x \le 1$$

and which deviates most from zero in the intervals $|x| \ge k^{-1}$, where k, $0 < k < 1$ is given.

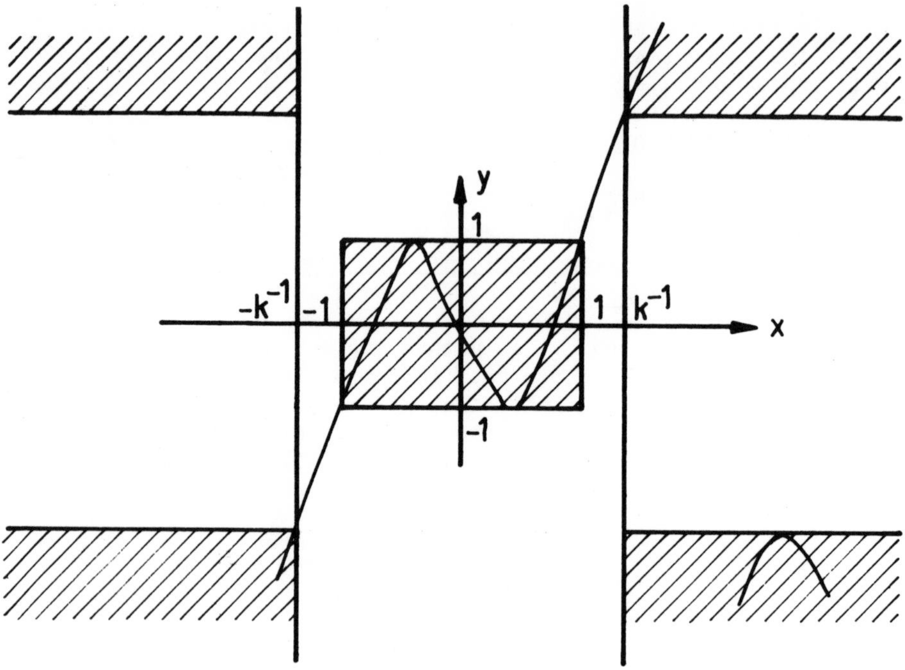

Figure 2

(Z4) Find the rational function $y = \phi(x)/\psi(x)$, where the degrees of the polynomials ϕ, ψ do not exceed n, such that

$$y(x) \geq 1 \quad \text{for} \quad 1 \leq x \leq k^{-1}, \; y(x) \leq -1 \quad \text{for} \quad -k^{-1} \leq x \leq -1$$

and which deviates least from zero in these intervals, where $0 < k < 1$.

We mention here a problem discussed by Achiezer:

(A1) Determine

$$\min_{(a)} \max |x^n + a_1 x^{n-1} + \dots + a_n|$$

where the max is over all x in the *two* intervals $-1 \leq x \leq -\lambda$, $\lambda \leq x \leq 1$ where λ, $0 < \lambda < 1$ is given.

§3. SOME APPLICATIONS OF ZOLOTAREV'S THIRD PROBLEM (Z3)

A. Design of Filters

An electrical filter is a "black box" with "knobs", containing variable components (condensers, resistances), which influences an input signal according to a response curve:

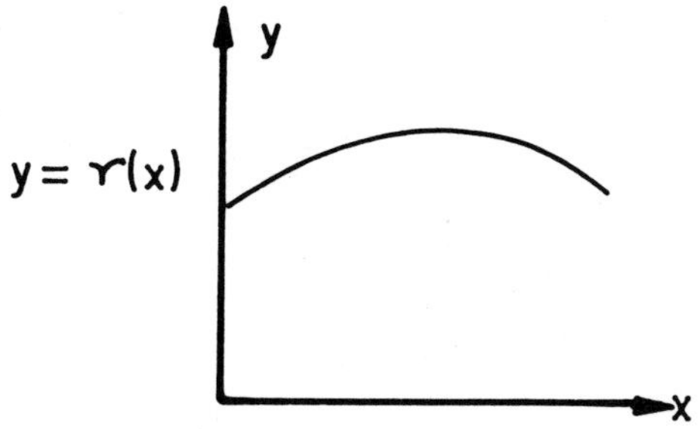

Figure 3

We have contact every day with filters: implicitly in telephone conversations and explictly in high-fidelity equipment. For a more detailed discussion see Melzak [1976].

To filter out the "high notes" requires a response of the form:

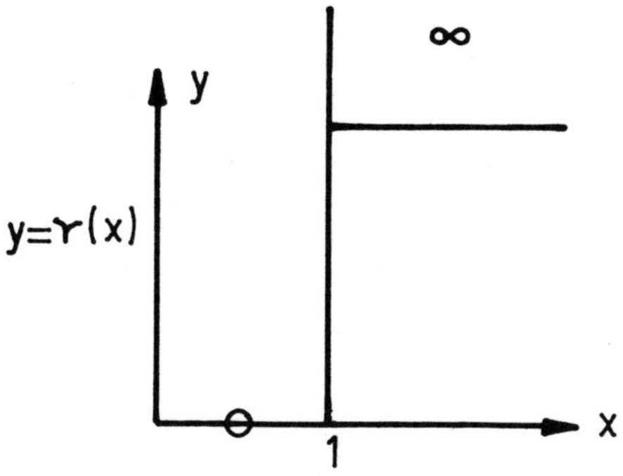

Figure 4

This is a "low pass" filter. This realises the truncation of a

Fourier series. This steep cut-off is not realisable and so we have to modify our demand to

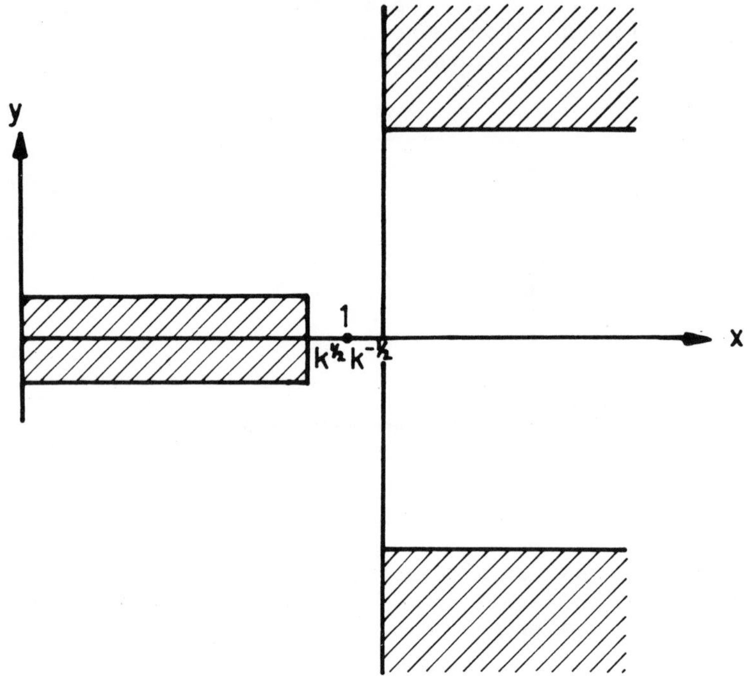

Figure 5

i.e. we want to adjust the parameters so that, given k, $0 < k < 1$

$$|r| \text{ is small in } [0, k^{1/2}], \quad |r| \text{ is large in } [k^{-1/2}, \infty).$$

For certain circuits the response is of the form

$$r(x) = \Pi |(a_j^2 - x^2)/(1 - a_j^2 x^2)|$$

and so our problem becomes

(Z3') <u>Determine</u>

$$\min_{(a)} \max_{0 \le x \le \sqrt{k}} |\Pi(a_j^2 - x^2)/(1 - a_j^2 x^2)|$$

which is very similar to (T1) when written as

(T1') Determine $\min_{(r)} \max_{-1\le x\le 1} \Pi(x - r_j)$.

[Because $r(x)\, r(x^{-1}) = 1$ we have only to consider one of the conditions (1).]

These problems were discussed in Germany, beginning with W. Cauer [1933] and in U.S.A. at Bell Telephone Laboratories [1939] by S. D. Darlington and E. L. Norton. Cauer was employed at the Mix and Genest organisation, later a subsidiary of the ITT Corporation.

The solution to (Z3') is given by

$$a_j = \sqrt{k}\ \mathrm{sn}(2jK/(2m+1),k) \quad j = 1, 2, \ldots, m$$

and the extremal value is

$$\sqrt{k_{2m}} = \sqrt{\frac{1-k'_m}{1+k'_m}}$$

where k_m corresponds to q^m as k corresponds to q.

The number n determines the size (cost) of the filter, the parameter k determines how sharp the cut-off is and the min-max gives the attenuation in the pass-band. To assist the designer, tables of the quantities involved were made -- now-a-days program packages would be written -- by e.g., Glowatzki [1955] U.S. National Bureau of Standards, (1956, unpublished), and A. R. Curtis [1964].

E. Stiefel [1961] contemplated extensions of this problem where we want to filter out a band (or several bands) of frequencies

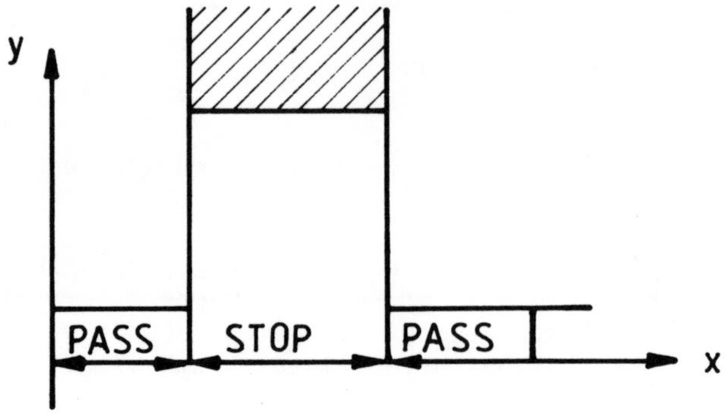

Figure 6

The solution to this problem involves hyperelliptic functions. Stiefel's associates, Amer and Schwarz [1964] have solved some practical problems, not in closed form, but by linear programming methods. This work was subsidized by the Hasler foundation.

We mention here that a problem concerning the optimal design of radio transmitters which produce a narrow principal beam and small subsidiary beams has been discussed by Pokrovskii [1962]. This leads to an extremal problem which generalizes (A1).

3B. ADI Parameters

The alternating direction implicit method for the iterative solution of the discrete approximations to elliptic partial differential equations was introduced in 1955 by Peaceman and Rachford [1955] in work supported by Humble Oil Company. In the case of the "model problem" the speed of convergence depends on

$$\max_{k \le x \le 1} \left| \prod_{j=1}^{m} \{(x-r_j)/(x+r_j)\} \right|$$

where the r_j are certain parameters to be chosen and where k, l are lower and upper bounds to the characteristic values of the (normalized) matrix of the system of linear equations approximating the differential equation $u_{xx} + u_{yy} = 0$ in a square $0 \le x, y \le 1$. The question of the optimal choice of the r_j

was answered when $m = 2^{\ell}$ by Gastinel [1962] and Wachspress [1963]. Indeed, more generally, from an optimal set of m parameters an optimal set of $2m$ can be found by the use of the arithmetic-geometric mean. The optimal parameters were found by W. B. Jordan (see Wachspress [1963]), based on Cauer's work. Jordan and Wachspress were employed at the Knolls Atomic Power Laboratory of the General Electric Company.

The actual result is the following:

$$r_j = \mathrm{dn}((2j-1)K/2m,k), \quad j = 1, 2, \ldots, m$$

$$L = L_m = (1 - \sqrt{k'_m})/(1 + \sqrt{k'_m})$$

where k_m corresponds to q^m as k corresponds to q.

The question of the optimal value of m arises. The proper question is about the behavior of

$$\eta_m = L_m^{1/m} .$$

Gaier and Todd [1967] showed that $\eta_m \uparrow$ and in fact that

$$\log \eta_m = \log q + [(\log 2)/m] + O(m^{-2}) \qquad (2)$$

which implies that the asymptotic value was attained (in the practical range of q) for moderate values of m: favourite values of m were 8 or 16 or 10. De Boor and Rice [1963] had obtained empirical results which were very close to (2).

Another question which was recently studied by V. I. Lebedev [1977], was: What is the best order to use the parameters?

3C. Square Roots

Consider the determination of N by the Newton process: "guess x_0 and improve by $x_{n+1} = \frac{1}{2}(x_n + (N/x_n))$". Convergence takes place for any $x_0 > 0$. What is the best x_0 to use? We have to make this question more precise. Let us only consider using floating point calculators. It is then natural to restrict N to lie between 10^{-2} and 1 (in the decimal case) or between $\frac{1}{4}$ and 1 (in the binary case). It is then appropriate to consider the relative error

$$r_n = |(x_n - \sqrt{N})/\sqrt{N}|.$$

Next we ask: What value of n? In virtue of a minor miracle, first observed by Moursund [1967], it does not matter, provided $n \geq 1$. It is natural to consider making x_0 a rational function of N of type (μ,ν), say, i.e. numerator with degree $\leq \mu$, denominator with degree $\leq \nu$,

$$x_0 = n_\mu(N)/d_\nu(N)$$

so that we consider

$$r_1 = |1 - (x_1/\sqrt{N})|.$$

In virtue of another minor miracle the extremal x_1 is a constant multiple of the extremal x_0 so we have to look at

$$|1 - \{n_\mu(N)/\sqrt{N}\ d_\nu(N)\}| .$$

For a collection of references to work in this area, see Todd [1977].

The problem of determining the optimal coefficients in n_μ, d_ν when $\mu = \nu$ or $\mu = \nu + 1$ was solved in general by Ninomiya [1970]. (For small μ, ν the coefficients were found algebraically by Maehly (see Cody [1964]).

A typical numerical result in the case of (2,1) approximation in $(\frac{1}{4}, 1)$ is

$$x_0 = \frac{0.3432201292\ N^2 + 1.071299971\ N + 0.085805032}{N + 0.5}$$

for which we have

$N = 0.25$	$x_0 = 0.50020\ 0044$	$N = 1$	$x_0 = 1.00090\ 088$
	$x_1 = 0.50000\ 0040$		$x_1 = 1.00000\ 0451$

The general solution in the (n,n) case is

$$x_0 = \frac{2}{1+\lambda'} \Pi \frac{(1 - c_{2r}) + c_{2r}x}{(1 - c_{2r-1}) + c_{2r-1}x} \qquad (3)$$

where

$$c_r = \mathrm{sn}^2(rK/2n,k), \qquad \lambda = k^{2n} \Pi c^2_{2r-1}$$

and the min-max is

$$L = (1-\lambda')/(1+\lambda').$$

We shall derive the general formula in §5 below and obtain this one by specialization.

In a memorandum of 25 June, 1962, E. L. Wachspress used the A.G.M. parameters to accelerate the convergence of the Newton process for the positive square root of a positive definite matrix.

§4. ZOLOTAREV'S FIRST PROBLEM (Z1)

The applications of this seem less interesting. For instance, we can "economize" polynomials approximating a polynomial of degree n by one of degree $n - 2$. These ideas have been exploited by C. Lanczos [1893-1974 and S. Paszkowski [1962].]

However the solution to the problem looks quite mysterious.

$$\text{(Z1)} \quad \min_{(a)} \max_{-1\le x\le 1} |x^n - n\sigma x^{n-1} + a_2 x^{n-2} + \dots + a_n| .$$

We may assume $\sigma \geq 0$. For $\sigma = 0$ we are back to (T1).

For small σ the solution is a distorted Chebyshev polynomial. Specifically, if $0 \leq \sigma \leq \tan^2(\pi/2n)$, the extremal polynomial is

$$2^{1-n}(1+\sigma)^n \, T_n((x-\sigma)/(\sigma+1)),$$

and the minmax is $2^{1-n}(1+\sigma)^n$.

For larger σ the solution is given in an extremely complicated form, involving various elliptic quantities. We use the standard notation of Whittaker and Watson [1927]. We are given n, σ and first solve the following equation for k, which is involved in K and in the elliptic and theta functions

$$* \quad 1 + \sigma = \frac{2\ \mathrm{sn}\ (K/n)}{\mathrm{cn}(K/n)\,\mathrm{dn}(K/n)} \ \{\mathrm{ns}(2K/n) \ - \frac{\tilde{I}_4'(\pi/2n)}{\tilde{I}_4(\pi/2n)}\} .$$

This k is unique and $0 < k < 1$ when $\sigma > \tan^2(\pi/2n)$.

Then the extremal polynomial $y = x^n - n\,\sigma\,x^{n-1} + \ldots$ is given parametrically by

$$y = \frac{1}{2} L[X^n + X^{-n}],$$

$$x = [\mathrm{sn}^2 u + \mathrm{sn}^2(K/n)]/[\mathrm{sn}^2 u - \mathrm{sn}^2(K/n)],$$

where

$$X = [\tilde{I}_1((\pi/2n) - (\pi u/2K))]/[\tilde{I}_1((\pi/2n) + (\pi u/2K))].$$

When u runs from 0 to iK' then x runs from -1 to 1. The corresponding min max is

$$L = 2^{1-n}[k^{1/2}\tilde{I}_3^2/\{\tilde{I}_2(\pi/2n)\tilde{I}_3(\pi/2n)\}]^{2n}.$$

A similar separation into cases occurs in the solution of (A1). Here the solution is a distorted Chebyshev polynomial for all α when n is even, and for small α (specifically, $0 \le \alpha \le \sin(\pi/2n)$) when n is odd; for n odd and $\alpha > \sin(\pi/2n)$ the solution is complicated. See, e.g.,Achiezer [1970, p. 209].

What we shall do is to discuss a simple problem, not directly relevant, but which illustrates the general method of solution and shows how to dispel some of the mysteries about [Z1 . The problem, which has been discussed by Hornecker [1958], Achiezer [1956, 1967], Bernstein [1926], Talbot [1962, 1964], is:

(B1) Determine

$$\min_{(a)} \max_{0 \le x \le 1} |(1+x)^{-1} - (a_0 x^n + \ldots + a_n)|.$$

For references to the literature on this problem, see Todd [1984b].

*$\tilde{I}$ denotes "curly theta".

A method of handling *all* the problems discussed is to guess the answer and then confirm it by appealing to the weighted rational equal ripple theorem, due essentially to Chebyshev but refined by de la Vallée Poussin and others. In the non-degenerate case we have:

Equal Ripple Theorem (ERT). Suppose $f(x)$ and $w(x)$ are continuous in $[a,b]$, and that $w(x) \neq 0$ in $[a,b]$. Then the extremal function for the problem

$$\min_{N,D} \max_{a \le x \le b} |E(x)|, \quad \text{where} \quad E(x) = f(x) - w(x)\frac{N_n(x)}{D_d(x)}$$

(where $N_n(x) = b_0x^n + \ldots + b_n$, $D_d(x) = a_0x^d + \ldots + a_d, a_0 \neq 0$) is characterized by $E(x)$ assuming its maximum absolute value with positive and negative signs alternately, $n + d + 2$ times in $[a,b]$.

The original application of this is to (T1) (with $n + 1$ for n) when $f(x) = x^{n+1}$ and $w(x) \equiv 1$, $d = 0$, $a_0 = -1$. Another instance is when $f(x) = 1$, $w(x) = x^{-1/2}$ which occurs in §3C. Detailed knowledge of the trigonometric and elliptic functions ensures quick but unmotivated proofs. Thus H. Lebesgue [1920], reviewing de la Vallée Poussin's [1919] Borel Tract, writes "..par une sorte de divination qui rappelle bien son illustre compatriote Tchebycheff, M. Bernstein trouve les polynomes d'approximation de $(z-a)^{-1}$, ...".

We now illustrate, by discussing (B1), a second approach to our problems which involves the use of ERT at the start to get a differential equation for the extremal error which may then be solved and lead to the result required. In this approach we refer to tables of (elliptic) integrals instead of to the properties of elliptic functions.

What we apply used to be called "Curve Tracing": From the qualitative behavior specified by ERT we can get a differential equation for the solution to (B1). The even and odd cases look slightly different -- we shall deal only with the case $n = 2$. Using ERT we show that the graph of the error function $y(x)$ must be of the form:

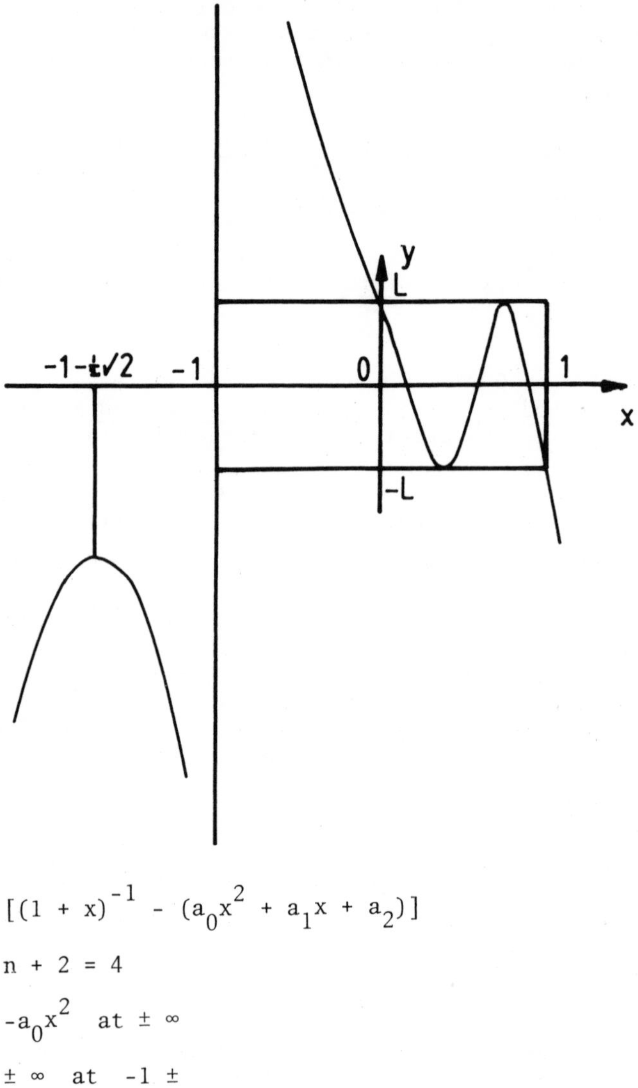

$$y = [(1 + x)^{-1} - (a_0x^2 + a_1x + a_2)]$$

$$n + 2 = 4$$

$y \sim -a_0x^2$ at $\pm \infty$

$y \sim \pm \infty$ at $-1 \pm$

y' three zeros

$y \pm L$ each three zeros

Figure 7

If the extremal values of $y(x)$ are $\pm L$ then $y^2 - L^2$ has simple zeros at $0,1$ and n double zeros in the interior of $[0,1]$ while y' has n simple zeros at these points and a single extraneous zero, α in $(-\infty, -1)$. It follows that

$$(1 + x)^2 x(1 - x) y'^2 = n^2(x - \alpha)^2(L^2 - y^2)$$

where α, L are yet to be found. Writing $y = L\eta$ and noting that $\eta(0) = 1$ we find

$$\int_1^\eta \frac{\pm dY}{\sqrt{1-Y^2}} = n \int_0^x \left[1 - \frac{1+\alpha}{1+X}\right] \frac{dX}{\sqrt{X(1-X)}}$$

where the ambiguous sign changes at each extrema, beginning with a negative sign.

The integrals involved here are elementary and we can solve explicitly for η. If we use the fact that $y(1) = -L$ we can determine

$$\alpha = \alpha_n = -1 - \sqrt{2}n^{-1}.$$

To determine L we use the fact that $(1+x)y(x) \sim 1$ as $x \sim -1$. This gives

$$L = L_n = \frac{1}{4}(3-2\sqrt{2})^n.$$

[The results for α_n, L_n for $n = 1, 2$ can be checked by elementary methods.]

The final result is that the best approximation is given by

$$\sqrt{2}\,\{[\tfrac{1}{2} - c\,T_1(2x - 1) + c\,T_2(2x - 1) + \dots +$$

$$+ (-1)^{n-1}c^{n-1}T_{n-1}(2x - 1)\} + (-1)^n(1 - c^2)^{-1}\,T_n\,(2x - 1)$$

where $c = 3 - 2\sqrt{2}$. This expression is remarkable because it is got by truncating the Fourier-Chebyshev expansion of $(1 + x)^{-1}$ and dividing the last term by $1 - c^2$. This was pointed out explicitly by Hornecker [1958] and Talbot [1962] and examined further by Rivlin [1962].

We now return to (Z1). From ERT it follows that the extremal $y(x) = x^n - n\sigma x^{n-1} + a_2 x^{n-2} + \dots$ has $(n-2) + 2 = n$ alternating extrema. Curve tracing arguments show that either all $n - 1$ zeros of y', or all but one of these lie in $[-1,1]$, and this is the cause for the separation into the trigonometric and elliptic cases. We sketch the behavior of y, when $n = 2$ and $n = 3$, in the two cases:

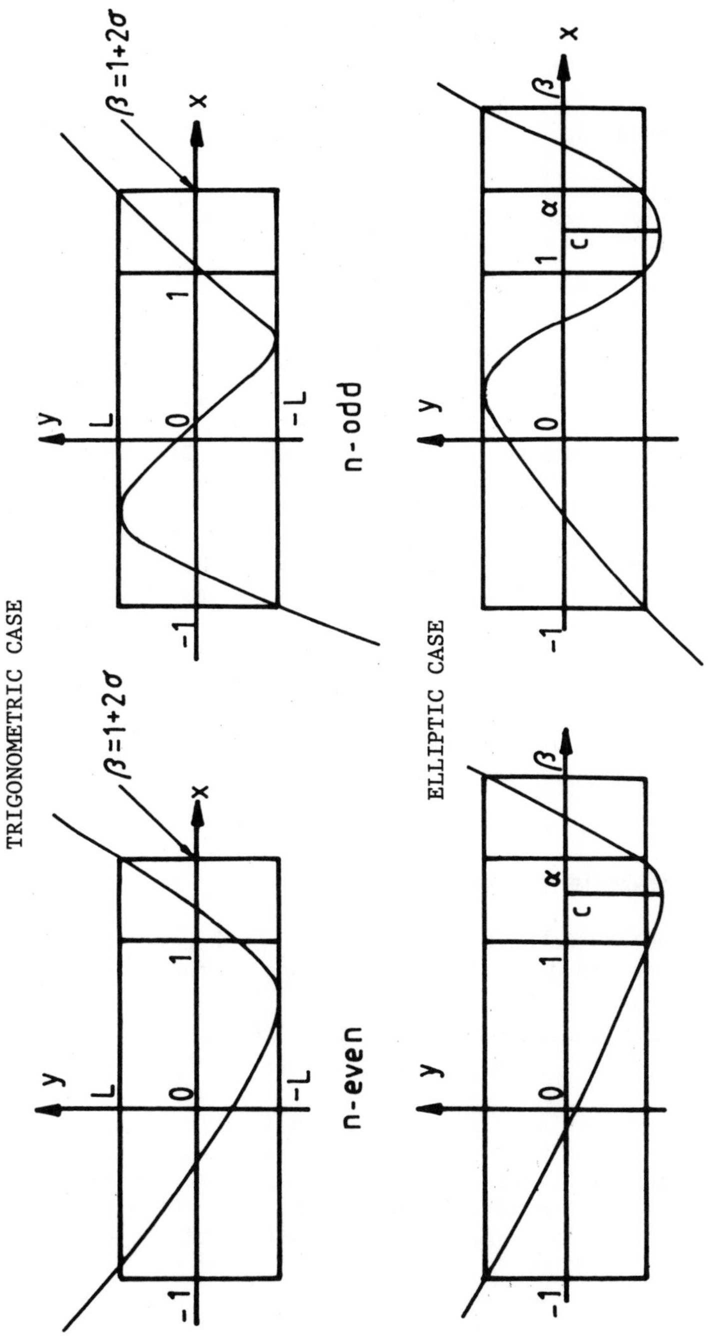

TRIGONOMETRIC CASE
n-even
n-odd
β=1+2σ
β=1+2σ
ELLIPTIC CASE
x
y
L
0
-L
1
-1
α
c
β

Figure 8

The differential equation in the trigonometric case is

$$\frac{\pm\, dy}{\sqrt{(L^2 - y^2)}} = \frac{n\,dx}{\sqrt{\{(1 + x)(\beta - x)\}}}$$

where the deviation L and the largest extrema, β, have to be determined.

The differential equation in the elliptic case is

$$\frac{\pm\, dy}{\sqrt{(y^2 - L^2)}} = \frac{n(c - x)\,dx}{\sqrt{\{(x^2 - 1)(\alpha - x)(\beta - x)\}}}$$

where the deviation L, the two largest extrema α, β and the extraneous turning point c have to be determined.

Alternative accounts of the solution of (Z1) are given by Erdös and Szegö [1942], by Achiezer [1953; 1967] and by Carlson and Todd [1983].

In the solution of (A1) the degenerate form of the ERT is required: if the polynomials N and/or D are truncated say

$$N_n = b_0 x^n + \ldots + b_{n-\mu} x^{n-\mu}, \quad D_d = a_0 x^d + \ldots + a_{d-\nu} x^{\nu}$$

then the number of extrema must be reduced by $\max(\mu, \nu)$. For a discussion of the general case of ERT see, e.g. Achiezer [1953, p. 55].

§5. THE GENERAL METHOD OF SOLUTION

The problems we have discussed led, by use of the Equal Ripple Theorem, to differential equations of the form

$$\frac{dx}{\sqrt{\{(1 - x^2)(1 - k^2x^2)\}}} = \frac{M\,dy}{\sqrt{\{(1 - y^2)(1 - \lambda^2y^2)\}}} \tag{1}$$

where y was to be an algebraic, rational or polynomial function of x and where k was given and M, λ to be determined. In some cases k, or k and λ were zero. In some cases elliptic differentials of the third kind or hyperelliptic differentials were involved instead of those of the first kind.

Thus our problems are those of the "transformation" of

elliptic objects (differentials, integrals, functions or $\mathcal{I}$-functions). This theory has been with us since the very beginning with formulation and solutions by Abel, Jacobi, Gauss, Legendre, Riemann, etc.

The recent literature on transformation is not very extensive: Riemann [1899], Tricomi [1948], Achiezer [1970], Lang [1973], Rauch-Lebowitz [1973], Robert [1973], Houzel [1978].

We begin with four elementary examples: the first three deal with integrals.

(I) Fagnano (1682-1766) pointed out essentially that

$$\int_0^T \frac{dt}{\sqrt{(t(1-t^2))}} = \int_Z^1 \frac{dz}{\sqrt{(z(1-z)))}}$$

if $T = (1 - Z)/(1 + Z)$ which is established by the (linear) transformation

$$t = (1 - z)/(1 + z).$$

Hence if $T = \sqrt{2} - 1$ then $Z = \sqrt{2} - 1 = T$. This means we have succeeded in bisecting a quadrant of a lemniscate ($r^2 = \cos 2\,\theta$, in polar coordinates). We can represent the curve parametrically as

$$x = (t(1 + t)/2)^{1/2},\ y = (t(1 - t)/2)^{1/2}$$

and, differentiating, we find

$$\dot{s}^2 = \dot{x}^2 + \dot{y}^2 = \{4t(1 - t^2)\}^{-1}.$$

(II) Gauss (1777-1855) and Landen (1719-1790) essentially used a quadratic transformation to determine the arithmetic-geometric mean M of non-negative a_0, b_0 where $a_0 > b_0$. The existence of $M = \lim a_n = \lim b_n$, where a_n, b_n are defined by

$$a_{n+1} = \frac{1}{2}(a_n + b_n),\ b_{n+1} = (a_n b_n)^{1/2},$$

is easily established by monotony. The transformation

$$y = \frac{1}{2}(x^2 - a_n b_n)/x$$

which runs from $-\infty$ to ∞ as x runs from 0 to ∞ gives, with a little algebra,

$$\int_{-\infty}^{\infty} \{(x^2 + a_n^2)(x^2 + b_n^2)\}^{-1/2}\, dx =$$

$$\int_{\infty}^{\infty} \{(y^2 + a_{n+1}^2)(y^2 + b_{n+1}^2)\}^{-1/2}\, dy.$$

Repetition of this gives

$$\int_{-\infty}^{\infty} \{(x^2 + a_0{}^2)(x^2 + b_0{}^2)\}^{-1/2}\, dx =$$

$$\int_{-\infty}^{\infty} \{(t^2 + M^2)(t^2 + M^2)\}^{-1/2}\, dt = \pi/M$$

so that

$$M = \frac{\pi}{\int_{-\infty}^{\infty} \{(x^2 + a_0{}^2)(x^2 + b_0{}^2)\}^{-1/2}\, dx} =$$

$$= \frac{\pi a_0}{2\, K\{(a_0{}^2 - b_0{}^2)/a_0{}^2)^{1/2}\}} .$$

This presentation is based on one of D. J. Newman [1982]; there are several others available, all depending on an invariant integral.

(III) Landen showed, by a geometric argument essentially equivalent to that in (II), that if

$$k_1 = (1 - k')/(1 + k')$$

then

$$K_1 = \frac{1}{2}(1 + k')\, K, \quad K_1' = (1 + k')\, K'.$$

(IV) The differential equation

$$n(1 - x^2)^{-1/2}\, dx = (1 - y^2)^{-1/2}\, dy, \quad y(1) = 1$$

is satisfied by the Chebyshev polynomial

$$y = T_n(x) = \cos(n \arccos x) .$$

It is not too surprising that the well-known approximation properties of this degenerate case of the transformation equation carry over in some measure to the general case.

The special case of (1), when $\lambda = k$, is called the multiplication problem. The solutions to this will either hold for all k and then $M = n$, a positive integer, or will hold only for special values of k, when the period ratio is a complex quadratic surd, as is the multiplier, M. The first type is called real multiplication, the second is called complex multiplication.

We pursue this question of transformation and multiplication further. The Weierstrassian case is trivial, since the periods of $\wp$ can be chosen arbitrarily (subject only to $(w_2/w_1) > 0$). Thus if $\wp$ has periods $2w_1$, $2w_2$ then

$$\wp(z) + (e_1 - e_2)(e_1 - e_3)(\wp(z) - e_1)^{-1}$$

has periods w_1, $2w_2$. [Cf. W. & W, 456, 444.] Also [W & W, 441]

$$\wp(2u) = \frac{\wp^4 + \frac{1}{2} g_2 \wp^2 + 2g_3 \wp + \frac{1}{16} g_2^3}{4\wp^3 - g_2 \wp - g_3}$$

where the arguments u of the $\wp$'s on the right have been omitted.

There is a very different state of affairs in the Jacobian case, for the quarter-periods K, K' are not independent, both being uniquely determined by k. We cannot construct a Jacobian function with say, quarter periods, K_1, K_1' with $K_1' = 2K'$, $K_1 = K$; the best we can do is to introduce a "multiplier" for the argument. We discuss a numerical example.

As k increases from 0 to 1, K increases from $\frac{1}{2}\pi$ to ∞ and K' decreases from ∞ to $\frac{1}{2}\pi$. Consequently, K'/K decreases from ∞ to 0. Here are two sets of numerical values:

$k = 0.8$, $k' = 0.6$, $K = 1.9953$, $K' = 1.7508$, $K'/K = 0.8774$.

$k = 0.25$, $k' = 0.9682$, $K = 1.5962$, $K' = 2.8012$, $K'/K = 1.749$.

These are special cases of the (complete) Landen transformations (W&W, §22.42) mentioned in Example III above.

If we begin with $\mathrm{sn}(u, k)$, with $k = 0.8$ and $K'/K = 0.8774$ it is clear from the graph of K'/K against k that there will be a unique λ, actually $\lambda = 0.25$, such that $\Lambda'/\Lambda = 2\,K'/K$ and that Λ/K has a specific value, actually 0.8, the multiplier, M. The elliptic function

$\mathrm{sn}(u/M, \lambda)$, actually $\mathrm{sn}(1.25\,u, 0.25)$

will have quarter periods K, $2K'$.

In general the multiplier will depend on n, the order of transformation and on the modulus, k.

We now outline the solution of the general transformation problem. There are some advantages in using the Riemann Normal Form. We shall, however, for simplicity use the Weierstrass form. But, for applications, we will return to the traditional Jacobi form, despite its complications.

Take the case

$$u = \int_0^X \frac{dX}{\sqrt{(4X^2 - g_2X - g_3)}}, \quad v = \int_0^Y \frac{dY}{\sqrt{(4Y^2 - \gamma_2Y - \gamma_3)}} \qquad (2)$$

with

$u = Mv$.

Inverting (2) we get

$x = (u; 2w_1, 2w_2)$, $y = (v; 2\Omega_1, 2\Omega_2)$

where x has periods $2w_1$, $2w_2$ and y has periods $2\Omega_1$, $2\Omega_2$. Because of the hypothesis that x, y are related by a polynomial equation we can conclude that the periods are not independent and that the elliptic functions x, y have a common period parallelogram and, say,

$$\begin{aligned} r\Omega_1 &= aw_1 + bw_2 \\ s\Omega_2 &= cw_1 + dw_2 \end{aligned} \tag{3}$$

where r, s, a, b, c, d are integers. An algebraic relation between x, y can be obtained by taking a polynomial of sufficiently high degrees in x, y, balancing the principal parts by choice of the coefficients, and then appealing to Liouville's Theorem. (Cf. e.g. Copson, [1935].) We can show that if the relation is $A(x, y) = 0$ then to a value of x corresponds r values of y and to a value of y corresponds $n = ad - bc$ values of x.

We shall now show that the algebraic relation between x, y can be replaced by two rational ones. In fact denote by z the γ function with half-periods $\pi_1 = r\,\Omega_1$, $\pi_2 = s\,\Omega_2$ and apply the last remark in the immediately preceding paragraph to the pair x, z and to the pair y, z. We conclude that x, z are connected by an equation

$$B(x,z) = 0$$

which is of degree $1 \times 1 = 1$ in x and degree rs in z and that y, z are connected by an equation

$$C(x, z) = 0$$

of degree 1 in y and degree n in z. So x and y are expressible rationally in terms of the new variable z.

Our problem now is to find in closed form the transformation between the two elliptic objects (integrals, functions (Weierstrass or Jacobian) or theta functions) with which we are concerned.

We have seen that it is enough to discuss the rational case when the half-periods are related by

$$\left.\begin{aligned} \Omega_1 &= aw_1 + bw_2 \\ \Omega_2 &= cw_1 + dw_2 \end{aligned}\right\}$$

with a, b, c, d integers (without common factor) and $ad - bc = n > 0$.

To do this we consider the factorization of the matrices

$$M = \begin{bmatrix} a & b \\ c & d \end{bmatrix} .$$

First of all, any such matrix M can be represented as a product of the form

$$U^{\cdot}S^{\cdot}\ U^{\cdot}S^{\cdot}\ \ldots\ N\ \ldots\ S^{\cdot}U^{\cdot}S^{\cdot}U^{\cdot}$$

where

$$N = \begin{bmatrix} n & 0 \\ 0 & 1 \end{bmatrix}, \quad S = \begin{bmatrix} 1 & 0 \\ 1 & 1 \end{bmatrix}, \quad U = \begin{bmatrix} 1 & 1 \\ 0 & 1 \end{bmatrix},$$

and where the powers to which U, S are raised are not indicated, except by dots. (This result is due to C. Cellitti [1914].) For instance,

$$\begin{bmatrix} 3 & 4 \\ 2 & 20 \end{bmatrix} = \begin{bmatrix} 1 & 0 \\ 1 & 1 \end{bmatrix}^3 \begin{bmatrix} 1 & 1 \\ 0 & 1 \end{bmatrix}^7 \begin{bmatrix} 52 & 0 \\ 0 & 1 \end{bmatrix} \begin{bmatrix} 1 & 0 \\ 1 & 1 \end{bmatrix}^{-7} \begin{bmatrix} 1 & 1 \\ 0 & 1 \end{bmatrix}^{-1} .$$

We can now factorize N into diagonal matrices using the decomposition of n into prime factors. Continuing our example

$$\begin{bmatrix} 52 & 0 \\ 0 & 1 \end{bmatrix} = \begin{bmatrix} 2 & 0 \\ 0 & 1 \end{bmatrix}^2 \begin{bmatrix} 13 & 0 \\ 0 & 1 \end{bmatrix} .$$

Thus M can be represented as a product of 2×2 matrices of determinant 1, 2 and odd primes. The transformation is similarly decomposed into transformations of order 1, order 2, and order an odd prime. The last two are called principal transformations of the first kind -- those of the second kind correspond to matrices

$$\begin{bmatrix} 1 & 0 \\ 0 & n \end{bmatrix}$$

which can however, be represented in terms of matrices of the first kind and the matrix of order 1

$$T = \begin{bmatrix} 0 & 1 \\ -1 & 0 \end{bmatrix}$$

since

$$T^{-1}\begin{bmatrix}1 & 0\\ 0 & n\end{bmatrix} T = \begin{bmatrix}n & 0\\ 0 & 1\end{bmatrix}.$$

Notice that a principal transformation means that one or other period is multiplied or divided by the order n and that a principal transformation of the first kind followed by one of the second kind, both of the same order produce a multiplication (division) of the periods by the order.

There are various ways to derive the relevant transformation formulas, e.g. the usual Liouville arguments or the use of the elementary multiple angle formulas and the definitions of the elliptic functions in terms of $\tilde{I}$-functions. We outline the latter approach.

Lemma. $x^{2n} - 2x^n \cos n\theta + 1 = \prod_{s=1}^{n} \{x^2 - 2x\cos(\theta+(s\pi/n))+1\}$.

Proof. The left hand side is $(x^n - \exp(in\theta))(x^n - \exp(-in\theta))$. Use Demoivre's Theorem for each factor and then combine conjugate factors.

If in the classical formula ($\phi(q)$ standing for $\Pi(1-q^{2r})$)

$$\tilde{I}_4(z,q) = \phi(q) \prod_{r=1}^{\infty} \{1 - 2q^{2r-1}\cos 2z + q^{4r-2}\}$$

we replace z by nz, and q by q^n and use the lemma in each factor on the right, change the order of multiplication, and recombine factors we find

$$\tilde{I}_4(nz,q^n) = \{\phi(q^n)/[\phi(q)]^n\} \prod_{s=0}^{n-1} \tilde{I}_4\,(z+(s\pi/n),q).$$

Similar results hold for the other $\tilde{I}$-functions.

We now use the representations of the Jacobian elliptic functions as quotients of theta functions, such as

$$\mathrm{sn}(u,k) = \frac{\tilde{I}_3}{\tilde{I}_2} \cdot \frac{I_1(u\tilde{I}_3^{-2})}{\tilde{I}_4(u\tilde{I}_3^{\,2})}$$

to get transformation formulas for the Jacobian functions, in particular the following result.

Theorem. If n is odd

$$\mathrm{sn}(uM^{-1},\lambda^2) = M^{-1}\mathrm{sn}(u,k^2)\ \Pi\ \frac{1-c_{2s}^{-1}\mathrm{sn}^2(u,k)}{1-k^2c_{2s}\mathrm{sn}^2(u,k)} \tag{9}$$

where $c_r = \mathrm{sn}^2(rKn^{-1},k)$, and

$$\lambda = k^n\ \Pi\ c_{2r-1}^2,\quad M = \Pi(c_{2r-1}/c_{2r})$$

(all products are over s, from $s = 1$ to $s = \frac{1}{2}(n-1)$).

The transformation (9) is a First (Principal) Transformation: The periods Λ, Λ' being connected with K, K' by the relations

$$\Lambda = K/nM,\quad \Lambda' = K'/M.$$

In the Second (Principal) Transformation the periods are connected by the relations

$$\Lambda = K/M,\quad \Lambda' = K'/nM.$$

For details of these see e.g. Cayley [1895] and Achiezer [1970, p. 284].

We note here that in Greenhill [1892] there is a proof of the basic formulas by means of an electromagnetic analogy (Kelvin's method of images).

This formula (9) is essentially that used in the discussion of the Cauer problem (see, e.g. Oberhettinger and Magnus [1949]). To solve the ADI problem we have to use the n even analog of (9). The analog of (9) for the dn function gives the solution to the Ninomiya problem given above; we continue this discussion as announced.

Theorem. If n is an integer then

$$\mathrm{dn}(u\,M^{-1},\ \lambda) = \mathrm{dn}(u,k)\ \Pi\ \frac{c_{2m-1} + s_{2m-1}\,\mathrm{dn}^2(u,k)}{c_{2m} + s_{2m}\,\mathrm{dn}^2(u,k)} \tag{10}$$

where

$$\lambda = k^n \pi s_{2m-1}^2, \quad M = \pi(s_{2m-1}/s_{2m}),$$

$$s_j = sn^2(jK/n, k), \quad c_j = cn^2(jk/n, k),$$

and where all products run from $m = 1$ to $m = [n/2]$.

The optimal starting value, for $\sqrt{x}$, when x is restricted to $[a, 1]$, is given by

$$y = \sqrt{x/\lambda'}\ dn(uM^{-1}, \lambda)$$

when

$$x = a\ nd^2(u,k).$$

[As u runs from 0 to K, dn runs from 1 to k' and x from a to a/k'^2 so that we take $k' = \sqrt{a}$.]

If we specialize this to $n = 4$, $a = 1/4$ we find

$$y = \frac{(c_1 x + \frac{1}{4} s_1)(c_3 x + \frac{1}{4} s_3)}{\frac{1}{2}\sqrt{\lambda'}\ (c_2 x + \frac{1}{4} s_2)}$$

since $c_4 = 0$, $s_4 = 1$. When $k' = \sqrt{a} = \frac{1}{2}$ since

$$s_2 = (1 + k')^{-1}, \quad c_2 = k'(1 + k')^{-1}$$

we have

$$y = \frac{(2/c_2\sqrt{\lambda'})\ [c_1c_3x^2 + \frac{1}{4}x(s_1c_3 + s_3c_1) + (s_1s_3/16)]}{x + \frac{1}{4}}.$$

From Ninomiya [1970] or Carlson and Todd [1983] we can compute the coefficients in the numerator. These are all surds and, e.g., we find the coefficient of x^2 to be

$$\alpha = 2(\sqrt{2} - 1)(24\sqrt{2})^{-1/4} = 0.34322\ 0129;$$

that of x is

$$\alpha(3 + \surd 2)/\surd 2 = 1.07129\ 9971$$

and the constant term is

$$\frac{1}{4}\alpha = 0.08580\ 5032.$$

These values agree with those in the continued fraction representation given by Ninomiya [1970, p. 403].

§6. A MODERN TREATMENT OF TRANSFORMATION

This is described in the books of Houzel [1978], Lang [1973] and Robert [1973]. It is somewhat sophisticated and the essential identity of elliptic functions, the tori which are the corresponding Riemann surfaces, elliptic curves and period lattices is assumed.

A singular cubic curve is rational, i.e. its points can be expressed rationally in terms of a parameter, e.g. the slope of a line through the singularity (such a line meeting the curve in one other point). To parameterize a nonsingular cubic, we need elliptic functions. For the curve

$$y^2 = 4x^3 - g_2x - g_3$$

or, in homogeneous form,

$$y^2z = 4x^3 - g_2xz^2 - g_3z^3$$

we can take

$$x = \wp(u),\ y = \wp'(u) \tag{1}$$

where the $\wp$ function has invariants g_2, g_3. We can define an abelian group on such a curve by the following geometrical construction, when a point 0 on the curve is taken arbitrarily as the zero element: to find the sum C of points A, B on the curve we denote by D the residual intersection of the line AB and then the sum "$A + B$" is the residual intersection of OD with the curve. It is easy to see that if 0 is taken as a point at infinity $(0,1,0)$ then if A has parameter a and B parameter b in representation (1) above then the parameter of $C =$ "$A + B$" is $a + b$.

It is natural to study mappings between two elliptic curves (or their lattices) which preserve the group operation -- these are called isogenies. It turns out that these are essentially the transformations which we have been studying.

§7. REMARKS

1. There have been discussions of the Zolotarev problems when the L_1 norm is used in place of the L_∞, when trigonometric polynomials, and when entire functions are used in place of polynomials. See e.g. Gončar [1969], Ryžakov [1965, 1969], Meiman [1960, 1962], Galeev [1975], Feherstorfer [1979].

2. The elegant solutions to the problems can be used to indicate the efficiency of algorithms for optimal parameters in cases where there are no theoretical results available. Of course, from general principles, near optimal parameters will give very near optimal results, in a smooth environment.

3. There has been a certain amount of activity in other aspects of rational approximation, e.g. by D. J. Newman [1978], A. R. Reddy [1977, 1978], E. B. Saff and R. S. Varga [1980]. Particulary relevant is recent work by Lorentz, Saff, Varga and others on approximation by incomplete polynomials.

---

* This study is a sketch for part of an extensive survey article "Applications of elliptic functions and elliptic integrals" which will appear elsewhere. A preliminary version was presented at a special session on "History of Contemporary Mathematics" at the Annual Meeting of the American Mathematical Society, 7 January 1981.

REFERENCES

1. Achiezer, N. I., "Vorlesungen über Approximations-theorie", Akademic Verlag, Berlin, 1953, translated as, "On the Theory of Approximation", Ungar, New York, 1956. (Second German edition, Berlin, 1967.) "Elements of the Theory of Elliptic Functions", 2nd edition (in Russian), 1970. For complete bibliographies of Achiezer see the articles on him in Uspehi Mat. Nauk ≡ Russian Math. Surveys, 16, 6, (102), 1961; p. 14; 26, 6, (162), 1971, pp. 257-261; 36; 4, (220), 1981, pp. 183-4.

2. Amer, R. A., Schwarz, H. R., "Contributions to the approximation problem of electrical filters", Mitt. Inst. ang. Math., ETH, Zürich, #9, 1964.

3. Bateman Project, "Higher Transcendental Functions, II", McGraw Hill, New York, 1953.

4. Bernstein, S., "Lecons sur les propriétés exctrémales et la meilleure approximation des fonctions analytiques d'une variable réelle", Gauthier-Villars, Paris, 1926. Chel., 1970.

5. Blum, E. K., "Numerical analysis and computation, theory and practice", Addison-Wesley, 1972.

6. Borwein, J. M., and Borwein, P. B., "Elliptic integrals and approximations to π", to appear; Cubic and Higher Order Algorithms for π, Canada Math. Bull. to appear; Explicit algebraic nth order approximations to π, this volume; The arithmetic geometric mean and fast computation of elementary functions, SIAM Rev., to appear.

7. Brent, R. P., "Fast multiple precision evaluation of elementary functions", J. ACM 23, 1976, pp. 212-251.

8. Carlson, B. C., and Todd, J., "The degenerating behavior of elliptic functions", SIAM J. Numerical Anal. 20, 1983, pp. 1120-1129; "Zolotarev's First Problem - the best approximation by polynomials of degree ≤ n - 2 to $x^n - n E x^{n-1}$ in [0, 1]σ], Aq. Math., 1984, 26 (1983, pp. 1-33.

9. Cauer, W., "Synthesis of linear communications networks", McGraw Hill, New York, 1958, German edition 1934 "Bemerkung über eine Extremalaufgabe von E. Zolotareff" ZAMM 20, 1940, 358.

10. Cayley, A., "Elliptic functions", Deighton Bell, Cambridge 1895.

11. Cellitti, C., "Nuova representazione della sostituzione lineare binaria primitua, Atti Reale Acc. deiLincei. Ser. V, 23, 1914, pp. 208-213.

12. Chebyshev, P. L., "Oeuvres, I, II", St. Petersburg, 1899, Chelsea, New York.

13. "Sur les fractions algébriques...", Bull. Soc. Math. France, 12, 1884, pp. 167-168 (≡II, 725).

14. Cody, W. J., "Double-precision square root for the CDC-300", Comm. ACM 7, 1964, pp. 715-718.

15. Copson, E. T., "Functions of a complex variable", Oxford, 1935.

16. Curtis, A. R., "Tables of Jacobian elliptic functions whose arguments are rational fractions of the quarter period", NPL Math. Tables, v. 7, H. M. Stationery Office, London, 1964.

17. De Boor, C., and Rice, J. R., "Chebyshev approximation by a $\pi[(x-x_i)/(x-s_i)]$ and application to ADI iteration", J. SIAM, 11, 1963, pp. 159-169.

18. de la Vallée Poussin, C. J., "Lecons sur l'approximation des fonctions d'une variable réelle",Gauthier-Villars, Paris, 1919, Chelsea, 1970.

19. Dixon, A. C., "The elementary properties of elliptic functions", Macmillan, London, 1894.

20. Dumas, S., "Sur le développement des functions elliptiques en fractions continues", Thesis, Zürich, 1908.

21. Darlington, S. D., "Synthesis of resistance 4-poles which produce prescribed insertion loss characteristics including special application to filter design", J. Math. Phys., 18, 1939, pp. 257-353.

22. Enneper, A., rev. F. Müller, "Elliptische Functionen, Theorie und Geschichte", Halle, 1890.

23. Erdös, P., and Szegö, G., "On a problem of I. Schur", Ann. of Math. {2}43, 1942, pp. 451-470.

24. Farhutcinova, R. F., "Approximate analytic formulas for coefficients of Zolotarev polynomials and the process of economization via Zolotarev polynomials", p. 126-135 in Uniform approximations and the moment problem, ed. M. Ja. Zinger, Dal´nevostocn. Naucn. Centr. Akad. Nauk, SSSR, Vladivostok., 1977.

25. Fricke, R., "Lehrbuch der Algebra", 1924.

26. Gaier, D., Todd, J., "On the rate of convergence of optimal ADI processes", Numer. Math. 9, 1967, pp. 452-459.

27. Galeev, E. M., "Zolotarev Problem in the Metric of $L_1(-1,1)$", Mat. Zametki 17, 1975, pp. 13-20 = Math. Notes, 1975, pp. 9-13.

28. Gastinel, N., "Sur le meilleur choix des paramètres de sur-relaxation", Chiffres 5, 1962, pp. 109-126.

29. Gautschi, W., "Computational methods in special functions", pp. 1-98 in Theory and Application of Special Functions, Academic Press, 1975.

30. Glowatski, E., "Sechstellige Tafel der Cauer-Parameter", Abh. Bayer. Akad. Wiss., Math. Nat. Klasse, 67, 1955.

31. Goncar, A. A., "The problems of E. I. Zolotarev which are connected with rational functions", Math. Sb. (N.S.), 78, 120, 1969, pp. 640-654 translated in USSR - Sb. 7, 1969, pp. 623-625.

32. Greenhill, G., "The Applications of Elliptic Functions", Macmillan, London, 1892.

33. Hornecker, G., "Évaluation approcheé de la meilleure approximation polynomiale d'ordre n de f(x) sur un segment fini [a,b]", Chiffres 1, 1958, pp. 157-169.

34. Houzel, C., "Fonctions elliptiques et intégrales abéliennes", in Abrégé d'histoire des mathématiques 1700-1900, J. Dieudonné, directeur de la publication, 2 vols., Hermann, Paris, 1978.

35. King, L. V., "On the direct numerical calculation of elliptic functions and integrals", Cambridge, 1924.

36. Kuznetsov, P. I., "D. F. Egorov (on the centenary of his birth)", Usp. mat. nauk. = Russian Math. Surveys 26, 1971, #5, pp. 124-164.

37. Lang, S., "Elliptic Functions", Addison-Wesley, Reading, Mass., 1973.

38. Lebedev, V. I., "On a Zolotarev problem in the method of alternating directions", Z. Vycisl. Mat. i Mat. Fiz., 17, 1977, pp. 349-366, translated in USSR Comp. Math. and Math. Phys., 17, 1977, pp. 58-76, 1978.

39. Lebesgue, H., "Review of de la Vallée Poussin [1919]", Bull. Sc. Math. {2}44, 1920, pp. 137-153 ≡ Oeuvres Scientifiques 5, 1973, pp. 322-338.

40. Magnus, W., Oberhettinger, F., and Soni, R. P., "Formulas and theorems for the special functions of mathematical physics", 3, Springer, Heidelberg, 1966.

41. Markushevich, A. I., "The remarkable sine functions", Elsevier, New York, 1966.

42. Meiman, N. N., "Solution of the fundamental problems of the theory of polynomials and entire functions which deviate least from zero", (in Russian), Trudy Moskov. Math. Obsc. 9, 1960, pp. 507-535, MR24A#2031.

43. Meiman, N. N., "On the theory of polynomials deviating least from zero", 1960, translated in Soviet Math. Dokl. 1, pp. 41-44.

44. Meiman, N. N., "Polynomials deviating least from zero with an arbitrary number of given coefficients", 1960, translated in Soviet. Math. Dokl. 1, pp. 72-75.

45. Meiman, N. N., "The zeros of a class of multiple valued-functions", translated in AMS Trans. 2, 19, 1962, pp. 167-171.

46. Meiman, N. N., "The theory of functions of class HB and B^x", translated in AMS Trans. (2), 19, 1962, pp. 173-178.

47. Meinardus, G., "Approximation von Funktionen und ihre numerische Behandlung", Springer, 1964, translated by L. L. Schumaker as "Approximation of Functions, Theory and Numerical Methods", Springer, 1967.

48. Meinardus, G., and Taylor, G. D., "Optimal partitioning of Newton's method for calculating roots", Math. Comp. 35, 1980, pp. 1231-1250.

49. Meinguet, J., and Belevitch, V., "On the realizability of ladder filters", IRE Trans. Prof. Groups on Circuit Theory, CT 5, 1958, pp. 253-255.

50. Melzak, Z. A., "Mathematical ideas, modeling and applications", Vol. II of Companion to concrete mathematics, Wiley, New York, 1976.

51. Milne-Thomson, L. M., "Jacobian elliptic function tables", Dover, New York, 1950.

52. Moursund, D. G., "Optimal starting values for Newton-Raphson calculations of $\sqrt{x}$", Comm. ACM 10, 1967, pp. 430-432.

53. Moursund, D. G., and Taylor, G. D., "Optimal starting values for the Newton-Raphson calculation of inverses of certain functions", SIAM J. Numer. Anal. 5, 1968, pp. 138-150.

54. NBS Handbook of Mathematical Functions, ed. M. Abramowitz - I. A. Stegun, U. S. Government Printing Office, Washington, D.C., 1964.

55. Neville, E. H., "Jacobian elliptic functions", Oxford, 1946.

56. Newman, D. J., "Approximation with rational functions", CBMS, Regional conference series in mathematics #41, 1978, American Math. Soc., Providence, R.I.

57. Newman, D. J., "Rational approximation versus fast computer methods", in Lectures on approximation and value distribution, Univ. de Montréal, 1982.

58. Ninomiya, I., "Generalized rational Chebyshev approximation", Math. Comp. 24, 1970, pp. 159-169. "Best rational starting approximations and improved Newton iteration for the square roots", Math. Comp. 24, 1970, pp. 391-404.

59. Oberhettinger, F., and Magnus, W., "Anwendung der elliptischen Funktionen in Physik und Technik", Springer-Verlag, Heidelberg, 1949.

60. Paszkowski, S., "The theory of uniform approximation", I, Rozprawy Matem., 25, Warsaw, 1962.

61. Piloty, H., "Zolotareffsche rationale Funktionen", ZAMM 34, 1954, pp. 175-189.

62. Ozigova, E. P.,"Zolotarev, E. I., 1847-1878",Nauka, Moscow, 1966 (in Russian).

63. Peaceman, D. W., and Rachford, H. H., Jr., "The numerical solution of parabolic and elliptic differential equations", J. SIAM 3, 1955, pp. 28-41.

64. Feherstorfer, F., "On the representation of extremal functions in the L^1-norm", J. Approx. Theory, 27, 1979, pp. 61-75.

65. Pokrovskii, V. L., "On a class of polynomials with extremal properties", Mat. Sb. (N.S.) 48, 90, pp. 257-276, translated in AMS Trans. (2), 19, 1962, pp. 199-219.

66. Rauch, H. E., and Leibowitz, A., "Elliptic functions, theta-functions and Riemann surfaces", Baltimore, 1973.

67. Rivlin, T. J., "Polynomials of best approximation to certain rational functions", Numer. Math. 4, 1962, pp. 345-369.

68. Robert, Alain, "Elliptic curves", Springer Lecture Notes in Mathematics, #326, 1973.

69. Reddy, A. R., "Rational approximation to $x^{n+1} - \sigma x^n$", Quart. J. Math. (Oxford), 2, 28, 1977, pp. 123-127.

70. Reddy, A. R., "A note on a result of Zolotarev and Bernstein", Manuscripta Math. 10, 1977, pp. 95-97.

71. Reddy, A. R., "On certain problems of Chebyshev, Zolotarev, Bernstein and Akhieser", Invent. Math. 45, 1978, pp. 83-110.

72. Rice, J. R., "The approximation of functions", 2 vols., Addison-Wesley, 1964, 1969.

73. Riemann, B., "Elliptische Functionen", ed. H. Stahl, Teubner, Leipzig, 1899.

74. Rivlin, T. J., "An introduction to the approximation of functions", 1969, 1981. Dover, N.Y.

75. Ryzakov, I. Ju., "An analog of a problem of E. I. Zolotarev", DAN USSR 160, 1965, 552-559 ≡ Soviet Math. Dokl. 6, 1965, pp. 157-159.

76. Ryzakov, I. Ju., "The trigonometrical analogue of a problem of E. I. Zolotarev", Izv. Vyss. Ucebn. Zaved. Mat. 8, 87, 1969, pp. 75-88.

77. Saff, E. B., and Varga, R. S., "Remarks on some conjectures of G. G. Lorentz", J. Approx. Theory 30, 1980, pp. 29-36.

78. Salamin, E., "Computation of π using arithmetic-geometric mean", Math. Comp. 30, 1976, pp. 565-570.

79. Stiefel, E. L., "Le problème d'approximation dans la théorie des filtres électriques", pp. 81-87, Belgian Colloquium, 1961.

80. Talbot, A., "On a class of Tchebysheffian approximation problems solvable algebraically", Proc. Cambridge Phil. Soc. 58, 1962, pp. 266-267.

81. Talbot, A., "The Tchebysheffian approximation of one rational function by another", Proc. Cambridge Phil. Soc. 60, 1964, pp. 877-890.

82. Talbot, A., Inaugural Lecture, "Approximation Theory or a miss is better than a mile", 18 November 1970, University of Lancaster Library, 1971.

83. Todd, J., "Introduction to the constructive theory of functions", Birkhäuser, Academic Press, 1963.

84. Todd, J., "Optimal ADI-parameters", in Functional Analysis, Approximationstheorie, Numerische Mathematik, ISNM, 7, Birkhäuser, Basel, 1967.

85. Todd, J., "Inequalities of Chebyshev, Zolotarev, Cauer and W. B. Jordan", pp. 321-328 in Inequalities, II., ed. O. Shisha, Academic Press, 1967.

86. Todd, J., "Optimal parameters in two programs", Bull. Inst. Math. Appl. 6, 1970, pp. 31-35.

87. Todd, J., "The lemniscate constants", Comm. ACM 18, 1975, pp. 16-19.

88. Todd, J., "The many values of mixed means", I., pp. 5-22, in General Inequalities, ed. E. F. Beckenbach, ISNM 41, Birkhäuser, Basel, 1978.

89. Todd, J., "Some applications of elliptic functions and integrals", Coll. Math. Soc. Janos Bolyai, vol. 22, Numerical Methods, ed. P. Rózsa, Keszthely (Hungary), 1977, pp. 591-618, North Holland, New York, and Amsterdam, 1980.

90. Todd, J., "The best polynomial approximation to $(1+x)^{-1}$ in [0,1]", to appear.

91. Tricomi, F. G., and Kraft, M., "Elliptische Funktionen", Leipzig, 1948.

92. Varga, R. S., "Matrix iterative analysis", Prentice-Hall, Englewood Cliffs, 1962.

93. Voronowskaja, E. V., "The functional method and its applications", Trans. Russian Math. Monographs, 28, 1970, American Math. Soc., Providence, R.I.

94. Wachspress, E. L., "Extended application of ADI iteration model problem theory", J. SIAM 11, 1963, pp. 994-1016.

95. Wachspress, E. L., "Optimum ADI iteration parameters for a model problem", J. SIAM 10, 1962, pp. 339-350.

96. Wachspress, E. L., "Iterative solution of elliptic systems", Prentice-Hall, Englewood Cliffs, 1966.

97. Whittaker, E. T., and Watson, G. N., "Modern analysis", Cambridge Univ. Press, 1927.

98. Zolotarev, E. I., "Complete collected works", I, 1931, II, 1932, Leningrad,

99. Zolotarev, E. I., "Sur l'application des functions elliptiques aux questions de maxima et minima", Bull. Acad. Sc. St. Petersburg 3, 24, 1878, 305-310, Melanges, 5, pp. 419-426 (≡I, pp. 369-376).

100. Zolotarev, E. I., "Anwendungen den elliptischen Functionen auf Probleme über Functionen die von Null am wenigsten oder am meisten abweichen"(in Russian), Abh. St. Petersburg 30, 1877, 5. (≡II, 1-59). See also remarks by N. G. Chebotarev and N. I. Achieser, II, pp. 357-361, Ob odnom voprose o naimehlshix velichnax, Diss. 1868, (≡II, 130-166).

EXPLICIT ALGEBRAIC NTH ORDER APPROXIMATIONS TO PI

J. M. Borwein and P. B. Borwein

Dalhousie University

ABSTRACT

We present a family of algorithms for computing pi which converge with order m (m any integer larger than one). Details are given for two, three and seven.

INTRODUCTION

In the course of a general study of elliptic integral transforms and their applications in the construction of good algebraic approximations to transcendental functions and natural constants [2], the authors discovered the following general multiplication formula which gives algebraic approximations of order m to π (for m any integer greater than 1). The formula is constructed as follows. Let

$$K(k) := \int_0^{\pi/2} \frac{1}{\sqrt{1-k^2\sin^2}} \tag{1.1}$$

and

$$E(k) := \int_0^{\pi/2} \sqrt{1 - k^2\sin^2} \tag{1.2}$$

denote the complete elliptic integrals of first and second kind respectively, for $0 \leq k \leq 1$. For each integer m there is an

S. P. Singh et al. (eds.), Approximation Theory and Spline Functions, 247–256.

integral polynomial in two variables u and v, ϕ_n, called the modular equation (of order m) and a rational function, M_m, of u and v, called the multiplier such that for any u in]0,1[

$$K(u^4) = mM_m(u,v)K(v^4) \tag{1.3}$$

whenever v is the (unique) solution in]0,u[to $\Phi_m(u,v) = 0$, [5]. When $u^8 + v^8 = 1$, $u := u(m)$ and $v := v(m)$ are said to be conjugate. Let K(m) and E(m) denote $K(v^4(m))$ and $E(v^4(m))$. In [2] the authors showed that there is a computable algebraic constant, $\alpha(m)$, such that

$$\frac{\pi}{4} = K^2(m)[\sqrt{m}\,(\frac{E(m)}{K(m)} - 1) + \alpha(m)] \tag{1.4}$$

and

$$0 \le \pi - \alpha(m)^{-1} = 0(10^{-\sqrt{m}}). \tag{1.5}$$

In fact $\alpha(m)^{-1}$ converges monotonically [2] to π. Heuristically this goes as follows. As m tends to infinity $\lambda(m) := v^4(m)$ tends to zero and so K(m) tends to $\pi/2$. Moreover, E(m)/K(m) decreases to one sufficiently fast to validate (1.5). When m = 1, $\alpha(m) = 1/2$ and (1.4) is Legendre's identity [5]. The multiplication formula, which allows one to compute π rapidly, is now able to be stated. Let p be any positive integer. Then for integral m

$$\alpha(p^2m) = p^2M_p^2\,\alpha(m) + p\sqrt{m}\,[\frac{v(1-v^8)}{4M_p}\frac{dM_p}{dv} + v^8 - pM_p^2u^8] \tag{1.6}$$

where $u := v(m)$ and v is the unique solution to $\Phi_p(v(m),v) = 0$ in]0,v(m)[. Also $\frac{dM_p}{dv}$ is the complete differential of M_p with respect to v. To compute this quantity it helps to know Jacobi's identity

$$\frac{du}{dv} = \frac{u(1-u^8)}{v(1-v^8)}\,pM_p^2 \tag{1.7}$$

whenever $\Phi_p(u,v) = 0$ and $0 \le v \le u$. For convenience we denote v as $T_p(u)$, and let $k := u^4$, $\lambda := v^4$.

Algebraic details for p a prime less than twenty are given in [6]. The general theory is nicely laid out in [5].

A GENERAL ITERATION

By iterating (1.5) we are led to the following algorithm. Let m be integral and let

$$\text{(i)}\quad \alpha_0 := \alpha(m), \quad v_0 := v(m). \tag{2.1}$$

For n in $\mathbb{N}$ we compute

$$\text{(ii)}\quad v_{n+1} := T_p(v_n) \tag{2.2}$$

$$\text{(iii)}\quad s_n := pM_p^2(v_n, v_{n+1}) \tag{2.3}$$

$$\text{(iv)}\quad d_n := \frac{v_{n+1}}{4} \frac{(1-v_{n+1}^8)}{M_p(v_n, v_{n+1})} \frac{dM_p(v_n, v_{n+1})}{dv} \tag{2.4}$$

and have

$$\alpha_{n+1} := ps_n\alpha_n + p^{n+1}\sqrt{m}\,[d_n + v_{n+1}^8 - s_n v_n^8]. \tag{2.5}$$

Moreover,

$$\alpha_n^{-1} - \pi = O(10^{-p^n\sqrt{m}}). \tag{2.6}$$

The larger m is, the better the initial approximation. This is illustrated below. For small m we have the following initial values [2].

Starting Values

m	$v^4(m) = \lambda(m)$	$\alpha(m)$
1	$2^{-1/2}$	$1/2$
2	$\sqrt{2} - 1$	$\sqrt{2} - 1$
3	$\sqrt{2}(\sqrt{3}-1)/4$	$(\sqrt{3}-1)/2$
4	$3 - 2\sqrt{2}$	$6 - 4\sqrt{2}$
5	$(\sqrt{\sqrt{5} - 1} - \sqrt{3 - \sqrt{5}})/2$	$(\sqrt{5} - \sqrt{2(\sqrt{5}-1)})/2$
7	$\sqrt{2}(3-\sqrt{7})/8$	$(\sqrt{7}-2)/2$

Other values are computed in [2]. We now specialize our algorithm for $m = 2, 3, 7$. The specializations are remarkably clean. Note also that (1.7) allows one to calculate (2.2), (2.3) and (2.4) as soon as

Φ_p is known since $\dfrac{du}{dv} = -\dfrac{\partial\Phi_p}{\partial v} \Big/ \dfrac{\partial\phi_p}{\partial u}$.

The Quadratic Case

In this case the multiplier is $M_2 := \frac{1+\lambda}{2}$ and the transformation T_2 is given by $\lambda := (1 - \sqrt{1-k^2})/(1 + \sqrt{1-k^2})$. The iteration becomes

$$\text{(i)} \quad x_{n+1} := (1 - \sqrt{1-x_n^2})/(1 + \sqrt{1-x_n^2}) \qquad (2.1)$$

$$\text{(ii)} \quad \alpha_{n+1} := (1+x_{n+1})^2\alpha_n - 2^{n+1}\sqrt{m}\, x_{n+1}, \qquad (2.2)$$

with $\alpha_0 := \alpha(m)$ and $x_0 := \lambda(m)$; and

$$\alpha_n^{-1} - \pi = O(10^{-2^n\sqrt{m}}).$$

A more exact asymptotic is given in [2]. The first few iterations behave as follows:

Digits Correct in Quadratic Algorithms

	n=1	2	3	4	5	6	7	8
m = 1	0	3	8	19	41	84	171	344
m = 2	2	5	13	28	56	120	242	> 400
m = 7	5	12	26	55	112	227	> 400	

If one replaces x_n by c_n/a_n (where $a_{n+1} := (a_n+b_n)/2$; $b_{n+1} := \sqrt{a_n b_n}$; $c_n := \sqrt{a_n^2 - b_n^2}$ is the AGM iteration [4], [7]) then we may replace (2.2) by

$$a_{n+1}^2 \alpha_{n+1} = a_n^2\alpha_n - 2^{n-1} \sqrt{m}\, c_n^2 , \tag{2.3}$$

which on summing yields

$$\pi = \frac{\lim a_{n+1}^2}{\alpha(m) - \sqrt{m} \sum_{n=0}^{\infty} 2^{n-1} c_n^2} . \tag{2.4}$$

When m=1 this is an identity known to Gauss [3] which forms the basis for the Salamin-Brent [4], [7] algorithm recently used by Tamura and Kanada to compute 2^{24} digits of π, ([8] and private communication). There is some advantage to (2.2) over (2.4) in that all root extractions in the former are of numbers converging rapidly to one.

THE CUBIC CASE

The modular equation is $u^4 - v^4 + 2uv(1-u^2v^2) = 0$. It is convenient, though, to use a form of the modular equation given in Cayley [5] which uses an auxiliary variable t. We have

$$M_3 := \frac{2t+1}{3} ; \; t := v^3/u$$

$$u^8 := t \left(\frac{t+2}{2t+1}\right)^3 ; v^8 := t^3 \left(\frac{t+2}{2t+1}\right) .$$

As in [1] we can explicitly compute the v_n. The algorithm becomes:

$$\text{(i)}\quad v_{n+1} := v_n^3 - \sqrt{v_n^6 + \sqrt[3]{4v_n^2(1-v_n^8)}} + v_{n-1} \qquad (3.1)$$

$$\text{(ii)}\quad t_n := \frac{v_{n+1}^3}{v_n} \qquad (3.2)$$

$$\text{(iii)}\quad \alpha_{n+1} := (2t_n+1)^2\alpha_n - 2\sqrt{m}\,3^n(t_n+2)t_n, \qquad (3.3)$$

with $\alpha_0 := \alpha(m)$, $v_0 := v(m)$, and v_1 calculated from the quartic formula as in [1]. When $m = 1$, we have $v_1 := ((\frac{1-\sqrt{3}}{\sqrt{2}} + 3^{1/4})\ 2^{-7/8}$. Then

$$\alpha_n^{-1} - \pi = O(10^{-3^n\sqrt{m}}).$$

Alternately, we can give (3.1) and (3.2) in terms of t_n directly. We get

$$t_n := \left(\frac{t_{n-1}+2}{2t_{n-1}+1}\right)\left[(t_{n-1}+1) - \sqrt{t_{n-1}^2 + (1+t_{n-1})\sqrt[3]{\frac{4(1-t_{n-1})(2t_{n-1}+1)}{(t_{n-1}+2)^2}}}\ \right]^3 .$$

For practical purposes it seems better to directly invert the modular equation. The first few iterations give:

Digits Correct in Cubic Algorithms

	n=1	2	3	4	5
m=1	2	10	34	107	327
m=7	8	30	93	288	873

In both the quadratic and cubic cases it is easy to directly establish the error estimate.

The Septic Case. The modular equation is

$(1-u^8)(1-v^8) = (1-uv)^8$. We use (1.7) and

$$7M_7^2 = v(u-v^7)/u(u^7-v) = b/a$$

where $b := uv/(u^8 - uv)$; $a := uv/(uv-v^8)$ to derive the following algorithm.

(i) Generate (v_n) decreasingly from

$$(1-v_n^8)(1-v_{n+1}^8) = (1-v_n v_{n+1})^8 \tag{4.1}$$

(ii) $$a_n := v_n v_{n+1}/(v_n v_{n+1}-v_{n+1}^8);\ b_n := v_n v_{n+1}/(v_n^8-v_n v_{n+1}) \tag{4.2}$$

(iii) $s_n := b_n/a_n$

(iv) $$t_n := 1/8[(1-v_{n+1}^8)(49a_n-b_n) + (1-v_n^8)(s_n-1)b_n] \tag{4.3}$$

(v) $$\alpha_{n+1} := s_n\alpha_n + 7^n\sqrt{m}\,(7-s_n-t_n) \tag{4.4}$$

with $\alpha_0 := \alpha(m)$, $v_0 := v(m)$ as before. Then

$$\alpha_n^{-1} - \pi = O(10^{-7^n\sqrt{m}}) .$$

The first few iterations are as follows:

Digits Correct in Septic Algorithms

	n=1	2	3
m=1	7	63	464
m=7	22	173	>1000

We finish by observing that while the rate of convergence improves as p increases the complexity remains unchanged [1] [3]. Also, it is possible, using the data given in [6], to write down an explicit iteration for p any odd number less than twenty. The case p = 5 can be handled almost as cleanly as 3 or 7.

REFERENCES

1. Borwein, J. M., and Borwein, P. B., "Cubic and higher order algorithms for π", to appear, Canad. Math. Bull.

2. Borwein, J. M., and Borwein, P. B., "Elliptic integral and approximation of π'. Preprint.

3. Borwein, J. M., and Borwein, P. B., "The arithmetic-geometric mean and fast computation of elementary functions", SIAM Review, 26, 1984.

4. Brent, R. P., "Fast multiple-precision evaluation of elementary functions", J. Assoc. Comput. Mach. 23, 1976, pp. 242-251.

5. Cayley, A., "An elementary treatise on elliptic functions", Bell and Sons, 1895, republished Dover 1961.

6. Cayley, A., "A memoir on the transformation of elliptic functions", Phil. Trans. T., 164, 1874, pp. 397-456.

7. Salamin, E., "Computation of π using arithmetic-geometric mean", Math. Comput. 135, 1976, pp. 565-570.

8. Tamura, Y., and Kanada, Y., "Calculation of π to 4,196,293 decimals based on Gauss-Legendre algorithm", preprint.

Research partially supported by NSERC Grants.

THREE ALGORITHMS FOR π

QUADRATIC: with $\alpha_0 := \alpha(m)$ $x_0 := \lambda(m)$; and

$$\text{(i)}\quad x_{n+1} := (1 - \sqrt{1-x_n^2})/(1 + \sqrt{1-x_n^2})$$

$$\text{(ii)}\quad \alpha_{n+1} := (1+x_{n+1})^2\alpha_n - \sqrt{m}\, 2^{n+1}x_{n+1},$$

$$\alpha_n^{-1} - \pi = 0(10^{-2^n\sqrt{m}}).$$

CUBIC: with $\alpha_0 := \alpha(m)$, $v_0 := v(m)$; and

(i) generate (v_n) decreasingly from

$$v_{n+1}^4 + 2v_nv_{n+1} = v_n^4 + 2(v_nv_{n+1})^3$$

$$\text{(ii)}\quad t_n := v_{n+1}^3/v_n$$

$$\text{(iii)}\quad \alpha_{n+1} := (2t_n+1)^2\alpha_n - 2\sqrt{m}\, 3^n(t_n+2)t_n ,$$

$$\alpha_n^{-1} - \pi = 0(10^{-3^n\sqrt{m}}).$$

SEPTIC: with $\alpha_0 := \alpha(m)$, $v_0 := v(m)$; and

(i) generate (v_n) decreasingly from

$$(1-v_n^8)(1-v_{n+1}^8) = (1-v_nv_{n+1})^8$$

$$\text{(ii)}\quad a_n := v_nv_{n+1}/(v_nv_{n+1}-v_{n+1}^8);\ b_n := v_nv_{n+1}/(v_n^8-v_nv_{n+1})$$

$$\text{(iii)}\quad s_n := b_n/a_n$$

$$\text{(iv)}\quad t_n := 1/8[(1-v_{n+1}^8)(49a_n-b_n) + (1-v_n^8)(s_n-1)b_n]$$

$$\text{(v)}\quad \alpha_{n+1} := s_n\alpha_n - \sqrt{m}\, 7^n(s_n+t_n-7)$$

$$\alpha_n^{-1} - \pi = 0(10^{-7^n\sqrt{m}}).$$

In cubic algorithms: (i) may be replaced by

$$v_{n+1} := v_n^3 - \sqrt{v_n^6 + \sqrt[3]{4v_n^2(1-v_n^8)} + v_{n-1}}$$

once v_1 is known. When $m = 1$

$$v_1 := \left(\left(\frac{1-\sqrt{3}}{\sqrt{2}}\right) + 3^{1/4}\right)2^{-7/8}.$$

Selected Starting Values

m	$v^4(m) = \lambda(m)$	$\alpha(m)$
1	$2^{-1/2}$	$1/2$
2	$\sqrt{2} - 1$	$\sqrt{2} - 1$
3	$\sqrt{2}(\sqrt{3}-1)/4$	$(\sqrt{3}-1)/2$
5	$(\sqrt{\sqrt{5}-1} - \sqrt{3-\sqrt{5}})/2$	$(\sqrt{5} - \sqrt{2(\sqrt{5}-1)})/2$
7	$\sqrt{2}(3-\sqrt{7})/8$	$(\sqrt{7}-2)/2$

SOLVING INTEGRAL EQUATIONS OF NUCLEAR SCATTERING BY SPLINES

M. Brannigan

University of Georgia

INTRODUCTION

In the definitive work of Faddeev [2] the three-body scattering problem is shown to be reducible to the solution of singular integral equations. We here consider spline approximation techniques to give reliable and computational methods for the numerical solution to these equations.

We consider the method of collocation and the Galerkin method both of which are classed as projection methods. For these techniques we can show that a convergence rate of order four can be achieved using cubic B-splines as basis functions.

SCATTERING EQUATIONS

Let us first consider the two-body scattering problem. The partial-wave off-shell amplitude, $M(q,\kappa;z)$, is given by a solution of the equation

$$M(q,\kappa;z) = v(q,\kappa) - \frac{2}{\pi} \int_0^\infty v(q,q') \frac{q'^2}{q'^2 - z} M(q',\kappa;z)dq', \quad q \in (0,\infty).$$

Here $v(q,q')$ is a two-body potential and z is a complex energy variable.

If κ is the incident two-particle momentum and setting $z = \kappa^2$, then we obtain the half-off-shell K-matrix

S. P. Singh et al. (eds.), Approximation Theory and Spline Functions, 257–264.

$M(q,\kappa) = M(q,\kappa;\kappa^2)$ given by

$$M(q,\kappa) = v(q,\kappa) - \frac{2}{\pi}\int_0^\infty v(q,q') \frac{q'^2}{q'^2-\kappa^2} M(q',\kappa)dq' .$$

It is convenient to map the momentum variable q onto a finite interval $[-1,1]$ using

$$q(x) = \left(\frac{x+1}{x-1}\right) .$$

We obtain

$$M(q(x),\kappa) = v(q(x),\kappa)$$

$$- \frac{2}{\pi}\int_{-1}^{1} v(q(x),q'(x')) \frac{x'+1}{x'-1} \frac{M(q'(x'),\kappa)}{[(x'+1)^2-\kappa^2(x'-1)^2]} dx .$$

We also consider a system of three identical particles each with spin zero and isospin zero interacting via a separable two-particle potential of the form

$$v(q,q') = \lambda\phi(q)\phi(q'),$$

where ϕ is a smooth vertex function. For this choice of potential, the Faddeev equations reduce to an integral equation in one momentum variable [3]. The partial-wave half-off-shelf K-matrix $X(p,k)$ is given by

$$X(p,k) = Z(p,k;k) - \frac{2}{\pi}\int_0^\infty Y(p,p';k) \frac{p'^2 X(p',k)}{p'^2-k^2} dp',$$

where k is the momentum of the incident particle in the three-body c.m. frame. The driving term Z is defined by the integral

$$Z(p,p';k) = -\int_{-1}^{1} \frac{\phi(q_1)\phi(q_2)\ dx}{p^2+p'^2+pp'x - \frac{3}{4}k^2+\varepsilon-i0}$$

where

$$q_1 = (\tfrac{1}{4}p^2+p'^2+pp'x)^{1/2} ,$$

$$q_2 = (p^2+\tfrac{1}{4}p'^2+pp'x)^{1/2} ,$$

and ϵ is the two-particle bound-state energy. The i0 denotes the direction we approach the real energy axis. The function Y in the kernal is given by the two-particle amplitude and the function Z.

We see that the three-body equation described above has the same basic structure as the two-body equation and hence can be solved in a similar fashion.

NUMERICAL PROCEDURES

As can be seen from the above equations we are dealing with equations of the general form

$$(I - K)f = y,$$

where $f, y \in C[a,b]$; the space of continuous functions; K, I are linear operators mapping $C[a,b]$ into itself, with I the identity operator and K an integral with a singular kernal.

Let π_n be a partition of the interval $[a,b]$ defined by the knots $a = t_1 < t_2 < \ldots < t_n = b$ with a mesh spacing $h_n = \max\{(t_{i+1}-t_i): 1 \le i \le n\}$. On this partition together with the extended knots $t_{-2} \le t_{-1} \le t_0 \le t_1$ and $t_n \le t_{n+1} \le t_{n+2} \le t_{n+3}$ we can construct the cubic-B-splines B_{ni}, which are non-zero over the respective intervals (t_{i-2}, t_{i+2}). It is known that $\{B_{ni} : i=0,\ldots,n+1\}$ forms a basis for the linear space of cubic splines with partition π_n, and which has continuity C_2 over $[a,b]$. Using this linear space we can approximate the function f by the linear combination

$$\sum_{i=0}^{n+1} \alpha_{ni} B_{ni}$$

and then we seek an appropriate set of coefficients $\{\alpha_{ni}\}$.

Consider now the function $r \in C[a,b]$ given by

$$r = \sum_{i=0}^{n+1} \alpha_{ni}(I - K)B_{ni} - y,$$

and define the usual inner product

$$(\psi,\phi) = \int_a^b \psi(x)\phi(x)\,dx.$$

Two ways in which we can now choose our coefficients α_{ni} are:

(a) take any appropriate $n+2$ abscissae values s_j, $j = 0,\ldots,n+1$ then solve the linear system

$$r(s_j) = 0 \quad j=0,\ldots,n+1$$

(b) set up and solve the linear system

$$(r,B_{ni}) = 0 \quad i=0,\ldots,n+1.$$

Choice (a) is the method of collocation, while (b) is the Galerkin technique.

To show convergence of these methods we write the singular kernal $K(\cdot,t)$ as

$$K(\cdot,t) = \frac{\tilde{K}(\cdot,t)}{t-u},$$

where $u \in [a,b]$.

To solve our equations we use a numerical quadrature, namely

$$\int_a^b F(x)\,dx = \sum_{i=0}^{p} w_i F(x_i)$$

for some suitabley chosen x_i, $i=1,\ldots,p$.

For method (a) we obtain for $j=0,\ldots,n+1$,

$$r(s_j) = \sum_{i=0}^{n+1} \alpha_{ni}\left[B_{ni}(s_j) - \int_a^b \frac{\tilde{k}(s_j,t)}{t-u} B_{ni}(t)dt\right] - y(s_j)$$

$$= \sum_{i=0}^{n+1} \alpha_{ni}\left[B_{ni}(s_j) - \sum_{q=1}^{p} w_q \frac{\tilde{K}(s_j,t_q)}{t_q-u} B_{ni}(t_q)\right] - y(s_j),$$

where we can choose out t_q such that $t_q \neq u$ for any $q=1,\ldots,p$.

For method (b) the s_j, $j=1,\dots,p$ are chosen from the relation

$$(r,B_{nk}) = \int_a^b r(s)\, B_{nk}(s)\, ds$$

$$= \sum_{j=1}^{p} w_j\, r(s_j)\, B_{nk}(s_j),$$

for $k=0,\dots,n+1$.

We thus see that our problem is defined by the grid of points (s_j,t_q). On this grid we are able to approximate $\bar{K}(s,t)$ by the function $(t-u)k(s,t)$ such that

$$(t_q-u)k(s_j,t_q) = \tilde{K}(s_j,t_q).$$

We have therefore an equivalent formulation for our evaluation of $r(s)$, namely

$$r(s_j) = \sum_{i=0}^{n+1} \alpha_{ni} \left[B_{ni}(s_j) - \sum_{q=1}^{p} w_q\, k(s_j,t_q)\, B_{ni}(t_q) \right] - y(s_j)$$

over the appropriate grid of points for (a) or (b).

This evaluation would result if we were solving the integral equation

$$f(s) - \int_a^b k(s,t) f(t) dt = y(s),$$

and the methods (a) or (b) cannot distinguish between this equation and the original singular equation. We shall write K_1 for the kernal in the operator form of this integral equation.

For both methods it can be shown that there exists a projection operator P such that the solution $g \in S$ of our procedure is the solution to the equation

$$P(I - K_1)g = Py.$$

If f_1 is the solution to the operator equation

$$(I - K_1)f = y$$

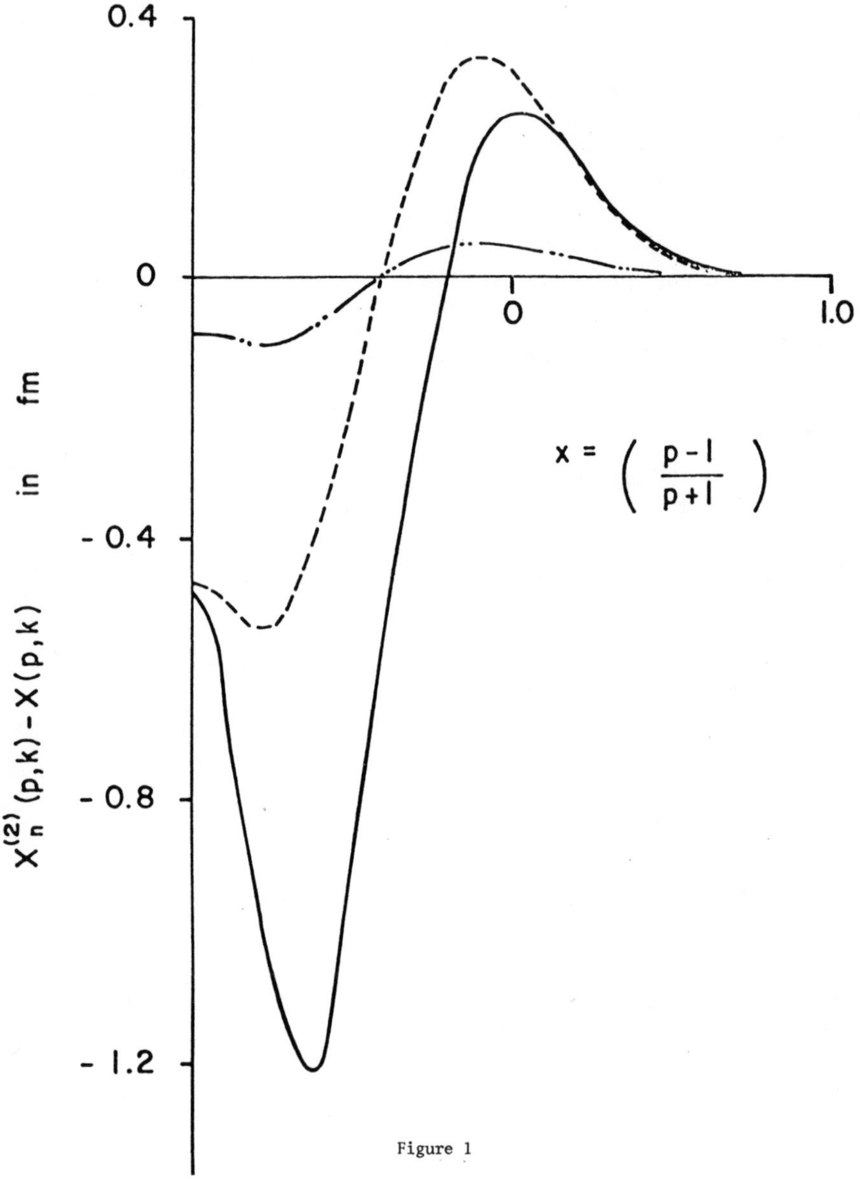

Figure 1

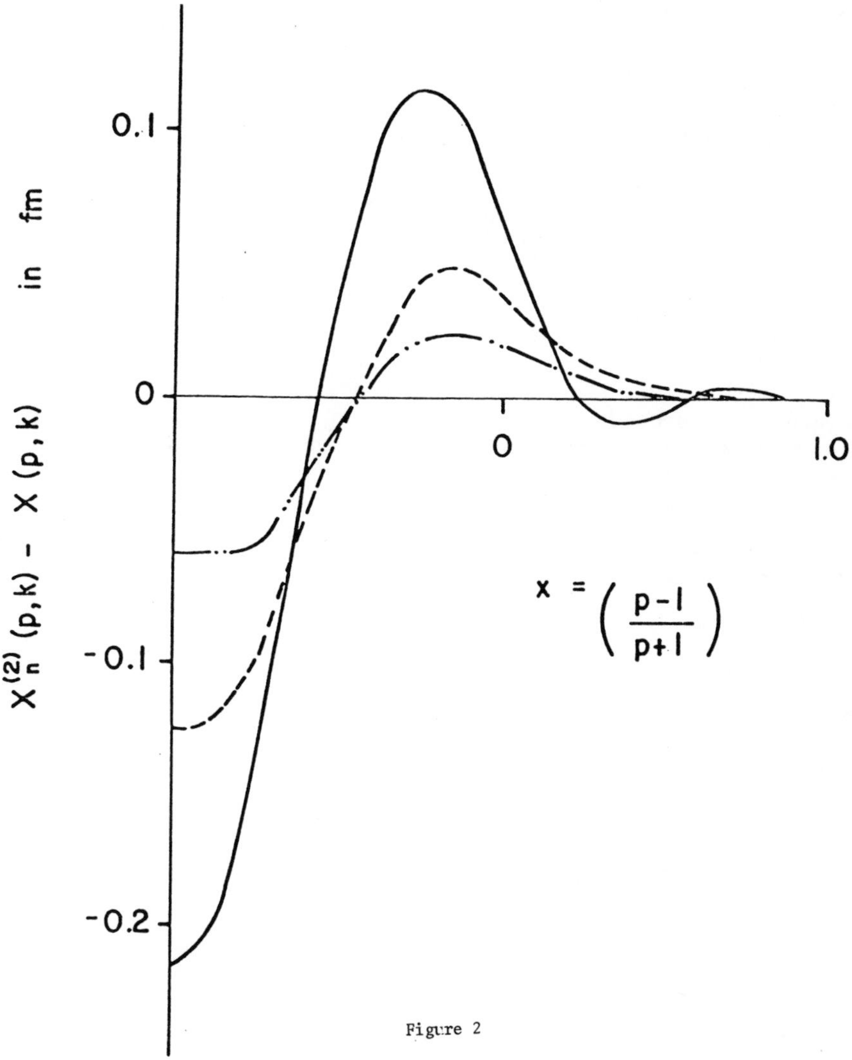

Figure 2

then, using the uniform norm, we obtain the bound, see [1]

$$\|f_1-g\| \le \|(I - PK_1)^{-1}\| \ \|f_1-Pf_1\|$$

where $f_1 \in C^4[a,b]$. Thus showing that the methods converge as $O(h_n^{\ 4})$ to some solution f_1.

NUMERICAL EXPERIMENTS

To compare the methods presented here we obtained solutions for the Amado-Lovelace equations given above for three identical bosons. We used uniformly spaced partitions π_n to obtain our results.

Our fig. 1 shows the error in the solution using the collocation method for $n = 4,6,8$, while fig. 2 shows the error using the Galerkin procedure. As is to be expected the Galerkin procedure gives better results. One can consider the Galerkin technique as a collocation procedure where the collocation points $\{s_i\}$ are optimal in the least squares sense. This advantage over the collocation procedure is offset by the fact that the Galerkin method requires the extra system of linear equations to obtain $\{s_i\}$. For this reason it is desirable to be able to obtain accurate solutions for small values of n. From the equations we have considered this has been shown to be true.

REFERENCES

1. Atkinson, K. E., "A survey of numerical methods for the solution of Fredholm integral equations", SIAM, Philadelphia, 1976.

2. Faddeev, L. D., "Mathematical aspects of the three-body problem in quantum scattering theory", Steklov Math. Institute 69, 1963, Davey, New York, 1965.

3. Lovelace, C., "Practical theory of three particle states", I. Nonrelativistic, Phys. Rev. 135B, 1964, pp. 1225-1249.

H-SETS FOR NON-LINEAR CONSTRAINED APPROXIMATION

M. Brannigan

University of Georgia

INTRODUCTION

The idea of an H-set was first proposed by Collatz [4], which was primarily introduced to give lower bounds for the numerical calculation of best linear approximation by sets not satisfying the Haar condition. The usefulness of H-sets for the characterisation and anlysis of best approximation in this non-Haar setting is set out in Brannigan [1]. A complete exposition in terms of functions in a normed linear setting is given in Brannigan [2]. For non-linear approximation an extension of H-sets is given in Collatz and Krales [5]. We here consider the non-linear approximation problem where constraints on the parameter set are present. This setting is of most practical use, and the analysis of numerical non-linear constrained approximation is needed. For the general characterisation of best approximation see Brannigan [3]. The usefulness of H-sets as developed here is in the ability to compute such sets in many cases. In the following the theorems are merely stated, those interested should consult Brannigan [3].

H-SETS

Let X be a compact subset of $\mathbb{R}^n$, and V a subset of $C(X)$ the space of continuous real-valued functions defined on X. Elements in V we denote by $h(p, x)$, where p is a parameter in $\mathbb{R}^n$. We use throughout the uniform norm on $C(X)$, namely

S. P. Singh et al. (eds.), Approximation Theory and Spline Functions, 265–270.

$$\|f(x)\| = \max\{|f(x)| : x \in X\}.$$

Let A be some index set, then the parameters p are to be constrained by $\psi_\alpha \in C(X)$, $\alpha \in A$ such that

$$\psi_\alpha(p, x) \leq 0, \quad x \in X, \alpha \in A$$

and for some $p \in \mathbb{R}^n$ we assume the constraint qualification that

$$\psi_\alpha(p, x) < 0 \quad x \in X, \alpha \in A.$$

In this setting we can pose

Problem 1. *Let* $f \in C(X)$ *find* $p \in \mathbb{R}^n$ *satisfying the constraints such that*

$$\|f - h(p, \cdot)\|$$

is minimised over all p.

Such a best approximation is referred to as 'global'. Also we have

Problem 2. *Let* $f \in C(X)$ *find* $p \in \mathbb{R}^n$ *satisfying the constraints such that*

$$\|f - h(p, \cdot)\|$$

is minimised over some neighbourhood of p *in* $\mathbb{R}^n$.

With such problems in mind we can consider H-sets. We shall assume throughout that the functions $h(p, \cdot)$ and $\psi_\alpha(p, \cdot)$ are differentiable with respect to the parameter p. Thus if we denote by $q_1, \ldots, q_n$ the components of vectors in $\mathbb{R}^n$ we are assuming that the partial derivatives

$$\frac{\partial h(p,\cdot)}{\partial q_i}, \quad \frac{\partial \psi_\alpha(p,\cdot)}{\partial q_i}; \quad \alpha \in A; \ i = 1, \ldots, n,$$

exist and are continuous functions in $C(X)$.

We are thus led to our definition of H-sets.

Definition 1. The set $[\{x_i, \ell_i, \lambda_i, k\}, \{y_i, \alpha_i, \mu_i, m\}]$ forms an H_1-set with respect to V at p if and only if

$$\sum_{i=1}^{k} \lambda_i \ell_i \frac{\partial h(p,x_i)}{\partial q_r} + \sum_{j=1}^{m} \mu_j \frac{\partial \psi_{\alpha_j}(p,y_j)}{\partial q_r} = 0, \quad r = 1,\dots,n,$$

$$\sum_{i=1}^{k} \lambda_i = 1;$$

with $\lambda_i, \mu_j > 0$, $i = 1, \dots, k$, $j = 1, \dots, m$; and $\ell_i = \pm 1$, $i = 1, \dots, k$.

An H_1-set is minimal if no subset of the points $\{x_i\} \cup \{y_j\}$ can form such an H_1-set. We note here the dependency on the chosen parameter p, in constrast to the linear case where no such dependency exists.

We shall denote by $\mathcal{H}_1(p)$ the set of all H_1-sets with respect to V at p. From this definition we obtain

Theorem 1. Let $[\{x_i, \ell_i, \lambda_i, k\}, \{y_j, \alpha_j, \mu_j, m\}] \in \mathcal{H}_1(p)$ then no $c = (c_1, \dots, c_n)$ exists such that

$$\sum_{i=1}^{n} c_i \ell_r \frac{\partial h(p,x_r)}{\partial q_i} \geq 0, \quad r = 1, \dots, k;$$

and

$$\sum_{i=1}^{n} \frac{\partial \psi_{\alpha_j}(p,y_j)}{\partial q_i} \geq 0, \quad j = 1, \dots, m;$$

with strict inequality for some r, j.

Conversely, given sets $M, N \subset X$, and for each $x \in M$ an $\ell_x = \pm 1$, and for each $y \in N$ an $\alpha_y \in A$, if no $c = (c_1, \dots, c_n)$ exists such that

$$\sum_{i=1}^{n} c_i \ell_x \frac{\partial h(p,x)}{\partial q_i} > 0, \quad x \in M,$$

and

$$\sum_{i=1}^{n} c_i \frac{\partial \psi_{\alpha_y}(p,y)}{\partial q_i} > 0, \quad y \in N,$$

then there exist finite sets $\{x_i\} \subset M$, $\{y_i\} \subset N$ such that $[\{x_i, \ell_{x_i}, \cdot, \cdot,\},\{y_j, \alpha_{y_j}, \cdot, \cdot\}] \in H_1(p)$.

We note that theorem 1 shows how our definition 1 relates to definitions given by Collatz.

For a non-differentiable definition we proceed as follows:

<u>Definition 2</u>. $[\{x_i, \ell_i, \lambda_i(q), k\}, \{y_j, \alpha_j, \mu_j(q), m\}]$ forms an H_2-set with respect to V at p if and only if for $q \in \mathbb{R}^n$

$$\sum_{i=1}^{q} \lambda_i(q)\ell_i(h(q, x_i) - h(p, x_i)) +$$

$$\sum_{j=1}^{m} \mu_j(q)(\psi_{\alpha_j}(q, y_i) - \psi_{\alpha_j}(p, y_i)) = 0$$

where $\lambda_i(q), \mu_j(q) > 0$, $i = 1, \ldots, k$; $j = 1, \ldots, m$, and $\ell_i = \pm 1$ $i = 1, \ldots, k$, also

$$\sum_{i=1}^{k} \lambda_i(q) = 0.$$

We shall denote by $H_2(p)$ the set of all H_2-sets with respect to V at p. Note here that the multipliers λ_i, μ_j are dependent on the q chosen. A Collatz type result is available as follows:

<u>Theorem 2</u>. Let $[\{x_i, \ell_i, \lambda_i(q), k\}, \{y_j, \alpha_j, \mu_j(q), m\}\}$ $\in H_2(p)$ then no $q \in \mathbb{R}^n$ exists such that

$$\ell_i(h(q, x_i) - h(p, x_i)) \geq 0, \quad i = 1, \ldots, k;$$

and

$$\psi_{\alpha_j}(q, y_j) - \psi_{\alpha_j}(p, y_j) \geq 0, \quad j = 1, \ldots, m,$$

with strict inequality for some i, j.

Conversely, given sets $M, N \subset X$ and for each x an $\ell_x = \pm 1$, and for each y and $\alpha_y \in A$, then if no $q \in \mathbb{R}^h$ exists such that

$$\ell_x(h(q, x) - h(p, x)), \quad x \in M,$$

and

$$\psi_{\alpha_y}(q, y) - \psi_{\alpha_y}(p, y), \quad y \in N,$$

are of the same sign then there exists finite sets $\{x_i\} \subset X$, $\{y_j\} \subset X$ which will form an H_2-set with respect to V at p.

To see a relationship between the two types of H-sets we are able to show

Theorem 3. $H_2(p) \subset H_1(p)$.

Also for neighbourhoods of p we have:

Theorem 4. If $G = [\{x_i, \ell_i, \lambda_i, k\}, \{y_j, \alpha_j, \mu_j, m\}] \in H_1(p)$ then for some neighbourhood U of V with centre $h(p, \cdot)$ G is an H_2-set with respect to U at p.

From this general setting of H-sets characterisation of approximations satisfying problems 1 and 2 can be made.

One essential difference between H_1-sets and H_2-sets is that H_1-sets can, in principle, be numerically evaluated using QU or SV decomposition of matrices. This cannot be said for H_2-sets.

REFERENCES

1. Brannigan, M., "H-sets in linear approximation", J. Approx. Theory, 20, 1977, pp. 153-161.

2. Brannigan, M., "Uniform linear approximation in normed linear spaces", Numer. Funct. Anal. and Optimization, 2(1), 1980, pp. 79-91.

3. Brannigan, M., "Uniform non-linear constrained approximation in normed linear spaces", Numer. Funct. Anal. and Optimisation, to appear.

4. Collatz, L., "Approximation von Funktionen bei einer und bei mehrenen unabhangigen Veranderlichen", Zamur, 36, 1956, pp. 198-211.

5. Collatz, L., and Krales, W., "Approximationstheorie", Teubner, Stuttgart, 1973.

OPERATOR PADE APPROXIMANTS:
Some ideas behind the theory and a numerical illustration

Annie A.M. Cuyt, aspirant N.F.W.O. Belgium

University of Antwerp U.I.A.

ABSTRACT

Section 1 will be devoted to the discussion of some generalizations of the concept of Pade-approximant for multivariate functions, based on the interpolation property of a Pade-approximant. Most of those generalizations preserve, under some conditions, a number of properties of the univariate Pade-approximant. In Section 2 we will repeat the recursive schemes used for the computation of the univariate Pade-approximant: the ε-algorithm and the qd-algorithm. We will also show that, if a generalizing definition is based on these recursive algorithms, then much more interesting properties remain valid for the generalization. In Section 3 we will illustrate the approximation power of this type of Pade approximants on a numerical example. Other applications are: the solution of nonlinear systems of equations [7], the solution of nonlinear differential and integral equations [3], the acceleration of convergence [4]. Since those applications have already been treated extensively, they will not be mentioned here; the interested reader is referred to the literature.

1. SOME DEFINITIONS FOR MULTIVARIATE PADE APPROXIMANTS

Let us first of all repeat the definition of univariate Pade approximant. Suppose we are given a function f(x) by its Taylor series expansion around the origin,

$$f(x) = \sum_{k=0}^{\infty} c_k x^k \quad \text{with} \quad c_k = \frac{1}{k!} f^{(k)}(0)$$

S. P. Singh et al. (eds.), Approximation Theory and Spline Functions, 271–288.

In the Pade approximation problem of order (n,m) we look for polynomials

$$p(x) = \sum_{i=0}^{n} a_i \, x^i$$

and

$$q(x) = \sum_{j=0}^{m} b_j x^j$$

such that in the power series $(f \cdot q - p)(x)$ the first $n + m + 1$ terms disappear, i.e.

$$(f - q - p)(x) = \sum_{k=n+m+1}^{\infty} d_k x^k$$

It is well-known that this problem had indeed a nontrivial solution for the a_i and b_i and that the rational functions $(p/q)(x)$ satisfy a number of beautiful properties.

The fact that all terms of degree up to and including $n + m$ vanish in $(f \cdot q - p)(x)$, can be represented by means of an "interpolationset" E describing the fulfilled equations:

$$d_k = 0 \quad \text{for} \quad k \in E = \{0, \ldots, n + m\} \subset N$$

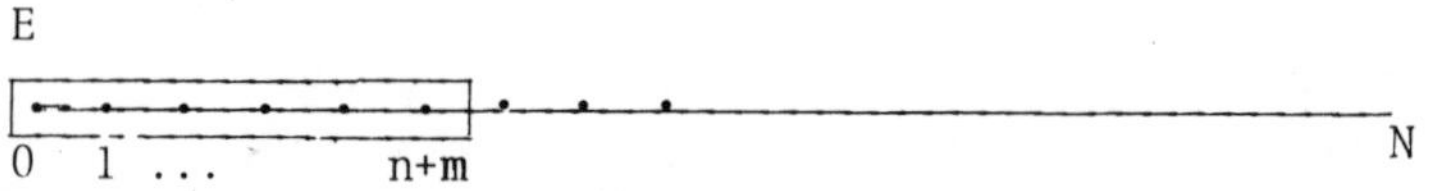

A natural generalization of this approximation problem to the multivariate case, is the following. We will describe the situation in the case of two variables, because the case of more than two variables is only notationally more difficult.

Suppose we know the Taylor series expansion (or at least part of it) of a bivariate function

$$f(x,y) = \sum_{i,j=0}^{\infty} c_{ij} \, x^i \, y^j \quad \text{with} \quad c_{ij} = \frac{1}{i!} \frac{1}{j!} \frac{\partial^{i+j} f}{\partial x^i \partial y^j} (0,0)$$

Let us try to calculate polynomials

$$p(x,y) = \sum_{i+j=0}^{n} a_{ij}\, x^i\, y^j$$

$$q(x,y) = \sum_{i+j=0}^{m} b_{ij}\, x^i\, y^j$$

of total degree n and m respectively (a term $x^i y^j$ is said to be of total degree $i+j$), such that

$$(f \cdot q - p)(x,y) = \sum_{i+j=n+m+1}^{\infty} d_{ij}\, x^i y^j$$

If we represent this demand by an interpolation set E in $\mathbb{N}^2$, we have

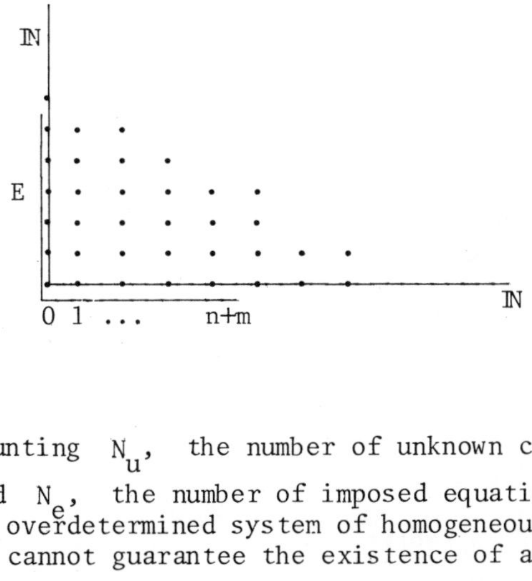

Counting N_u, the number of unknown coefficients a_{ij} and b_{ij}, and N_e, the number of imposed equations, we remark that we have an overdetermined system of homogeneous equations and thus that we cannot guarantee the existence of a nontrivial solution:

$$N_u = \frac{1}{2}(n+1)(n+2) + \frac{1}{2}(m+1)(m+2)$$

$$N_e = \frac{1}{2}(n+m+1)(n+m+2)$$

The ideal situation would be

$$N_e = N_u - 1$$

because one unknown can always be determined by a normalization of the denominator $q(x,y)$.

Several authors have created this ideal situation by altering the form of the polynomials $p(x,y)$ and $q(x,y)$ and/or by choosing another interpolation set. The definitions which we shall compare are those of Chisholm and his group at the University of Kent in Canterbury, Lutterodt who introduces two types of approximants, Karlsson and Wallin who are primarily interested in convergence properties.

All those definitions are of the following kind. This very general setting has also been given by Levin [12].

Take a "numerator index set" N and a "denominator index set" D describing the form of the polynomials

$$p(x,y) = \sum_{(i,j)\in N\subseteq \mathbb{N}^2} a_{ij}\, x^i y^j$$

$$q(x,y) = \sum_{(i,j)\in D\subseteq \mathbb{N}^2} b_{ij}\, x^i y^j$$

and construct on interpolationset E such that

$$N \subset E$$

$$\#\,(E\backslash N) = \#\,D - 1$$

where # denotes the number of elements in a set.

Now one can assure the existence of nontrivial a_{ij} and b_{ij} when solving

$$(f\cdot q - p)(x,y) = \sum_{(i,j)\in \mathbb{N}^2\backslash E} d_{ij}\, x^i y^j$$

For Chisholm's bivariate Pade approximants

$$N = ([0,n_1]\times[0,n_2])\cap \mathbb{N}^2 \quad \text{with} \quad n_1,n_2\in\mathbb{N}$$

$$D = ([0,m_1]\times[0,m_2])\cap \mathbb{N}^2 \quad \text{with} \quad m_1,m_2\in\mathbb{N}$$

$$E = \{(i,j)\,|\,0\le i\le \max(n_1,m_1),\ 0\le j\le \min(n_2,m_2)\}$$
$$\cup\ \{(i,j)\,|\,0\le i\le \min(m_1,m_1),\ 0\le j\le \max(n_2,m_2)\}$$
$$\cup\ \{(i,j)\,|\max(n_2,m_2) < j\le n_2+m_2,\ \max(n_2,m_2) < i+j\le n_2+m_2,$$
$$0\le i\le \min(n_1,m_1)\}$$

$$\cup\{(i,j)\,|\max(n_1,m_1) < i \le n_1+m_1,\ \max(n_1,m_1) < i+j \le n_1+m_1,$$
$$0 \le j \le \min(n_2,m_2)\}$$

$$d_{ij} = 0 \quad \text{for} \quad (i,j) \in E$$

$$d_{n_1+m_1+1-\ell,\ell}+d_{\ell,n_2+m_2+1-\ell} = 0 \text{ for } \ell=n,_{-1}\min(n_1,m_1,n_2,m_2)$$

For Lutterodt's approximants we have

$$N = ([0,n_1] \times [0,n_2]) \cap \mathbb{N}^2 \quad \text{with} \quad n_1,n_2 \in \mathbb{N}$$

$$D = ([0,m_1] \times [0,m_2]) \cap \mathbb{N}^2 \quad \text{with} \quad m_1,m_2 \quad \mathbb{N}$$

$$E \supseteq N$$

E satisfies the inclusion property, i.e. if $(i,j) \in E$ then $([0,i] \times [0,j]) \cap \mathbb{N}^2 \subset E$

$$\#E = (n_1+1)(n_2+1) + (m_1+1)(m_2+1) - 1$$

$$d_{ij} = 0 \quad \text{for} \quad (i,j) \in E$$

and for his Pade approximants of type B^1

$$N = ([0,n_1] \times [0,n_2]) \cap \mathbb{N}^2 \quad \text{with} \quad n_1,n_2 \in \mathbb{N}$$

$$D = ([0,m_1] \times [0,m_2]) \cap \mathbb{N}^2 \quad \text{with} \quad m_1,m_2 \in \mathbb{N}$$

$$E = N\cup\{(i,0)\,|\,n_1+1 \le i \le n_1+m_1\}\cup\{(0,j)\,|\,n_2+1 \le j \le n_2+m_2\}$$
$$\cup\{(i,j)\,|\,n_1+1 \le i \le n_1+m_1,\ n_2+1 \le j \le n_2+m_2\}$$

$$d_{ij} = 0 \quad \text{for} \quad (i,j) \in E$$

For the Karlsson-Wallin approximants

$$N = \{(i,j)\,|\,0 \le i+j \le n\} \quad \text{with} \quad n \in \mathbb{N}$$

$$D = \{(i,j)\,|\,0 \le i+j \le m\} \quad \text{with} \quad m \in \mathbb{N}$$

$$E \supseteq N$$

E satisfies the inclusion property

$$\#E \ge \frac{1}{2}(n+1)(n+2) + \frac{1}{2}(m+1)(m+2) - 1$$

None of those types of bivariate Pade approximants can be calculated recursively, and unicity of the approximant itself can only be guaranteed under certain conditions. By imposing restrictions on N,D and E one can preserve some of the univariate properties of Pade approximants. We will discuss this in detail at the end of the next section. Those who want to know more about some of these multivariate Pade approximants are referred to [1, 9, 14, 15, 11].

2. OPERATOR PADE APPROXIMANTS

Since the univariate ε- and qd-algorithm will serve as a motivation for the introduction of operator Pade approximants (multivariate Pade approximants turn out to be a special case), we will first repeat some facts about those recursive computation schemes.

Consider a series $\sum_{i=0}^{\infty} t_i$ in $\mathbb{R}$ and also the sequence $(s_i)_{i\in \mathbb{N}}$ of its partial sums; so $s_i = t_o + \ldots + t_i$.

Input of the ε-algorithm are the elements s_i. We perform the following computations:

a) $\varepsilon_{-1}^{(i)} = 0 \qquad i = 0,1, \ldots$

$\varepsilon_{0}^{(i)} = s_i$

b) $\varepsilon_{2j}^{(-j-1)} = 0 \qquad j = 0,1, \ldots$

c) $$\varepsilon_{j+1}^{(i)} = \varepsilon_{j-1}^{(i+1)} + \frac{1}{\varepsilon_{j}^{(i+1)} - \varepsilon_{j}^{(i)}} \qquad \begin{array}{l} j = 0,1, \ldots \\ i = -j,-j+1, \ldots \end{array}$$

The index j refers to a column while i refers to a diagonal in the ε-table. If the algorithm does not break down the following property can be proved for the ε-algorithm. We denote by $\Delta s_k = s_{k+1} - s_k$.

Theorem 2.1: With $s_i = 0$ for $i < 0$

$$\varepsilon_{2j}^{(i)} = \frac{\begin{vmatrix} s_{i+j} & \cdots & & s_i \\ \Delta s_{i+j} & \cdots & \Delta s_{i+1} & \Delta s_i \\ \vdots & & & \vdots \\ \Delta s_{i+2j-1} & & \Delta s_{i+j} & \Delta s_{i+j-1} \end{vmatrix}}{\begin{vmatrix} 1 & \cdots & 1 \\ \Delta s_{i+j} & \cdots & \Delta s_i \\ \vdots & & \vdots \\ \Delta s_{i+2j-1} & \cdots & \Delta s_{i+j-1} \end{vmatrix}}$$

Input of the qd-algorithm are the terms t_i. One performs the following calculations:

a) $e_0^{(i)} = 0 \;,\; q_1^{(i)} = \dfrac{t_{i+1}}{t_i}$ $\qquad i = 0,1, \ldots$

b) $e_j^{(i)} = q_j^{(i+1)} + e_{j-1}^{(i+1)} - q_j^{(i)}$ $\qquad i = 0,1,2, \ldots \; j = 1,2, \ldots$

c) $q_{j+1}^{(i)} = q_j^{(i+1)} \cdot e_j^{(i+1)} / e_j^{(i)}$ $\qquad i = 0,1,2, \ldots \; j = 1,2, \ldots$

Again the index j refers to a column while i refers to a diagonal. If all the $q_j^{(i)}$ and $e_j^{(i)}$ exist, one can prove the following property.

Theorem 2.2

For $s_i = t_o + \ldots + t_i$,

$$\varepsilon_{2j}^{(i)} = s_i + \frac{t_{i+1}|}{|1} - \frac{q_1^{(i+1)}|}{|1} - \frac{e_1^{(i+1)}|}{|1} - \frac{q_2^{(i+1)}|}{|1} - \frac{e_2^{(i+1)}|}{|1} - \ldots - \frac{q_j^{(i+1)}|}{|1}$$

For $t_i = c_i x^i$ (the ith term in the Taylor series of f), thus for $s_i = \sum_{k=0}^{i} c_k x^k$, it is well-known that $\varepsilon_{2m}^{(n-m)}$ is the Pade approximant of order (n,m) for f. This can easily be seen as follows. The imposed conditions

$$(f \cdot q - p)(x) = \sum_{k=n+m+1} d_k x^k$$

result in two linear systems of equations in the unknown coefficients a_i and b_j of the polynomials p and q:

$$\begin{cases} c_0 b_0 = a_0 \\ c_1 b_0 + c_0 b_1 = a_1 \\ \vdots \\ c_n b_0 + \ldots + c_0 b_n = a_n \end{cases} \qquad \begin{cases} c_{n+1} b_0 + \ldots + c_{n+1-m} b_m = 0 \\ \vdots \\ c_{n+m} b_0 + \ldots + c_n b_m = 0 \end{cases}$$

with $c_k = 0$ for $k < 0$ and $b_j = 0$ for $j > m$.

A solution of the homogeneous system of equations is given by the following determinants:

$$b_0 = \begin{vmatrix} c_n & c_{n-1} & \ldots & c_{n+1-m} \\ c_{n+1} & c_{n-1} & & \\ \vdots & & & \\ c_{n+m-1} & c_{n+1} & c_n & \end{vmatrix} \qquad b_j = \begin{vmatrix} c_n & \boxed{\begin{matrix} -c_{n+1} \\ \vdots \\ -c_{n+m} \end{matrix}} & \begin{matrix} c_{n+1-m} \\ \vdots \\ c_n \end{matrix} \\ c_{n+1} & & \\ \vdots & & \\ c_{n+m-1} & & \end{vmatrix}$$

↑ j^{th} column in b_0 replaced by this column

Hence an explicit formula for $\frac{p(x)}{q(x)}$ is

$$\frac{\begin{vmatrix} \sum_{k=0}^{n} c_k x^k & \sum_{k=0}^{n-1} c_k x^k & \ldots & \sum_{k=0}^{n-m} c_k x_k \\ c_{n+1} & c_n & \ldots & c_{n+1-m} \\ \vdots & \vdots & & \vdots \\ c_{n+m} & c_{n+m-1} & \ldots & c_n \end{vmatrix}}{\begin{vmatrix} 1 & x & \ldots & x^m \\ c_{n+1} & c_n & \ldots & c_{n+1-m} \\ \vdots & \vdots & & \vdots \\ c_{n+m} & c_{n+m-1} & & c_n \end{vmatrix}}$$

$$\frac{\begin{vmatrix} \sum_{k=0}^{n} c_k x^k & \sum_{k=0}^{n-1} c_k x^k & \cdots & \sum_{k=0}^{n-m} c_k x^k \\ c_{n+1}x^{n+1} & c_n x^n & \cdots & c_{n+1-m}x^{n+1-m} \\ \vdots & \vdots & & \\ c_{n+m}x^{n+m} & c_{n+m-1}x^{n+m-1} & \cdots & c_n x^n \end{vmatrix}}{\begin{vmatrix} 1 & x & \cdots & x^m \\ c_{n+1}x^{n+1} & c_n x^n & & c_{n+1-m}x^{n+1-m} \\ \vdots & \vdots & & \vdots \\ c_{n+m}x^{n+m} & c_{n+m-1}x^{n+m-1} & & c_n x^n \end{vmatrix}} \qquad (1)$$

$$= \varepsilon_{2m}^{(n-m)}$$

Remark that formula (1) is a rational function of the form

$$\frac{\sum_{i=0}^{n} a_i x^{i+nm}}{\sum_{j=0}^{m} b_j x^{j+nm}}$$

where

$$f(x) \sum_{j=0}^{m} b_j x^{j+nm} - \sum_{i=0}^{n} a_i x^{i+nm} = \sum_{k=n+m+1} d_k x^{k+nm}$$

This shift of the degrees over nm, which results from the determinantal expression for $\varepsilon_{2m}^{(n-m)}$, does not bother us in the univariate case because we can divide it out and it will serve as a useful tool in the operator case.

Let us now turn to the definition of Pade approximants for nonlinear operators $F: X \to Y$ where X is a Banach space and Y is a commutative Banach algebra. We shall denote elements in the space X by the symbol z.

Suppose the nonlinear operator F is abstract analytic in 0, i.e.

$$F(z) = \sum_{k=0}^{\infty} c_k z^k$$

where now $c_k = \frac{1}{k!} F^{(k)}(0)$ with $F^{(k)}(0)$ the kth Frechet derivative of F at the origin, and thus a k-linear bounded operator.

In the case $X = \mathbb{R}^2$ and $Y = \mathbb{R}$, $F(z)$ is merely a bivariate Taylor series expansion with $c_k z^k = \sum_{i+j=k} c_{ij} x^i y^j$.

For a given operator F we can now construct $\frac{P(z)}{Q(z)}$, the operator Pade approximant of order (n,m), as

$$\frac{P(z)}{Q(z)} = \frac{\begin{vmatrix} \sum_{k=0}^{n} c_k z^k & \sum_{k=0}^{n-1} c_k z^k & \cdots & \sum_{k=0}^{n-m} c_k z^k \\ c_{n+1} z^{n+1} & c_n z^n & \cdots & c_{m+1-m} z^{n+1-m} \\ \vdots & \vdots & & \vdots \\ c_{n+m} z^{n+m} & c_{n+m-1} z^{n+m-1} & \cdots & c_n z^n \end{vmatrix}}{\begin{vmatrix} I & I & \cdots & I \\ c_{n+1} z^{n+1} & c_n z^n & \cdots & c_{n+1-m} z^{n+1-m} \\ \vdots & \vdots & & \vdots \\ c_{n+m} z^{n+m} & c_{n+m-1} z^{n+m-1} & \cdots & c_n z^n \end{vmatrix}}$$

where I is the unit element for the multiplication in Y and division in Y is multiplication by the inverse element for the multiplicative operator defined in the Banach algebra Y.

It is easy to see [5] that $P(z)$ and $Q(z)$ are respectively of the form

$$P(z) = \sum_{i=0}^{n} A_i z^{i+nm}$$

$$Q(z) = \sum_{j=0}^{m} B_j z^{j+nm}$$

where A_i and B_j are (i+nm)- and (j+nm)- linear operators, and that

$$(F \cdot Q - P)(z) = \sum_{k=n+m+1} D_k \, z^{k+nm} \tag{2}$$

We want to emphasize here that originally these operator Pade approximants were not introduced in this way [2]. They were defined by means of the set of equations (2) and the validity of the ε-algorithm was only proved afterwards.

One of the immediate results of this definition is that the operator Pade approximants can be calculated recursively by means of the ε-algorithm and that there is also a connection with the theory of continued fractions by means of the qd-algorithm (see [6]).

But many other properties are satisfied. We give here a list of desirable properties and will compare the multivariate Pade approximants introduced by means of the ε-algorithm with the multivariate Pade approximants introduced via the interpolationsets:

a) unicity of the solution

b) reciprocal covariance: if $f(x,y)$ is replaced by $\frac{1}{f}(x,y)$ and if $\frac{p}{q}(x,y)$ is the Pade approximant of order (n,m) for $f(x,y)$, is then $\frac{q}{p}(x,y)$ the Pade approximant of order (m,n) for $\frac{1}{f}(x,y)$?

c) homografic covariance: if $f(x,y)$ is replaced by $\frac{a \cdot f + b}{c \cdot f + d}(x,y)$ with a,b,c,d in $\mathbb{R}$, and if $\frac{p}{q}(x,y)$ is the Pade approximant of order (n,n) for $f(x,y)$, is then $\frac{a \cdot p + b \cdot q}{c \cdot p + d \cdot q}(x,y)$ the Pade approximant of order (n,n) for $\frac{a \cdot f + b}{c \cdot f + d}(x,y)$?

d) projection property: if $\frac{p}{q}(x,y)$ is the Pade approximant of order (n,m) for $f(x,y)$, are then $\frac{p}{q}(x,0)$ and $\frac{p}{q}(0,y)$ the Pade approximants of order (n,m) for $f(x,0)$ and $f(0,y)$ respectively?

e) symmetry: if $f(x,y) = f(y,x)$, is then the Pade approximant symmetric too?

f) consistency property: if $f(x,y)$ is a rational function itself, do we then find f back as its Pade approximant by choosing the degrees of p and q appropriately?

g) recursive computation

h) block-structure: if the multivariate Pade approximants of order (n,m) are ordered in a two-dimensional array for increasing n and m, does the Pade-table then consist of square blocks containing equal Pade-approximants?

i) continued fractions: can the multivariate Pade approximant also be obtained as the convergent of a multivariate continued fraction?

The following review gives an answer to those questions. Chisholm's and Lutterodt's approximants shall be denoted by $(n_1,n_2)/(m_1,m_2)$ while the Karlsson-Wallin and the operator Pade approximants are indicated by n/m.

	Chisholm	Lutterodt		Karlsson-Wallin	Operator
Unicity	only under certain conditions on c_{ij}	only with respect to a given E, if the homogeneous system has a unique solution		only if E contains as many points as possible	yes
Recipr. Cov.	yes	yes	no	yes	yes
Homogr. Cov.	yes	yes	no	yes	yes
Projection Pr.	yes	only if E contains the univariate interpolationsets	yes	only if E contains the univariate interpolationsets	yes

	Chisholm	Lutterodt		Karlsson-Wallin	Operator
Symmetry	only for (n,n)/(m,m)	only for symmetric E and for (n,n)/(m,m)	only for (n,n)/(m,m)	only for symmetric E	yes
Consistency Pr.	no	no	no	no	yes
ε-algorithm	no	no	no	no	yes
Block-structure	no	no	no	no	yes
qd-algorithm	no	no	no	no	yes

3 NUMERICAL APPROXIMATION OF THE BETA FUNCTION

The Beta function is an example which has also been studied by the Canterbury group [8] and by Levin [13]. We will compare our results with theirs. The Beta function may be defined by

$$B(x,y) = \frac{\Gamma(x)\Gamma(y)}{\Gamma(x+y)}$$

where Γ is the Gamma function. Singularities occur for $x = -k$ and $y = -k$ $(k = 0,1,2,...)$ and zeros for $y = -x-k$ $(k=0,1,2,...)$. We write

$$B(x,y) = \frac{A(x-1,y-1)}{xy}$$

with

$$A(u,v) = 1 + uv\, f(u,v)$$

The coefficients in the Taylor series expansion of $f(u,v)$ have been calculated by the first method suggested in [8].

We will calculate some bivariate Pade approximants $\frac{p}{q}(u,v)$ of order (n,m) for $f(u,v)$ and compute

$$\frac{1 + (x-1)(y-1)\frac{p}{q}(x-1,y-1)}{xy}$$

as an approximation for $B(x,y)$.

Let us first take a look at the computational effort it takes for the calculation of a certain approximant. We denote by N_f the number of coefficients in the Taylor series of $f(x,y)$ which we shall need for the computation of the approximant. Since the coefficients a_{ij} can be calculated by substitution of the b_{ij} in the left hand sides of the equations, N_u will now denote the number of unknown coefficients b_{ij} in the homogeneous system.

For Chisholm's approximants with $n_1 = n_2 = n$ and $m_1 = m_2 = m$, we have

$$N_u = (m+1)^2$$

$$N_f = (m+1)^2 + (n+1)^2 + 2\min(n,m) - 1$$

For the operator Pade approximants

$$N_u = \frac{1}{2}[(nm + m+1)(nm+m+2) - nm(nm+1)] \quad \text{if} \quad nm > 0$$

$$\frac{1}{2}(m+1)(m+2) \quad \text{if} \quad nm = 0$$

$$N_f = \frac{1}{2}(n+m+1)(n+m+2)$$

The rational functions which Levin used for the approximation of the Beta function, were of the following type

$$\frac{\sum_{j=0}^{n_1} x^j \frac{\sum_{i=0}^{n_2} \alpha_{ij} y^i}{\sum_{i=0}^{n_2} \beta_{ij} y^i} + \sum_{j=0}^{n_1} y^j \frac{\sum_{i=0}^{n_2} p_{ij} x^i}{\sum_{i=0}^{n_2} q_{ij} x^i}}{\sum_{i=0}^{m} \sum_{j=0}^{m} \alpha_{ij} x^i y^j}$$

and we shall denote them by $[(n_1;n_2)/m]_r$ because for their computation:

$$N_u = (m+1)^2 + (n_2+1)(n_1+1)$$

$$N_f = 2(2n_2+1)(n_1+1) - (n_1+1)^2 + [\max(0,m+r-n_1)]^2 - 1$$

(for more details see [13]).

Using the prong method [10] the homogeneous system of equations for the calculations of Chisholm's approximants can be solved in $O[m^2(2m^2+2m-1)]$ operations. The calculation of a function value of an operator Pade approximant can be performed via the ε-algorithm in $O[m^2(n+m)^2]$ operations and we prefer this method to the solution of the system of equations (2).

The solution of the homogeneous system for the calculation of $[n_1;n_2)/m]_r$ involves $O[(m+1)^6 + (n_2+1)^2(n_1+1)]$ operations.

After comparison of the N_f, N_u and the computational effort, we decided to compare the numerical values of the bivariate Pade approximants given in the table below. Chisholm's approximants are of the type (n,n/(m,m); the operator Pade approximants are still indicated by n/m.

For the different groups (I), (II) and (III) we have N_f approximately equal to 87, 40 and 71 respectively.

It is easy to see that the operator Pade approximants can produce better results than the Chisholm approximants, e.g. for (x,y) = (-0.75),-0.75), and that they also can be better than the approximants Levin used, e.g. for (x,y) = (0.50,0.50). They are most accurate for (x-1,y-1) not too far from the origin.

(x,y)	(-0.75,-0.75)	(-0.50,-0.50)	(-0.25,-0.25)	(0.25,0.25)	(0.50,0.50)	(0.75,0.75)	(-1.75,1.75)
B(x,y)	9.88829829	0.	-6.77770467	7.41629871	3.14159265	1.694426166	0.
$[(4;5)/2]_3$	9.888	-0.00021	-6.777755	7.41629594	3.14159248	1.69442616	0.0186
(7;7)/(3,3)	9.820	-0.0010	-6.77774	7.41629871	3.14159265	1.69442617	0.0016
8/4	9.884	-0.00006	-6.777705	7.41629871	3.14159265	1.69442617	-0.0351

(x,y)	(-0.75,-0.75)	(-0.50,-0.50)	(-0.25,-0.25)	(0.25,0.25)	(0.50,0.50)	(0.75,0.75)	(0.75,0.25)
B(x,y)	9.88839829	0.	-6.77770467	7.41629871	3.14159265	1.694426166	4.44288293
$[(3;3)/1]_1$	9.94	-0.03	-6.794	7.416229	3.14159242	1.69442617	4.442883
(3;3)/(3,3)	7.0	-0.14	-6.787	7.416310	3.14159269	1.69442617	4.442883
4/4	8.38	-0.13	-6.802	7.416281	3.14159263	1.69442617	4.442883

(x,y)	(-0.75,-0.75)	(-0.50,-0.50)	(-0.25,-0.25)	(0.25,0.25)	(0.50,0.50)	(0.75,0.75)	(1.75,-.75)
B(x,y)	9.88839829	0.	-6.7777-467	7.41629871	3.14159265	1.694426166	-4.44288293
$[2;3)/2]_2$	9.86	-0.003	-6.7783	7.41629639	3.14159252	1.69442617	-4.4428
(7,7)/(2,2)	9.3	-0.014	-6.7783	7.41629881	3.14159265	1.69442617	-4.4421
8/3	9.74	-0.006	-6.7783	7.41629862	3.14159265	1.69442617	-4.4442

REFERENCES

1. Chisholm, J.S.R., N-variable rational approximants in [16], pp. 23-42.

2. Cuyt, A., Regularity and normality of abstract Pade approximants. Projection property and product property. Journ. Approx. Theory 35(1), 1981, pp. 1-11.

3. Cuyt, A., Pade approximants in operator theory for the solution of nonlinear differential and integral equations. Comps and Maths with Applcs *(6), 1982, pp. 445-466.

4. Cuyt, A., Accelerating the convergence of a table with multiple entry. Num. Math, 41, 1983, pp. 281-286.

5. Cuyt, A., The epsilon-algorithm and Pade-approximants in operator theory. Siam Journ. Math. Anal. 14, 1983, pp. 1009-1014.

6. Cuyt, A., The QD-algorithm and Pade-approximants in operator theory. To appear in Siam Journ. Math. Anal.

7. Cuyt, A., and Van der Cruyssen, P., Abstract Pade Approximants for the solution of a system of nonlinear equations. Comps and Maths with Applcs 9(4), 1983, pp. 617-624.

8. Graves-Morris, P., and Hughes Jones, R., and Makinson, G., The calculation of some rational approximants in two variables. Journ. Inst. Maths Applcs 13, 1974, pp. 311-320.

9. Hughes Jones, R., General rational approximants in N variables. Journ. Approx. Theory 16, 1976, pp. 207-233.

10. Hughes Jones, R., and Makinson, G., The generation of Chisholm rational polynomial approximants to power series in two variables. Journ. Inst. Maths Applcs 13, 1974, pp. 299-310.

11. Karlsson, J., and Wallin, H., Rational approximation by an approximation by an interpolation procedure in several variables. In [16], pp. 83-100.

12. Levin, D., General order Pade-type rational approximants defined from double power series. Journ. Inst. Maths Applcs 18, 1976, pp. 1-8.

13. Levin D., On accelerating the convergence of infinite double series and integrals. Math. Comp. 35(152), 1980, pp. 1331-1345.

14. Lutterodt, C.H., A two-dimensional analogue of Pade-approximant theory. Journ. Phys. A: Maths 7(9), 1974, pp. 1027-1037.

15. Lutterodt, C.H., Rational approximants to holomorphic functions in n dimensions. Journ. Math. Anal. Applcs 53, 1976, pp. 89-98.

16. Saff, E.B., and Varga, R.S., Pade and rational approximation theory and applications. Academic Press, London, 1977.

HARMONIC APPROXIMATION

Myron Goldstein

Arizona State University

The theory of harmonic approximation dates at least back to the last century when Runge proved that if K is a compact subset of R^2 such that R^2 - K is connected, then every harmonic function on K can be uniformly approximated on K by harmonic polynomials. In 1929, Walsh [12] proved that if K is a compact subset of R^2 such that R^2 - K is connected, then every function continuous on K and harmonic in the interior of K can be uniformly approximated on K by harmonic polynomials. In the 1940's, these theorems were generalized by Brelot [2] and Deny [3]. Thus they proved in particular that if K is a compact subset of R^n, $n \geq 2$, such that R^n - K is connected, then every harmonic function on K can be uniformly approximated on K by harmonic polynomials.

To generalize the theorem of Walsh alluded to above, we need to introduce the concept of thinness. To do this, we need to define the fine topology on R^n. The fine topology on R^n is by definition the smallest topology on R^n for which all superharmonic functions are continuous in the extended sense. This definition is due to H. Cartan. We now say that a subset E of R^n is thin at x if and only if x is not a fine limit point of E. It is interesting here to list some of the properties of the fine topology. Fuglede [7] has proved that the fine topology for R^2 is locally connected topology. Thus any finely open set 0 in R^2 breaks up into a number of disjoint, finely connected components, each of which is finely open. Furthermore,

S. P. Singh et al. (eds.), Approximation Theory and Spline Functions, 289–292.

it is known that every fine component of 0 has finite area and there are at most countably many fine components of a finely open set. In addition, it is known that 0 contains circles of arbitrarily small radii centred at each of its points. Further interesting facts about the fine topology on R^2 are that a finely open subset 0 of R^2 is finely connected if and only if 0 is connected with respect to the usual Euclidean topology of R^2 and that again if 0 is a finely connected open subset of R^2, then the finely connected components of 0 coincide with the Euclidean connected components of 0 and these in turn coincide with the path-connected components of 0. These facts are due to Gamelin and Lyons [8]. The situation in R^n, $n > 2$, is more complicated. Thus in R^n, $n > 2$, it is no longer true that any finely open subset of R^n, $n > 2$, includes circles of arbitrarily small radii centred at each of its points.

If E is a subset of R^n and x is a limit point of E, then it can be proved that E is thin at x if and only if there is a superharmonic function u on a neighbourhood of x such that

$$u(x) < \liminf_{\substack{y \to x \\ y \in E - \{x\}}} u(y).$$

Thus it follows that if E is an open subset of R^n having a Green's function, then a point $x \in \alpha E$ is regular for the Dirichlet problem if and only if $R^n - E$ is not thin at x.

Brelot [2] and Deny [3] also in the 1940's proved that if K is a compact subset of R^n such that $R^n - K$ is connected, then every function continuous on K and harmonic in K^0 (the interior of K) can be uniformly approximated on K by harmonic polynomials if and only if $R^n - K$ and $R^n - K^0$ are thin at the same points. It is interesting to note the similarities between this theorem of Brelot and Deny and Mergelyan's theorem which states that if K is a compact subset of the complex plane C, then a necessary and sufficient condition that every function continuous on K and holomorphic in K^0 can be uniformly approximated on K by polynomials is that $C - K$ be connected.

In a 1960 paper [10], Saginjan investigated the problem of L^p approximation of functions harmonic in K^0 and belonging to $L^p(K)$, K a compact subset of R^2, and $p \geq 1$. He proved that if K is a closed Caratheodory set, i.e., if the boundary of K coincides with the boundary of the unbounded component of $R^2 - K$, then every function harmonic in K^o and belonging to $L^p(K)$, $p \geq 1$, can be approximated in the L^p norm by harmonic polynomials. Sinanjan [11] later proved the holomorphic version of this theorem. He proved that if K is a closed Caratheodory set in R^2, then every function holomorphic in K^o and belonging to $L^p(K)$, $p \geq 1$, can be approximated in the L^p norm by polynomials. Hedberg [9] later investigated the harmonic L^p approximation problem in R^n for $n \geq 3$.

In a 1980 paper [4] Gauthier, Goldstein, and Ow, investigated the uniform harmonic approximation problem for arbitrary closed subsets (not assumed to be bounded) of R^2. Let $\hat{R}^2$ denote the Riemann sphere. Then Gauthier, Goldstein, and Ow, proved that if F is a closed subset of R^2 such that the complement of F with respect to the Riemann sphere is connected and locally connected, then every function harmonic on F can be uniformly approximated on F by harmonic functions in R^2. Actually, they stated and proved their theorem for open Riemann surfaces. In 1983 [5], Gauthier, Goldstein, and Ow, generalized this result to R^n for all $n \geq 3$. In 1982, Gauthier, Hengartner, and Lebrache [6] proved that if F is a closed subset of R^n such that $R^n - F$ and $R^n - F^o$ are thin at the same points and if the complement of F with respect to the one point compactification of R^n is connected and locally connected, and $n \geq 2$, then every function continuous on F and harmonic in F^o can be uniformly approximated on F by harmonic functions in R^n. Actually, their result holds for a relatively closed subset of a domain in $\mathbb{R}_n$. This theorem is the harmonic analogue of Arakelian's theorem [1] which is that if F is a closed subset of C such that the complement of F with respect to the Riemann sphere is connected and locally connected, then every function continuous on F and holomorphic in F^o can be uniformly approximated on F by entire functions.

When F is assumed only to be a relatively closed subset of R^n, $n \geq 2$, the problem of L^p approximation does not seem to have received any attention. At least, as far as the author is aware, there do not seem to have been any papers published in this area. However, it would be very interesting to see whether the results in the compact case generalize.

Finally, it should be mentioned that many of the results mentioned in this paper concerning harmonic approximation have been generalized to the case where harmonic functions are replaced by solutions of a second order linear elliptic partial differential equation.

REFERENCES

1. Arakelyan, N. U., "Uniform Approximation on Closed Sets by Entire Functions" (Russian), Akad. Nauk SSSR, Izvestia, ser. mat. 28, 1964, pp. 1187-1206.

2. Brelot, M., "Sur L'Approximation et La Convergence Dans La Theorie Des Fonctions Harmoniques Ou Holomorphes", Bull. Soc. Math. France, 73, 1945, pp. 55-70.

3. Deny, J., "Systemes Totaux De Fonctions Harmoniques", Ann. Inst. Fourier (Grenoble), 1, 1949, pp. 103-113.

4. Gauthier, P. M., Goldstein, M. and Ow, W. H., "Uniform Approximation on Unbounded Sets by Harmonic Functions with Logarithmic Singularities", Trans. Amer. Math. Soc., vol. 261, no. 1, 1980, pp. 169-183.

5. Gauthier, P. M., Goldstein, M., and Ow, W. H., "Uniform Approximation on Closed Sets by Harmonic Functions with Newtonian Singularities", J. London Math. Soc., 2, 28, 1983, pp. 71-82.

6. Gauthier, P. M., Hengartner, W., and Labreche, M., "Une Caracterisation des Ensembles D'Approximation Harmonique de R^n ou d'une Surface de Riemann", to appear, Canad. J. of Math.

7. Fuglede, B., "Finely Harmonic Functions", Lecture Notes in Mathematics, 289, Springer, Berlin, 1972.

8. Gamelin, T., and Lyons, T. J., "Jensen Measures for R(K)", J. London Math. Soc., 2, 27, 1983, pp. 317-330.

9. Hedberg, L. I., "Approximation in the Mean by Solutions of Elliptic Equations", Duke Math. J. 40, 1973, pp. 9-16.

10. Saginjan, A. L., "Theory of Approximation in the Complex Domain", Erevan, 1960, (Russian).

11. Sinjanan, S. O., "Approximation by Polynomials and Analytic Functions in the Areal Mean", Amer. Math. Soc. Transl. Series 2., vol. 74, 1968, pp. 91-123.

12. Walsh, J. L., "The Approximation of Harmonic Functions by Harmonic Polynomials and by Harmonic Rational Functions", Bull. Amer. Math. Soc., Vol. 35, pp. 499-544, 1929.

BEST HARMONIC L^1 APPROXIMATION TO SUBHARMONIC FUNCTIONS

M. Goldstein, W. Haussman and K. Jetter

Arizona State University and University of Duisberg

S. N. Bernstein [1, p. 130] proved the following. Let $h^{(m)}$ be positive in the open interval $(-1, 1)$. Then the algebraic polynomial of degree less than or equal to $m - 1$ of best approximation to h in the L^1 norm is the polynomial q_{m-1} which interpolates h in the m points $x_k = \cos(k/(m + 1))$, $k = 1, 2, \ldots m$. For the case $m = 2$, this says that the best linear polynomial approximant q^* to convex $C^2(-1, 1)$-function h in the L^1 norm can be computed by interpolating in the canonical set $(-\frac{1}{2}, \frac{1}{2})$ of interpolation points.

The two dimensional analogue of the convex $C^2(-1, 1)$ functions are the subharmonic functions in the open unit disk $D = \{(x, y) \in R^2: x^2 + y^2 < 1\}$ and the two dimensional analogue of the linear functions restricted to $(-1, 1)$ are the harmonic functions in D. Thus in view of the theorem of S. N. Bernstein mentioned above, one is led naturally to the following problem. Characterize those functions in $C(\overline{D})$ which are real analytic and subharmonic in D and which have best $L^1(\lambda)$ approximants from the set of functions in $C(\overline{D})$ which are harmonic in D. Here λ denotes two dimensional Lebesgue measure, $L^1(\lambda)$ denotes those real valued functions in F which are integrable with respect to λ, and $C(\overline{D})$ denotes those continuous real valued functions on $\overline{D}$ provided with the L^1 norm

S. P. Singh et al. (eds.), Approximation Theory and Spline Functions, 293–295.

$$||u|| = \frac{1}{\pi} \int_{\overline{D}} u d\lambda, \qquad u \in C(\overline{D}). \tag{1}$$

We have obtained a complete solution of this problem which is strikingly reminiscent of the theorem of S. N. Bernstein for the case $m = 2$ alluded to earlier. The characterization is as follows.

Let $H(D)$ denote the Banach space of all real-valued functions on $\overline{D}$ which are harmonic in D and which are provided with the L^1 norm as given in (1).

Theorem. Let $G = H(D) \cap C(\overline{D})$, and let h be subharmonic and real analytic in D, continuous on $\overline{D}$, and $h \notin G$. Then the following are equivalent:

(a) h has a best harmonic L^1 approximant q^* in G.

(b) There exists a function $q^* \in G$ with the following properties:

(i) q^* solves the Dirichlet problem on

$$D_o = \{(x, y) \in R^2: x^2 + y^2 < \tfrac{1}{2}\}, \quad \text{i.e. } q^*\big|_{\partial D_o} = h\big|_{\partial D_o}$$

(ii) $h - q^* > 0$ a.e. on $\overline{D} \setminus \overline{D}_o$.

Morover, if either (a) or (b) holds, then q^* is the unique best harmonic L^1 approximant to h in G.

As a consequence of our theorem, it follows that there exist functions h subharmonic and real analytic in D and continuous on $\overline{D}$ which do not have a best harmonic L^1 approximant in $H(D) \cap C(\overline{D})$. An example is $h = x^4y^4$. On the other hand, it follows from our theorem that $h = \frac{1}{4}(11x^4 - 18x^2y^2 + 3y^4)$ has a best harmonic L^1 approximant. Of course, by the theorem, best L^1 harmonic approximants when they exist are necessarily unique.

It is interesting to note the similarity of our theorem to that of S. N. Bernstein for the case of convex $C^2(-1. 1)$ functions. Thus the length of $(-\frac{1}{2}, \frac{1}{2})$ is half the length of

(-1. 1) whilst the area of D_o is half the area of D. In Bernstein's theorem, the change of sign occurs at $-\frac{1}{2}$ and $\frac{1}{2}$ whilst in our theorem, the change of sign occurs at ∂D_o. However, unlike the theorem of Bernstein, H(D) is infinite dimensional.

The L^∞ case has been treated by W. K. Hayman, D. Kershaw, and T. J. Lyons [2].

Full details and proofs will appear elsewhere.

REFERENCES

1. Bernstein, S. N., "Collected works", Vol. I (Acad. Nauk SSSR, Moscow, 1952).

2. Hayman, W. K., Kershaw, D., and Lyons, T. J., "The Best Harmonic Approximant to a Continuous Function", preprint.

B-SPLINES ON THE CIRCLE AND TRIGONOMETRIC B-SPLINES

T. N. T. Goodman and S. L. Lee*

1. INTRODUCTION

We shall first introduce the notion of circle splines. Denote by Π_n the space of polynomials of degree at most n on the unit circle $U = \{z \in \mathbb{C} : |z| = 1\}$. Take a non-decreasing sequence $\underset{\sim}{t} = (t_j)_{-\infty}^{\infty}$ with $t_j < t_{j+n+1}$, $t_{j+k} = t_j + 2\pi$ $(j \in Z)$ and write $z_j = e^{it_j}$ $(j \in Z)$. We now define $S_n(\underset{\sim}{t})$ to be the class of all functions on U which coincide with an element of Π_n on each non-trivial arc $[z_j, z_{j+1})$ and which are $C^{n-\mu}$ at z_j if t_j has multiplicity μ in $\underset{\sim}{t}$. (Arcs will be taken in an anti-clockwise direction, so that $[z_j, z_{j+1}) = \{e^{ix} : t_j \le x < t_{j+1}\}$. We say an element of $S_n(\underset{\sim}{t})$ is a <u>circle spline of degree n with knots at</u> $z_1, \ldots, z_k$.

Such splines were studied first by Ahlberg, Nilson and Walsh [1] and, slightly later, by Schoenberg [16]. Both papers studied interpolation properties and these were later extended by Tzimbalario [19] and the present authors [6]. Approximation by circle splines has recently been considered by Chen [5].

We now turn our attention to trigonometric splines and denote by T_n the linear span (over $\mathbb{C}$) of the functions

S. P. Singh et al. (eds.), Approximation Theory and Spline Functions, 297–325.

$\cos^\nu(\frac{1}{2}x)\sin^{n-\nu}(\frac{1}{2}x)$ $(x \in \mathbb{R}, \nu = 0, \ldots, n)$. With $\underset{\sim}{t}$ as before we define a class $T_n(\underset{\sim}{t})$ of trigonometric splines as follows. A function $S(x)$ on $\mathbb{R}$ is in $T_n(\underset{\sim}{t})$ if it coincides with an element of T_n on each non-trivial interval $[t_j, t_{j+1})$, it is $C^{n-\mu}$ at an element of $\underset{\sim}{t}$ with multiplicity μ, and it satisfies

$$S(x + 2\pi) = (-1)^n S(x) \quad (x \in \mathbb{R}). \tag{1.1}$$

There are other spaces of trigonometric splines that can be considered. We choose to study the space $T_n(\underset{\sim}{t})$ because it has the simple connection with circle splines (which can be easily seen):

$$S(z) \in S_n(t) \iff e^{-inx/2} S(e^{ix}) \in T_n(\underset{\sim}{t}). \tag{1.2}$$

We shall also consider briefly the following space of trigonometric splines with compact support. Take a non-decreasing sequence $\underset{\sim}{x} = \{x_j : j = 1, \ldots, n + k + 1\}$ with $x_j < x_{j+n+1}$ $(j = 1, \ldots, k)$. We define $T_n(\underset{\sim}{x})$ to be the class of all functions on $\mathbb{R}$ which vanish outside $[x_o, x_{n+k+1})$, which coincide with an element of T_n on each non-trivial interval $[x_j, x_{j+1})$ and which are $C^{n-\mu}$ at x_j if x_j has multiplicity μ in $\underset{\sim}{x}$.

Trigonometric splines were introduced by Schoenberg [15]. More recently they have been studied by Lyche and others [8], [10], [11], as well as the present authors [6]. Many of their properties can be extended to L-splines and for an account of these, as well as a summary of results on trigonometric splines, the reader is referred to [18].

In this paper we survey some properties of circle splines and trigonometric splines and also give a number of new results. We do not attempt to give a comprehensive survey. Instead, we concentrate on results related to B-splines for which we can exploi the relation (1.2) between circle splines and trigonometric splines. We study B-splines in §2 and §3, interpolation in §4 and §5 and approximation in §6 and §7. In §2 and §6 we derive results for circle splines by direct analogy with results for real polynomial splines. Using (1.2) we then deduce, in §3 and

§7, corresponding results for trigonometric splines. In §4 we study interpolation by trigonometric splines using real-variable methods such as zero-counting and total positivity. In §5 we then use (1.2) again to deduce interpolation results for circle splines which go far beyond those obtained in other papers which use essentially complex-variable techniques.

Finally, we mention that in [13] Micchelli and Sharma study interpolation by Λ-splines, which are the same as circle splines except that Π_n is replaced by the linear span of the functions z^{λ_j} $(j = 0, \ldots, n)$ where $\Lambda = \{\lambda_0, \ldots, \lambda_n\}$ is a prescribed set of distinct integers. (Thus when $\Lambda = \{0, 1, \ldots, n\}$, Λ-splines reduce to circle splines). In [7] it is shown that provided $\underset{\sim}{t}$ is uniform, most of the results of §2 and §6 in this paper generalise from circle splines to Λ-splines. Moreover by considering the case when Λ is symmetric about zero, one can deduce results for a corresponding generalisation of trigonometric splines.

2. B-SPLINES IN THE CIRCLE

We first establish the dimension of the space $S_n(\underset{\sim}{t})$ of circle splines. We shall need the analogue for the circle of the truncated power function: for a, b in U we define

$$\phi(z; a, b) = \begin{cases} 0, & z \in [a, b), \\ z - b, & \text{elsewhere.} \end{cases} \tag{2.1}$$

This cannot be expressed in terms of a function of one variable, as in the real case, though we could express it in terms of two variables by using the relation

$$\phi(\alpha z; \alpha a, \alpha b) = \alpha\phi(z; a, b) \quad (\alpha \in U).$$

Curiously we have not seen proofs of the next two results and so we include proofs here.

<u>Lemma 2.1.</u> The dimension of $S_n(\underset{\sim}{t})$ equals the maximum of k and n + 1.

<u>Proof.</u> Since $\Pi_n \subseteq S_n(\underset{\sim}{t})$, the dimension of $S_n(\underset{\sim}{t})$ must be at least n + 1. Furthermore dim $S_n(\underset{\sim}{t})$ cannot decrease on adding extra elements to $\underset{\sim}{t}$ and so it is sufficient to prove the

result for $k \geq n + 1$. Without loss of generality we may assume $t_{-1} < t_0$.

Now any function S in $S_n(\underset{\sim}{t})$ can be written in the form

$$S(z) = P(z) + \sum_{j=0}^{k-1} C_j \phi(z; z_0, z_j)^{n-\mu_j}, \tag{2.2}$$

where $\mu_j = |\{\ell < j : t_\ell = t_j\}|$ and S coincides with P in in Π_n on arc $[z_{-1}, z_0)$. Moreover, a function S of form (2.2) is in $S_n(\underset{\sim}{t})$ if and only if for z in arc $[z_{-1}, z_0)$ we have

$$S(z) - P(z) = \sum_{j=0}^{k-1} C_j (z - z_j)^{n-\mu_j} = 0.$$

Equating coefficients of z^ℓ $(\ell = 0, \ldots, n)$ gives a system of $n + 1$ equations in k unknowns where matrix is

$$\binom{n-\mu_j}{\ell} (-z_j)^{n-\mu_j-\ell} \qquad (\ell = 0, \ldots, n;\ j = 0, \ldots, k - 1),$$

(with the usual convention that $\binom{n-\mu_j}{\ell} = 0$ for $\ell > n-\mu_j$). Since this matrix is of rank $n + 1$, $S_n(\underset{\sim}{t})$ is of dimension $n + 1 + k - (n + 1) = k$. □

We note that if $k \leq n + 1$, then $S_n(\underset{\sim}{t}) = \Pi_n$. Henceforward, we shall assume that $k \geq n + 2$ and that each element of $\underset{\sim}{t}$ has multiplicity at most $k - n - 1$, so that

$$t_j < t_{j+n+1} < t_j + 2\pi \quad (j \in Z). \tag{2.3}$$

We can now define the B-splines for $S_n(\underset{\sim}{t})$, keeping throughout the normalisation which (as we shall see later) makes the B-splines into a partition of unity. Using the usual notation for divided differences we define, for j in Z,

$$N_j^n(z) = (-1)^{n+1} (z_{j+n+1} - z_j)$$
$$[x_j, \ldots, z_{j+n+1}] \{\phi(z; z_j, .)\}^n. \tag{2.4}$$

Clearly N_j^n is a circle spline of degree n with knots at $z_j, \ldots, z_{j+n+1}$ and $N_j^n(z)$ vanishes outside $[z_j, z_{j+n+1})$. We see from Lemma 2.1 that if any circle spline of degree n vanishes outside $[z_j, z_{j+n+1})$ and has knots at a strict subset of $z_j, \ldots, z_{j+n+1}$, then it must be in Π_n and hence vanish identically. This tells us that N_j^n has minimal support in the sense that no non-trivial element of $S_n(\underset{\sim}{t})$ can have support contained strictly within arc (z_j, z_{j+n+1}). It also tells us that N_j^n has an 'active knot' at each element w of $\{z_j, \ldots, z_{j+n+1}\}$, i.e. if w has multiplicity μ in $\{z_j, \ldots, z_{j+n+1}\}$ then $N_j^{n(n-\mu+1)}$ is discontinuous at w. Note that when $n = 0$,

$$N_j^0(z) = \begin{cases} 1, & z \in \operatorname{arc}[z_j, z_{j+1}), \\ 0, & \text{elsewhere.} \end{cases} \tag{2.5}$$

The function N_j^n is called a B-spline as abbreviation for 'basic spline'. The terminology is justified by

<u>Theorem 2.1</u>. The B-splines $N_0^n, \ldots, N_{k-1}^n$ form a basis for $S_n(\underset{\sim}{t})$.

<u>Proof</u>. We have only to show that $N_0^n, \ldots, N_{k-1}^n$ are linearly independent. So suppose $S = c_0N_0^n + \ldots + c_{k-1}N_{k-1}^n \equiv 0$. As before, we assume $t_{-1} < t_0$. Now define T by

$$T(z) = \begin{cases} 0, & z \in \operatorname{arc}\ [z_0 z_{k-n-1}), \\ \sum_{\ell=k-n-1}^{k-1} c_\ell N_\ell^n(z), & z \in \operatorname{arc}\ [z_{k-n-1}, z_k). \end{cases}$$

For z in $[z_{k-1}, z_k)$, $T(z) = S(z) = 0$. Thus T is a circle spline with knots at $z_{k-n-1}, \ldots, z_{k-1}$, which vanishes outside arc $[z_{k-n-1}, z_{k-1})$ and so, by our previous discussion,

T vanishes identically. If μ denotes the multiplicity of z_{k-n-1} in $\{z_{k-n-1}, \ldots, z_{k-1}\}$, then for $j = k-n-1, \ldots, k-1$, $N_j^{n(n-\mu+1)}$ is discontinuous at z_{k-n-1} if and only if $j = k-n-1$. Thus $c_{k-n-1} = 0$. Continuing in this way gives $c_{k-n} = \ldots = c_{k-1} = 0$. So $S = c_0 N_0^n + \ldots + c_{k-n-2} N_{k-n-2}^n \equiv 0$ and applying a similar argument gives $c_0 = \ldots = c_{k-n-2} = 0$. □

The following properties of B-splines on the circle can be derived in a very similar manner to the corresponding properties of real B-splines. For proofs of the latter we refer to [2].

Theorem 2.2. [5] *For $n \geq 1$ and z in U we have the identities:*

$$\frac{1}{n} N_j^{n\prime}(z) = \frac{N_j^{n-1}(z)}{z_{j+n} - z_j} - \frac{N_{j+1}^{n-1}(z)}{z_{j+n+1} - z_{j+1}} \tag{2.6}$$

$$N_j^n(z) = \frac{z - z_j}{z_{j+n} - z_j} N_j^{n-1}(z) + \frac{z_{j+n+1} - z}{z_{j+n+1} - z_{j+1}} N_{j+1}^{n-1}(z) \tag{2.7}$$

with the convention that terms involving N_ℓ^{n-1} vanish if $z_{\ell+n} = z_\ell$ $(\ell = j, j+1)$.

Theorem 2.3. [5] *Suppose f is in $L_1^{n+1}(U)$, i.e. $f^{(\nu)}$ is absolutely continuous for $\nu = 0, \ldots, n$ and $f^{(n+1)}$ is in $L_1(U)$. Then*

$$[z_j, \ldots, z_{j+n+1}]\, f = \frac{1}{n!(z_{j+n+1} - z_j)} \int_U N_j^n(z) f^{(n+1)}(z)\, dz. \tag{2.8}$$

Corollary 2.1. *For $j = 1, \ldots, k$.*

$$\int_U N_j^n(z)\, dz = \frac{z_{j+n+1} - z_j}{n+1}. \tag{2.9}$$

Theorem 2.4. [5] For w, z in U,

$$(w - z)^n = \sum_{j=1}^{k} (w - z_{j+1})\dots(w - z_{j+n})\, N_j^n(z). \tag{2.10}$$

Corollary 2.2. For $\ell = 0, \dots, n$, $z \in U$,

$$z^\ell = \sum_{j=1}^{k} z_j^{(\ell)} N_j^n(z) \tag{2.11}$$

where

$$\binom{n}{\ell} z_j^{(\ell)} = \sum\{z_{j+1}^{\alpha_1}\dots z_{j+n}^{\alpha_n} : \alpha_1 + \dots + \alpha_n = \ell,\ \alpha_i = 0 \text{ or } 1\ (i = 1, \dots, n)\}. \tag{2.12}$$

In particular, we have

$$\sum_{j=1}^{k} N_j^n(z) = 1 \quad (z \in U). \tag{2.13}$$

3. TRIGONOMETRIC B-SPLINES

We keep $\underset{\sim}{t}$ as in §2 and define now the 'trigonometric B-splines' P_j^n in the space $T_n(\underset{\sim}{t})$. For $n = 0$ we write

$$P_j^0(x) = N_j^0(e^{ix}) \quad (x \in \mathbb{R},\ j \in Z). \tag{3.1}$$

For $n \geq 1$ we put

$$\tau_j = \frac{1}{n}(t_{j+1} + \dots + t_{j+n}) \tag{3.2}$$

and define

$$P_j^n(x) = e^{in(\tau_j - x)/2}\, N_j^n(e^{ix}) \quad (x \in \mathbb{R},\ j \in Z). \tag{3.3}$$

Note that for any integer j,

$$P^n_{j+k} = (-1)^n P^n_j. \tag{3.4}$$

Now recalling (1.2) we can easily deduce the following results from Theorems 1.1, 1.2 and 1.4.

Theorem 3.1. *The B-splines* $P^n_0, \ldots, P^n_{k-1}$ *form a basis for* $\mathcal{T}_n(\underset{\sim}{t})$.

Theorem 3.2. [11] *For* $n \geq 1$ *and* $x \in \mathbf{R}$ *we have the identities*

$$\frac{2}{n} P^{n'}_j(x) = \frac{\cos 1/2\ (x - t_j)}{\sin 1/2\ (t_{j+n} - t_j)} P^{n-1}_j(x) - \frac{\cos 1/2\ (t_{j+n+1} - x)}{\sin 1/2\ (t_{j+n+1} - t_{j+1})} P^{n-1}_{j+1}(x), \tag{3.5}$$

$$P^n_j(x) = \frac{\sin 1/2\ (x - t_j)}{\sin 1/2\ (t_{j+n} - t_j)} P^{n-1}_j(x) + \frac{\sin 1/2\ (t_{j+n+1} - x)}{\sin 1/2\ (t_{j+n+1} - t_{j+1})} P^{n-1}_{j+1}(x), \tag{3.6}$$

with the convention that terms involving P^{n-1}_ℓ *vanish if* $t_{\ell+n} = t_\ell (\ell = j, j + 1)$.

Theorem 3.3. [11] *For* x, y *in* $\mathbb{R}$,

$$[\sin \frac{1}{2} (y - x)]^n = \sum_{j=1}^{k} \sin \frac{1}{2} (y - t_{j+1}) \ldots \sin \frac{1}{2} (y - t_{j+n}) P^n_j(x). \tag{3.7}$$

From (3.6) we can easily deduce, by induction on n,

Corollary 3.1. *For* $j \in Z$, *the function* P^n_j *is real-valued and*

$$P_j^n(x) > 0 \quad \text{for} \quad t_j > x \quad t_{j+n+1}. \tag{3.8}$$

From (2.11) for $\ell = 0$ and n we can see that

$$\cos \frac{1}{2} nx = \sum_{j=1}^{k} \cos \frac{1}{2} n\tau_j P_j^n(x), \tag{3.9}$$

$$\sin \frac{1}{2} nx = \sum_{j=1}^{k} \sin \frac{1}{2} n\tau_j P_j^n(x). \tag{3.10}$$

We can now derive a bound on the magnitude of $P_j^n(x)$ which will be useful later.

<u>Corollary 3.2</u>. <u>If</u> $t_{j+n+1} - t_j < \pi/n$, <u>then</u>

$$P_j^n(x) < \cos \frac{1}{2} n(x - \tau_j) \quad \text{for} \quad t_j < x < t_{j+n+1}. \tag{3.11}$$

<u>Proof</u>. From (3.9) and (3.10) we see that

$$\cos \frac{1}{2} n(x - \tau_j) = P_j^n(x) + \sum_{\ell=j-n}^{j+n} \cos \frac{1}{2} n(\tau_\ell - \tau_j) P_\ell^n(x). \tag{3.12}$$

Since $P_j^n(x)$ depends only on the values of $t_j, \ldots, t_{j+n+1}$, we may choose $t_{j-n}, \ldots, t_{j-1}$ to be arbitrarily close to t_j, and $t_{j+n+2}, \ldots, t_{j+2n}$ to be arbitrarily close to t_{j+n+1}. So recalling that $t_{j+n+1} - t_j < \pi/n$, we see that

$$\frac{1}{2} n|\tau_\ell - \tau_j| = \frac{1}{2} \sum_{p=1}^{n} |t_{\ell+p} - t_{j+p}| < \frac{1}{2}\pi .$$

Thus all the terms on the right side of (3.12) are positive and (3.11) follows. □

From Corollaries 3.1 and 3.2 we can now deduce properties of B-splines on the circle.

<u>Corollary 3.3</u>. <u>For</u> $t_j < x < t_{j+n+1}$, $\arg N_j^n(e^{ix}) = \frac{1}{2} n(x-\tau_j)$.

Furthermore, if $t_{j+n+1} - t_j < \pi/n$, then $|N_j^n(e^{jx})| < \cos \frac{1}{2} n(x - \tau_j)$.

We can also deduce from Theorem 2.3 a corresponding formula involving $P_j^n(x)$. However, this involves defining a 'trigonometric divided difference' and we refer the reader to [18, p. 458].

We now turn our attention to the space $T_n(\underset{\sim}{x})$ for a finite sequence $\underset{\sim}{x} = (x_j)_1^{n+k+1}$. Henceforward, we assume that

$$x_j < x_{j+n+1} < x_j + 2\pi \quad (j = 1, \ldots, k).$$

To define an element of $T_n(\underset{\sim}{x})$ with support in $[x_j, x_{j+n+1})$ we simply take a piece of the corresponding B-spline P_j^n. To be precise, for any j with $1 \le j \le k$ we define a sequence $\underset{\sim}{t} = (t_\ell)_{-\infty}^{\infty}$ by

$$t_\ell = x_\ell \quad (\ell = j, \ldots, j + n + 1), \; t_{\ell+n+2} = t_\ell + 2\pi (\ell \in Z).$$

With P_j^n defined as above we define the trigonometric B-spline Q_j^n by

$$Q_j^n(x) = \begin{cases} P_j^n(x), & x_j \le x < x_{j+n+1}, \\ 0 \quad , & \text{elsewhere.} \end{cases}$$

It is easily seen that $Q_1^n, \ldots, Q_k^n$ form a basis for $T_n(\underset{\sim}{x})$.

4. TRIGONOMETRIC INTERPOLATION

Since the case $n = 0$ is essentially trivial we shall henceforward assume $n \ge 1$. We first consider interpolation by functions in $T_n(\underset{\sim}{x})$. Take a non-decreasing sequence

$\underset{\sim}{\tau} = (\tau_1, \ldots, \tau_k)$ with multiplicities at most $n + 1$. We shall say $(\mathcal{T}_n(\underset{\sim}{x}), \tau)$ <u>is solvable</u> if for every sequence $\underset{\sim}{y} = (y_1, \ldots, y_k)$ there is a unique function S in $\mathcal{T}_n(\underset{\sim}{x})$ which interpolates $\underset{\sim}{y}$ at $\underset{\sim}{\tau}$, i.e.

$$S^{(\nu_i)}(\tau_i) = y_i \quad (i = 1, \ldots, k) \tag{4.1}$$

where

$$\nu_i = |\{j < i : \tau_j = \tau_i\}|. \tag{4.2}$$

For simplicity we assume we do not interpolate at a possible discontinuity, i.e. $S^{(\nu_i)}$ is continuous at τ_i for all S in $\mathcal{T}_n(\underset{\sim}{x})$ $(i = 1, \ldots, k)$. The properties we have derived for $Q_1^n, \ldots, Q_k^n$ show that they are 'generalized B-splines' in the sense of [6] and so we have

<u>Theorem 4.1.</u> [6] <u>The problem $(\mathcal{T}_n(\underset{\sim}{x}), \underset{\sim}{\tau})$ is solvable if and only if</u>

$$x_i < \tau_i < x_{i+n+1} \quad (i = 1, \ldots, k).$$

We now take $\underset{\sim}{t}$ as before and consider interpolation by elements of $\mathcal{T}_n(\underset{\sim}{t})$. Take a non-decreasing sequence $\underset{\sim}{\tau} = (\tau_j)_{-\infty}^{\infty}$ satisfying

$$\tau_j < \tau_{j+n+1}, \; \tau_{j+k} = \tau_j + 2\pi \quad (j \in Z). \tag{4.3}$$

We shall say $(\mathcal{T}_n(\underset{\sim}{t}), \underset{\sim}{\tau})$ <u>is solvable</u> if for every sequence $\underset{\sim}{y} = (y_1, \ldots, y_k)$ there is a unique function in $\mathcal{T}_n(\underset{\sim}{t})$ which interpolates $\underset{\sim}{y}$ at $(\tau_1, \ldots, \tau_k)$ (in the sense of (4.1), (4.2)). From the general theory of [6] we can deduce the following weaker analogue of Theorem 4.1.

<u>Theorem 4.2.</u> [6] <u>If $(\mathcal{T}_n(\underset{\sim}{t}), \underset{\sim}{\tau})$ is solvable, then</u>

$$\exists \ell \in Z \text{ with } t_i < \tau_{i+\ell} < t_{i+n+1} \quad (i \in Z). \tag{4.4}$$

If $n + k$ is odd and (4.4) holds, then $(T_n(\underset{\sim}{t}), \underset{\sim}{\tau})$ is solvable.

It therefore remains only to consider the case when (4.4) holds and $n + k$ is even. This case appears to be much more difficult. It will be useful to introduce the concept of 'eigenvalue', a notion introduced by Schoenberg [17] in relation to cardinal spline interpolation. We call a non-trivial function S on $\mathbb{R}$ an eigenspline for $(T_n(\underset{\sim}{t}), \underset{\sim}{\tau})$ with eigenvalue λ it is satisfies the following conditions.

1. If $t_j < t_{j+1}$, $S|[t_j, t_{j+1}) \in T_n$.

2. S is $C^{n-\mu}$ at an element of $\underset{\sim}{t}$ with multiplicity μ.

3. S interpolates zero data at $\underset{\sim}{\tau}$.

4. $S(x + 2\pi) = \lambda S(x)$ $(x \in \mathbb{R})$.

Recalling (1.1) we see that $(T_n(\underset{\sim}{t}), \underset{\sim}{\tau})$ is solvable if and only if $(-1)^n$ is not an eigenvalue.

Now assume that $\underset{\sim}{\tau}$ (like $\underset{\sim}{t}$) has multiplicities at most $k - n - 1$. Then to any problem $(T_n(\underset{\sim}{t}), \underset{\sim}{\tau})$ there corresponds a 'dual problem' $(T_n(\underset{\sim}{\tau}), \underset{\sim}{t})$.

Lemma 4.1. The problem $(T_n(\underset{\sim}{t}), \underset{\sim}{\tau})$ has an eigenvalue λ if and only if the dual problem $(T_n(\underset{\sim}{\tau}), \underset{\sim}{t})$ has an eigenvalue λ^{-1}.

Proof. We introduce the 'trigonometric truncated power function' ϕ_+ defined by

$$\phi_+(x) = \begin{cases} 0, & x < 0, \\ (\sin \frac{x}{2})^n, & x \geq 0. \end{cases}$$

Now choose any point α which is in neither $\underset{\sim}{t}$ nor $\underset{\sim}{\tau}$, and suppose S is an eigenspline for $(T_n(\underset{\sim}{t}), \underset{\sim}{\tau})$ with eigenvalue λ. We may choose the numbering of $\underset{\sim}{t}$ and $\underset{\sim}{\tau}$ so that

$$\alpha < t_1 < t_k < \alpha + 2\pi, \quad a < \tau_1 < \tau_k < \alpha + 2\pi.$$

Then the restriction of S to $[\alpha, \alpha + 2\pi)$ is of the form

$$S(x) = \sum_{j=0}^{n} a_j \phi_+^{(j)}(x - \alpha) + \sum_{j=1}^{k} b_j \phi_+^{(\mu_j)}(x - t_j), \qquad (4.5)$$

where

$$\mu_j = |\{\ell < j : t_\ell = t_j\}| .$$

Moreover S satisfies

$$S^{(\nu)}(\alpha + 2\pi) = \lambda S^{(\nu)}(\alpha) \quad (\nu = 0, \ldots, n), \qquad (4.6)$$

$$S^{(\nu_i)}(\tau_i) = 0 \, (i = 1, \ldots, k), \qquad (4.7)$$

where

$$\nu_i = |\{\ell < i : \tau_\ell = \tau_i\}| .$$

Conversely any function S on $[\alpha, \alpha + 2\pi)$ of the form (4.5) which satisfies (4.6) and (4.7) can be extended uniquely to an eigenspline with eigenvalue λ. Thus λ is an eigenvalue for $(T_n(\underset{\sim}{t}), \underset{\sim}{\tau})$ if and only if the system of $n + 1 + k$ equations in $\tilde{n} + \tilde{1} + k$ unknowns given by (4.5), (4.6) and (4.7) is singular.

Next suppose T is an eigenspline for the dual problem $(T_n(\underset{\sim}{\tau}), \underset{\sim}{t})$ with eigenvalue λ^{-1}. Then the restriction of T to $(\alpha, \alpha + 2\pi]$ is of the form

$$T(x) = \sum_{j=0}^{n} a_j \phi_+^{(j)}(\alpha + 2\pi - x) + \sum_{j=0}^{k} b_j \phi_+^{(\nu_j)}(\tau_j - x) \quad (4.8)$$

and satisfies

$$T^{(\nu)}(\alpha) = \lambda T^{(\nu)}(\alpha + 2\pi) \quad (\nu = 0, \ldots, n), \qquad (4.9)$$

$$T^{(\mu_i)}(t_i) = 0 \quad (i = 1, \ldots, k). \qquad (4.10)$$

As before we see that λ^{-1} is an eigenvalue for $(T_n(\underset{\sim}{\tau}), \underset{\sim}{t})$

if and only if the system given by (4.8), (4.9) and (4.10) is singular. But is is easily seen that this system is singular if and only if the system given by (4.5), (4.6) and (4.7) is singular. Thus the result is proved. □

We can immediately deduce

Theorem 4.3. *The problem* $(T_n(\underset{\sim}{t}), \underset{\sim}{\tau})$ *is solvable if and only if the dual problem* $(T_n(\underset{\sim}{\tau}), \underset{\sim}{t})$ *is solvable*.

This result is not of much value unless we can find some solvable problems. To this end we shall use the following Lemma which is proved by applying Theorem 6 of [4] and the fact, proved in [6], that 'the B-spline collocation matrix is totally positive'.

Lemma 4.2. [6] *Suppose one of the following conditions is satisfied.*

$$\underset{\sim}{t} \text{ is strictly increasing and } \tau_i = t_i (i \in Z). \tag{4.11}$$

$$t_i < \tau_i < t_{i+1} \; (i \in Z). \tag{4.12}$$

Then $(T_n(t), \tau)$ *has* m *distinct eigenvalues of sign* $(-1)^k$, *where* $m = n - 1$ *under condition* (4.11) *and* $m = n$ *under condition* (4.12).

We can now deduce

Theorem 4.4. *Suppose* $n + k$ *is even.*

a) *If (4.11) is satisfied, then* $(T_n(\underset{\sim}{t}), \underset{\sim}{\tau})$ *is solvable if and only if* n *is odd.*

b) *Suppose (4.12) is satisfied and for some* ℓ, $\tau_{i+\ell} = t_i + \pi$ $(i \in Z)$. *Then* $(T_n(\underset{\sim}{t}), \underset{\sim}{\tau})$ *is solvable if and only if* n *is even.*

c) *Suppose (4.12) is satisfied and* $\underset{\sim}{t}$ *and* $\underset{\sim}{\tau}$ *are symmetric about the same point. Then* $(T_n(\underset{\sim}{t}), \underset{\sim}{\tau})$ *is solvable if and only if* n *is even.*

Proof. We claim that in all three cases, λ is an eigenvalue of $(T_n(\underset{\sim}{t}), \underset{\sim}{\tau})$ if and only if λ^{-1} is also an eigenvalue.

For a) and b) this follows from Lemma 4.1, while for c) it follows from the symmetry of the problem. We can now apply Lemma 4.2 to show that $(-1)^k$ is an eigenvalue if and only if m is odd. But $(\mathcal{T}_n(\underset{\sim}{t}), \underset{\sim}{\tau})$ is solvable if and only if $(-1)^n$ is not an eigenvalue and, since $(-1)^k = (-1)^n$, the result follows. □

The range of problems which we can classify as solvable or unsolvable can be greatly extended by the following technique. We keep fixed all but one of the interpolation conditions per period. Then as the remaining interpolation condition is varied, either all the resulting problems are unsolvable or at most one of them is unsolvable. The precise details are given in

<u>Theorem 4.5.</u> [6] <u>Suppose</u> $\underset{\sim}{t}$ <u>satisfies</u> $t_{j+p} = t_j + 2\pi p/k$ $(j \in \mathbb{Z})$ <u>for some integer</u> p <u>which divides</u> k. <u>Fix numbers</u> $\sigma_1 < \dots < \sigma_{p-1} < \sigma_1 + 2\pi p/k$ <u>and let</u> T <u>denote the set of all strictly increasing sequences</u> $\underset{\sim}{\tau}$ <u>which include</u> $\sigma_1, \dots, \sigma_{p-1}$ <u>and satisfy</u>

$$t_j < \tau_j < t_{j+n+1}, \quad \tau_{j+p} = \tau_j + 2\pi p/k \quad (j \in \mathbb{Z}).$$

<u>Then either</u> $(\mathcal{T}_n(\underset{\sim}{t}), \underset{\sim}{\tau})$ <u>is unsolvable for all</u> $\underset{\sim}{\tau}$ <u>in</u> T <u>or it is unsolvable for at most one</u> $\underset{\sim}{\tau}$ <u>in</u> T <u>(apart from a renumbering of the index)</u>.

From Theorems 4.4 and 4.5 we can classify the solvability of a large variety of problems. In particular by taking $p = 1$ we can easily deduce

<u>Corollary 4.1.</u> [6] <u>Suppose</u> $\underset{\sim}{t}$ <u>and</u> $\underset{\sim}{\tau}$ <u>are uniform, i.e.</u> $t_{j+1} - t_j = \tau_{j+1} - \tau_j = 2\pi/k$ $(j \in \mathbb{Z})$. <u>Then</u> $(\mathcal{T}_n(\underset{\sim}{t}), \underset{\sim}{\tau})$ <u>is solvable except in the following cases.</u>

a) <u>n and k are even and for some</u> ℓ, $\tau_{i+1} = t_i$ $(i \in \mathbb{Z})$.

b) <u>n and k are odd and for some</u> ℓ, $\tau_{i+1} = \frac{1}{2}(t_i + t_{i+1})$ $(i \in \mathbb{Z})$.

We close this section by stressing that we have considered the space $\mathcal{T}_n(\underset{\sim}{t})$ only because of its relation to circle splines. There are corresponding results for spaces of periodic trigonometric splines with arbitrary period. There are also similar results for interpolation of an infinite sequence of bounded data by bounded trigonometric splines with periodically spaced knots (see [16]).

5. INTERPOLATION ON THE CIRCLE

As in the previous section we take a non-decreasing sequence $\underset{\sim}{\tau} = (\tau_i)_{-\infty}^{\infty}$ satisfying (4.3). This time we put $w_j = e^{i\tau_j} (j \in Z)$ and say that $(\mathcal{S}_n(\underset{\sim}{t}), \underset{\sim}{\tau})$ is solvable if for every sequence $y = (y_1, \ldots, y_k)$ there is a unique function S in $\mathcal{S}_n(\underset{\sim}{t})$ which interpolates $\underset{\sim}{y}$ at $(w_1, \ldots, w_k)$, i.e.

$$S^{(\nu_i)}(w_i) = y_i \quad (i=1,\ldots,k) \tag{5.1}$$

where

$$\nu_i = |\{j < i : \tau_j = \tau_i\}| \ . \tag{5.2}$$

Using (1.2) we can immediately deduce the following results from the results of §4.

Theorem 5.1. [6] If $(\mathcal{S}_n(\underset{\sim}{t}), \underset{\sim}{\tau})$ is solvable, then (4.4) holds. If $n + k$ is odd and (4.4) holds, then $(\mathcal{S}_n(\underset{\sim}{t}), \underset{\sim}{\tau})$ is solvable.

Theorem 5.2. If $\underset{\sim}{\tau}$ has multiplicities at most $k - n - 1$, then $(\mathcal{S}_n(\underset{\sim}{\tau}), \underset{\sim}{t})$ is solvable if and only if $(\mathcal{S}_n(\underset{\sim}{\tau}), \underset{\sim}{t})$ is solvable.

Theorem 5.3. Suppose $n + k$ is even.

a) If (4.11) is satisfied, then $(\mathcal{S}_n(\underset{\sim}{t}), \underset{\sim}{\tau})$ is solvable if and only if n is odd.

b) Suppose (4.12) is satisfied and for some $\ell, \tau_{i+\ell} = t_i + \pi$ $(i \in Z)$. Then $(\mathcal{S}_n(\underset{\sim}{t}), \underset{\sim}{\tau})$ is solvable if and only if n is even.

c) Suppose (4.12) is satisfied and $\underset{\sim}{t}$ and $\underset{\sim}{\tau}$ are symmetric about the same point. Then $(S_n(\underset{\sim}{t}), \underset{\sim}{\tau})$ is solvable if and only if n is even.

Theorem 5.4. [6] With $\underset{\sim}{t}$, T as in Theorem 4.5, either $(S_n(\underset{\sim}{t}), \underset{\sim}{\tau})$ is unsolvable for all $\underset{\sim}{\tau}$ in T or it is unsolvable for at most one $\underset{\sim}{\tau}$ in $\underset{\sim}{t}$ (apart from a renumbering of the above index).

Corollary 5.1. [19] If $\underset{\sim}{t}$ and $\underset{\sim}{\tau}$ are uniform, then $(S_n(\underset{\sim}{t}), \underset{\sim}{\tau})$ is solvable except for cases a) and b) of Corollary (4.1).

6. APPROXIMATION ON THE CIRCLE

In this section we consider approximation operators of the form

$$(Lg)(z) = \sum_{j=1}^{k} T_j(g) N_j^n(z) \quad (z \in U), \tag{6.1}$$

where $T_j(g)$ is a linear combination of $g^{(r)}(\sigma_j)$ $(r = 0, 1, \ldots, n)$ for some point σ_j. In [5] Chen considers the case when for $j = 1, \ldots, k$ the point σ_j is prescribed in the arc (z_j, z_{j+n+1}). He shows that (for strictly increasing $\underset{\sim}{t}$) there is a unique such operator L which reproduces $S_n(\underset{\sim}{t})$ and he studies the order of converges of $(Lg)^{(s)}$ to $g^{(s)}$ as the mesh size $\max |z_{j+1} - z_j|$ goes to zero. These results and their proofs are in direct analogy with results for the 'quasi-interpolant' of de Boor and Fix [3] for real polynomial splines. We shall not give the details here. Corresponding results for Λ-splines and their trigonometric analogues are proved in [7].

We now restrict out attention to the case when the linear functionals $T_j(g)$ in (6.1) depend only on the value of g at some point σ_j. In this case we denote the approximation operator by S, so that for some numbers A_j $(j = 1, \ldots, k)$,

$$(Sg)(z) = \sum_{j=1}^{k} A_j g(\sigma_j) N_j^n(z) \quad (z \in U). \tag{6.2}$$

We cannot expect such an operator S to reproduce $S_n(\underset{\sim}{t})$. Instead we require that S reproduces the functions z^α and z^β for some distinct α, β in $\{0, \ldots, n\}$, i.e. for $z \in U$,

$$z^\alpha = \sum_{j=1}^{k} A_j \sigma_j^\alpha N_j^n(z), \tag{6.3}$$

$$z^\beta = \sum_{j=1}^{k} A_j \sigma_j^\beta N_j^n(z). \tag{6.4}$$

Comparing with (2.11) for $\ell = \alpha$ and β we see that

$$A_j \sigma_j = z_j^{(\alpha)}, \quad A_j \sigma_j = z_j^{(\beta)} \quad (j = 1, \ldots, n)$$

which is equivalent to

$$\sigma_j = (z_j^{(\beta)}/z_j^{(\alpha)})^{1/(\beta-\alpha)}, \qquad A_j = (z_j^{(\alpha)})^{\beta/(\beta-\alpha)} (z_j^{(\beta)})^{\alpha/(\alpha-\beta)}. \tag{6.5}$$

We distinguish this operator (if necessary) by $S_n^{\alpha,\beta}$. Note that $S_n^{0,1}$ reproduces linear functions and in this case

$$\sigma_j = \frac{1}{n}(z_{j+1} + \ldots + z_{j+n}), \; A_j = 1 \; (j = 1, \ldots, n).$$

Thus $S_n^{0,1}$ is an analogue of the Bernstein-Schoenberg operator (see [14]).

<u>Theorem 6.1</u>. <u>Let</u> $\rho_n(\underset{\sim}{t}) = \max\{|t_{j+n+1} - t_j| : j = 1, \ldots, k\}$. <u>If</u> g <u>is continuous on some neighbourhood of</u> U, <u>then for fixed</u> α, β <u>as</u> $n\rho_n(\underset{\sim}{t}) \to 0$,

$$|(S_n^{\alpha,\beta} g)(z) - g(z)| \le Cnw(z^{-\alpha} f(z); \rho_n(t))$$

<u>where</u> C <u>is a constant and</u> w <u>denotes the modulus of continuity</u>.

Proof. From (6.2) and (6.3) we see that for z in U,

$$(Sg)(z) - g(z) = \sum_{j=1}^{k} A_j \sigma_j^\alpha \{\sigma_j^{-\alpha} g(\sigma_j) - z^{-\alpha} g(z)\} N_j^n(z). \tag{6.6}$$

Fixing z in U, we see that the terms in the summation (6.6) are non-zero only when arc (z_j, z_{j+n+1}) contains z. For such j it is easily seen from (6.5) that as $\rho_n(\underset{\sim}{t}) \to 0$, $|z_j^{(\alpha)} - z^\alpha|$, $|z_j^{(\beta)} - z^\beta|$, $|\sigma_j - z|$ and $|A_j - 1|$ are all of order $\rho_n(\underset{\sim}{t})$. We also see from Corollary 3.3 that as $n\rho_n(\underset{\sim}{t}) \to 0$, $|N_j^n(z)| \leq 1$. The result then follows from (6.6) on noting that the summation contains at most $n + 1$ non-zero terms. □

Henceforward we shall assume that the points $z_1, \ldots, z_k$ are uniformly spaced around U. Without loss of generality, we can assume that for $j = 1, \ldots, k$,

$$z_j = \xi^j, \quad \text{where} \quad \xi = e^{2\pi i/k}. \tag{6.7}$$

After some calculation we see from (2.12) that

$$z_j^{(\ell)} = c_\ell \xi^{1/2\ \ell(n+1)+\ell j} \quad (j = 1, \ldots, k) \tag{6.8}$$

where $c_0 = 1$ and for $\ell = 1, \ldots, n$,

$$c_\ell = \frac{\prod_{p=1}^{n} \frac{\sin p\pi/k}{p\pi/k}}{\prod_{p=1}^{\ell} \frac{\sin p\pi/k}{p\pi/k} \prod_{p=1}^{n-\ell} \frac{\sin p\pi/k}{p\pi/k}} \tag{6.9}$$

So from (6.5) we have for $j = 1, \ldots, k$,

$$\begin{cases} \sigma_j = R\xi^{1/2\ (n+1)+j}, & \text{where } R = (c_\beta/c_\alpha)^{1/(\beta-\alpha)}, \\ A_j = c_\alpha^{\beta/(\beta-\alpha)}\, c_\beta^{\alpha/(\alpha-\beta)} = A, & \text{say.} \end{cases} \tag{6.10}$$

We note that the points σ_j are uniformly spaced around a circle of radius R. Since R does not in general equal 1, we require the function g in (6.2) to be defined on some neighbourhood of U. However, given a continuous function f on U we can extend it to a continuous function g on a neighbourhood of U in a number of ways. Perhaps the simplest is by

$$g(\eta z) = f(z), \; z \in U, \; \eta > 0. \tag{6.11}$$

In this case the operator S takes the form

$$(Sf)(z) = A \sum_{j=1}^{k} f(\xi^{1/2\,(n+1)+j}) N_j^n(z). \tag{6.12}$$

We also note that when $\alpha + \beta = n$, we have $c_\alpha = c_\beta$ and so $R = 1$, $A = c_\alpha$ and the operator becomes

$$(S_n^{\alpha, n-\alpha} f)(z) = c_\alpha \sum_{j=1}^{k} f(\xi^{1/2(n+1)+j}) N_j^n(z). \tag{6.13}$$

In [7] more general formulae than (6.10) are derived for Λ-splines. As a special case of a result in [7] we have the following asymptotic formula which is reminiscent of a result of Voronovskaja (see [9, p.22]) and is analogous to a result of Marsden [12] for real polynomial splines.

Theorem 6.7. [7] *If g is holomorphic on a neighbourhood of U, then*

$$\lim_{k \to \infty} \frac{k^2}{n+1} \{ (S_n^{\alpha,\beta} g)(z) - g(z) \} = -\frac{1}{6}\pi^2 \{ z^2 g''(z) + (1 - \alpha - \beta) z g'(z) + \alpha\beta g(z) \} \tag{6.14}$$

uniformly on U.

Remark 6.1. A study of the proof in [7] shows that the result still holds if we fix α and β and let n vary, provided $kn^{-3} \to \infty$.

Remark 6.2. If $\alpha + \beta = n$, then we see from (6.13) that $S_n^{\alpha,\beta} f$ is defined for functions f defined only in U. In this case the proof of Theorem 6.2 is easily adapted to derive (6.14) for any function f in $C^3(U)$.

To conclude this section we shall consider the convergence of derivatives of Sg to derivatives of g. First we derive a formula for the derivatives of Sg.

Lemma 6.1. For $\nu = 0, 1, \ldots, n-1$, we have the identity

$$(Sg)^{(\nu)}(z) = \frac{\nu! R^{\nu} A}{c_{\nu}} \sum_{j=1}^{k} [\sigma_{j-\nu}, \ldots, \sigma_j] g N_j^{n-\nu}(z). \qquad (6.15)$$

Proof. We shall prove the identity by induction on ν. For $\nu = 0$ (6.15) reduces to (6.2). We assume then that (6.15) is true for some $\nu < n - 1$. Then from (2.6),

$$\begin{aligned}(Sg)^{(\nu+1)}(z) &= \frac{\nu! R^{\nu} A}{c_{\nu}} \sum_{j=1}^{k} [\sigma_{j-\nu}, \ldots, \sigma_j] g \\ &\quad \times (n-\nu) \left\{ \frac{N_j^{R-\nu-1}(z)}{z_{j+n-\nu} - z_j} - \frac{N_{j+1}^{n-\nu-1}(z)}{z_{j+n-\nu+1} - z_{j+1}} \right\} \\ &= \frac{\nu!(n-\nu) R^{\nu} A}{c_{\nu}} \sum_{j=1}^{k} \{[\sigma_{j-\nu}, \ldots, \sigma_j] g \\ &\quad - [\sigma_{j-\nu-1}, \ldots, \sigma_{j-1}] g\} \frac{N_j^{n-\nu-1}(z)}{z_{j+n-\nu} - z_j} \\ &= \frac{\nu!(n-\nu) R^{\nu} A}{c_{\nu}} \sum_{j=1}^{k} \frac{\sigma_j - \sigma_{j-\nu-1}}{z_{j+n-\nu} - z_j} [\sigma_{j-\nu-1}, \ldots, \sigma_j] g \, N_j^{n-\nu-1}(z).\end{aligned}$$

$$(6.16)$$

But from (6.7) and (6.10) we see that

$$\begin{aligned}\frac{\sigma_j - \sigma_{j-\nu-1}}{z_{j+n-\nu} - z_j} &= R\, \xi^{(n+1)/2} \frac{(\xi^j - \xi^{j-\nu-1})}{\xi^{j+n-\nu} - \xi^j} \\ &= R \frac{\sin(\nu+1)\pi/k}{\sin(n-\nu)\pi/k}.\end{aligned}$$

Substituting this into (6.16) and recalling (6.9) gives (6.15) with ν replaced by $\nu + 1$, which completes the proof. □

We can now prove

Theorem 6.3. *Fix non-negative integers* α, β *and* ν. *Suppose* g *is defined on some neighbourhood of* U *and for all* η *sufficiently close to* 1 *the function* $g(\eta z)(z \in U)$ *lies in* $C^\nu(U)$. *Moreover let* $h_\nu(\eta z) := \frac{d^\nu}{dz^\nu} g(\eta z)$ *be a continuous function for* z *in* U *and* η *close enough to* 1. *Then as* $kn^{-2} \to \infty$, $n \geq \max(\alpha, \beta, \nu+1)$, *there is a constant* C *such that for* $z \in U$,

$$|(S_n^{\alpha,\beta} g)^{(\nu)}(z) - g^{(\nu)}(z)| \leq Cn \left\{\frac{n}{k} \| g^{(\nu)} \| + w(h_\nu; \frac{n}{k})\right\}, \tag{6.17}$$

where

$$\| g^{(\nu)} \| = \max \{|g^{(\nu)}(z)| : z \in U\}.$$

Proof. From (6.15) and (2.13) we get

$$(Sg)^{(\nu)}(z) - g^{(\nu)}(z) = \sum_{j=1}^{k} \left\{\frac{\nu! R^\nu A}{c_\nu} [\sigma_{j-\nu}, \ldots, \sigma_j] g \right. \tag{6.18}$$

$$\left. - g^{(\nu)}(z)\right\} N_j^{n-\nu}(z).$$

Now putting $f(z) = g(Rz)$ $(z \in U)$, we see that

$$R^\nu [\sigma_{j-\nu}, \ldots, \sigma_j] g = [\xi^{1/2(n+1)+j-\nu}, \ldots, \xi^{1/2(n+1)+j}] f$$

$$= \frac{1}{(\nu-1)!(\xi^{1/2(n+1)+j} - \xi^{1/2(n+1)+j-\nu})} \int_U N(v) f^{(\nu)}(v) dv$$

by Theorem 2.3, where N is the B-spline with knots at $\xi^{1/2(n+1)+j-\nu}, \ldots, \xi^{1/2(n+1)+j}$. So applying Corollary 2.1 we get

$$\frac{\nu! R^\nu A}{c_\nu} [\sigma_{j-\nu}, \ldots, \sigma_j] g - g^{(\nu)}(z) \tag{6.19}$$

$$= \frac{\nu}{\xi^{1/2(n+1)+j} - \xi^{1/2(n+1)+j-\nu}} \int_U \{\frac{A}{c_\nu}\frac{d^\nu}{dv^\nu} g(Rv) - g^{(\nu)}(z)\}$$

$N(v)dv.$

Fixing z in U we see that the terms in the summation in (6.18) are non-zero only when $\mathrm{arc}(z_j, z_{j+n-\nu+1})$ contains z. Also the integrand in (6.19) is non-zero only for v in $\mathrm{arc}(\xi^{1/2(n+1)+j-\nu}, \xi^{1/2(n+1)+j})$. So as $k/n \to \infty$, $|Rv - z|$ is also of order n/k. Recalling Corollary 3.3 and noting that the summation in (6.18) has at most $n + 1$ non-zero terms, we obtain the result. □

Remark 6.3. If f is in $C^\nu(U)$ and we extend f as in (6.11) then $h_\nu(\eta z) = f^{(\nu)}(z)$. So if we define the operator S by (6.12), then the estimate (6.17) takes the simpler form

$$|(Sf)^{(\nu)}(z) - f^{(\nu)}(z)|Cn\{\frac{n}{k}\|f^{(\nu)}\| + w(f^{(\nu)}; \frac{n}{k})\}. \qquad (6.20)$$

Remark 6.4. If $\alpha + \beta = n$, we recall from (6.13) that $S_n^{\alpha,\beta}f$ is defined for functions f defined only on U. In this case we also have the estimate (6.20) for f in $C^\nu(U)$.

7. VARIATION DIMINISHING TRIGONOMETRIC APPROXIMATION

In this final section we shall consider the trigonometic analogues of the approximation operators in §6. As in §3 we set

$$\tau_j = \frac{1}{n}(t_{j+1} + \ldots + t_{j+n}) \qquad (3.2)$$

We first suppose $\underset{\sim}{t}$ is uniform, i.e. $t_j = 2\pi j/k (j \in Z)$. Then for any integer α with $0 \le \alpha < \frac{1}{2}n$ we define the operator $T = T_n^\alpha$ by

$$(T_n^\alpha f)(x) = c_\alpha \sum_{j=\ell-k+1}^{\ell} f(\tau_j)P_j^n(x), \quad \text{whenever} \quad t_\ell \le x < t_{\ell+1}, \qquad (7.1)$$

where c_α is given by (6.9). If the function f satisfies

$$f(x + 2\pi) = (-1)^n f(x) \quad (x \in \mathbb{R}), \tag{7.2}$$

then we see from (3.4) that (7.1) can be rewritten as

$$(T_n^\alpha f)(x) = c_\alpha \sum_{j=1}^{k} f(\tau_j) P_j^n(x) \quad (x \in \mathbb{R}). \tag{7.3}$$

Now for any function f in $C(\mathbb{R})$ satisfying (7.2) we can define a function g in $C(U)$ by

$$g(e^{ix}) = e^{inx/2} f(x). \tag{7.4}$$

Then recalling (3.3) we see that

$$(T_n^\alpha f)(x) = c_\alpha \sum_{j=1}^{k} e^{-in\tau_j/2} g(e^{i\tau_j}) e^{in(\tau_j - x)/2} N_j^n(e^{ix})$$

$$= e^{-inx/2} c_\alpha \sum_{j=1}^{k} g(e^{i\tau_j}) N_j^n(e^{ix}).$$

Recalling (6.13) and noting that $e^{i\tau_j} = \xi^{1/2(n+1)+j}$, we have

$$e^{inx/2} (T_n^\alpha f)(x) = (S_n^{\alpha,n-\alpha} g)(e^{ix}) . \tag{7.5}$$

Equations (7.4) and (7.5) now allow us to deduce results for T_n^α from the corresponding results for $S_n^{\alpha,n-\alpha}$. Firstly we recall that $S_n^{\alpha,n-\alpha}$ reproduces the functions z^α and $z^{n-\alpha}$. Putting $g(z) = z^\alpha$ in (7.4) gives $f(x) = e^{ix(\alpha-n/2}$. Then (7.5) gives

$$e^{inx/2} (T_n^\alpha f)(x) = e^{i\alpha x}$$

and so T_n^α reproduces the function $e^{ix(\alpha-n/2)}$. By putting $g(z) = z^{n-\alpha}$ we can similarly see that T_n^α also reproduces

$e^{ix(n/2-\alpha)}$. Thus we have

Theorem 7.1. *The operator* T_n^α *given by (7.1) reproduces the functions* $\cos(\frac{1}{2}n - \alpha)x$ *and* $\sin(\frac{1}{2}n - \alpha)x$.

From Theorem 6.2 and Remark 6.2 we can deduce

Theorem 7.2. *If* f *in* $C^3(\mathbb{R})$ *satisfies (7.2), then*

$$\lim_{k\to\infty} \frac{k^2}{n+1} \{(T_n^\alpha f)(x) - f(x)\} = \frac{1}{6}\pi^2 \{f''(x) + (\frac{1}{2}n - \alpha)^2 f(x)\}$$

uniformly on $\mathbb{R}$.

From Theorem 6.3 and Remark 6.4 we can deduce

Theorem 7.3. *Fix integers* α, ν *and* n *with* $0 \le \alpha < \frac{1}{2}n$, $0 \le \nu \le n-1$, *and suppose* f *in* $C^\nu(\mathbb{R})$ *satisfies (7.2). Then as* $k \to \infty$ *we have for* x *in* $\mathbb{R}$,

$$|(T_n^\alpha f)^{(\nu)}(x) - f^{(\nu)}(x)| \le C\{\frac{1}{k} \sum_{\ell=0}^{\nu} \|f^{(\ell)}\| + w(f^{(\nu)}; \frac{1}{k})\}$$

where C *is independent of* f *and* k, *and* $\|f^{(\ell)}\| = \sup\{|f^{(\ell)}(x)| : x \in \mathbb{R}\}$.

So far we have assumed that $\underset{\sim}{t}$ is uniform. However, we can allow $\underset{\sim}{t}$ to be non-uniform in the special case $\alpha = 0$, when we write

$$(T_n^0 f)(x) = \sum_{j=\ell-k+1}^{\ell} f(\tau_j)P_j^n(x), \quad \text{whenever} \quad t_\ell \le x < t_{\ell+1}. \tag{7.6}$$

We see from (3.9) and (3.10) that T_n^0 reproduces the functions $\cos\frac{1}{2}nx$ and $\sin\frac{1}{2}nx$. Corresponding to Theorem 6.1 we have

Theorem 7.4. *If* f *is uniformly continuous and bounded on* $\mathbb{R}$, *then as* $n\rho_n(\underset{\sim}{t}) \to 0$, *there is a constant* C *such that*

$$|(T_n^0 f)(x) - f(x)| \le Cn\{\rho_n(\underset{\sim}{t})^2 \|f\| + w(f; \rho_n(\underset{\sim}{t}))\} . \quad (7.7)$$

Proof. From (3.9) and (3.10) we see that for $x \in \mathbf{R}$,

$$1 = \sum_{j=1}^{k} \cos \frac{1}{2} n(\tau_j - x) P_j^n(x)$$

$$= \sum_{j=\ell-k+1}^{\ell} \cos \frac{1}{2} n(\tau_j - x) P_j^n(x)$$

for any ℓ. Thus if $t_\ell \le x < t_{\ell+1}$, then

$$(T_n^0 f)(x) - f(x) = \sum_{j=\ell-k+1}^{\ell} \{f(\tau_j) - \cos \frac{1}{2} n(\tau_j - x) f(x)\} P_j^n(x). \quad (7.8)$$

The terms in the summation in (7.8) are non-zero only for $j = \ell - n, \ldots, \ell$ and for such j, $|x - \tau_j| \le \rho_n(\underset{\sim}{t})$. The result then follows on recalling (3.11). □

Now from (3.8) we see that the operator T_n^α given by (7.1) or (7.6) is a positive operator, i.e. $f(x) \ge 0 (x \in \mathbf{R})$ implies $(T_n^\alpha f)(x) \ge 0 (x \in \mathbf{R})$. We shall show further that T_n^α is variation diminishing on the space $C_{2\pi}$ of all continuous, 2π-periodic functions on $\mathbf{R}$. It is clear that T_n^α maps $C_{2\pi}$ into itself. By 'variation diminishing' we mean in the sense of Schoenberg [14], and we proceed to describe this precisely.

For a real, non-trivial sequence $\underset{\sim}{a} = (a_j)_{-\infty}^{\infty}$ with period p we let $V_p(\underset{\sim}{a})$ denote the number of strong sign changes in the sequence $a_\ell, \ldots, a_{\ell+p}$, where ℓ is any integer with $a_\ell \ne 0$. (Clearly this is independent of ℓ). For a non-trivial function f in $C_{2\pi}$ we define its periodic variation $V_{2\pi}(f)$ to be the supremum of $V_p(f(\underset{\sim}{x}))$ over all $p \ge 1$ and all increasing sequences $\underset{\sim}{x} = (x_j)_{-\infty}^{\infty}$ satisfying $x_{j+p} = x_j + 2\pi$ $(j \in Z)$. To say T_n^α is variation diminishing simply means that the periodic variation of $T_n^\alpha f$ cannot exceed that of f, i.e.

<u>Theorem 7.5.</u> <u>For any function</u> f <u>in</u> $C_{2\pi}$, $V_{2\pi}(T_n^\alpha f) \leq V_{2\pi}(f)$.

<u>Proof.</u> For f in $C_{2\pi}$ we see from (7.1), (3.8) and (3.4) that

$$(T_n^\alpha f)(x) = C_\alpha \sum_{j=1}^{k} f(\tau_j) |P_j^n(x)| \quad (x \in \mathbb{R}).$$

Now it is shown in [6] that for any real, non-trivial sequence $\underset{\sim}{a} = (a_j)_{-\infty}^{\infty}$ with period k,

$$V_{2\pi}(a_1 |P_1^n| + \ldots + a_k |P_k^n|) \leq V_k(\underset{\sim}{a}),$$

where for $j = 1, \ldots, k$, $|P_j^n|(x) = |P_j^n(x)|$ $(x \in \mathbb{R})$. Thus

$$V_{2\pi}(T_n^\alpha f) \leq V_k(f(\tau_j)) \leq V_{2\pi}(f). \qquad \square$$

REFERENCES

1. Ahlberg, J. H., Nilson, E. N., and Walsh, J. L., "Properties of analytic splines, I. complex polynomial splines", J. of Analysis and Appl. 33, 1971, pp. 234-257.

2. de Boor, C., "Splines as linear combinations of B-splines", in Approximation Theory II, (G. G. Lorentz, C. K. Chui and L. L. Schumaker, Eds.), pp. 1-47, Academic Press, New York, 1976.

3. _____ and Fix, G. J., "Spline approximation by quasi-interpolants", J. Approx. Theory 8, 1973, pp. 19-45.

4. Cavaretta, A. S. Jr., Dahmen, W., Micchelli, C. A., and Smith, P. W., "A factorisation theorem for banded matrices", Lin. Alg. Appli. 39, 1981, pp. 229-245.

5. Chen, H. L., "Quasiinterpolant splines on the unit circle", J. Approx. Theory 38, 1983, pp. 312-318.

6. Goodman, T. N. T., and Lee, S. L., "Interpolatory and variation-diminishing properties of generalised B-splines", to appear in Proc. Royal Soc. Edinburgh.

7. _____, Lee, S. L., and Sharma, A., "Approximation by Λ-splines on the circle", to appear.

8. Koch, P. E., and Lyche, T., "Bounds for the error in trigonometric Hermite interpolation", in Quantative Approximation, (R. A. DeVore and K. Scherer, Eds.), pp. 185-196, Academic Press, New York, 1980.

9. Lorentz, G. G., "Bernstein polynomials", Univ. of Toronto Press, 1953.

10. Lyche, T., "A Newton form for trigonometric Hermite interpolation", BIT 19, 1979, pp. 229-235.

11. _____ and Winther, R., "A stable recurrence relation for trigonometric B-splines", J. Approximation Theory 25, 1979, pp. 266-279.

12. Marsden, M. J., "An identity for spline functions with applications to variation diminishing spline approximation", J. Approx. Theory 3, 1970, pp. 7-49.

13. Micchelli, C. A., and Sharma, A., "Spline functions on the circle: cardinal L-splines revisited", Canad. J. Math., 32, 1980, pp. 1459-1473.

14. Schoenberg, I. J., "On variation diminishing approximation methods", in On Numerical Approximation, (R. E. Langer, Ed.), pp. 240-274, Univ. of Wisconsin Press, Madison, 1959.

15. _____, "On trigonometric spline interpolation", J. Math. Mech. 13, 1964, pp. 795-826.

16. _____, "On polynomial spline functions on the circle I, II", in Proceedings of the Conference on Constructive Theory of Functions, pp. 403-433, Budapest, 1972.

17. ____, "Cardinal spline interpolation", CBMS12, SIAM, Philadelphia, 1973.

18. Schumaker, L. L., "Spline functions: basic theory", Wiley, New York, 1981.

19. Tzimbalario, J., "Interpolation by complex splines", Trans. Amer. Math. Soc. 243, 1978, pp. 213-222.

* S. L. Lee wishes to thank the Science and Engineering Research Council for a grant to support this research at the University of Dundee.

ON REDUCING THE COMPUTATIONAL ERROR IN THE SUCCESSIVE APPROXIMATIONS METHOD

Francois B. Guénard

Université de Poitiers

Let f be a continuous map from a compact interval $I = [a, b] \subset \mathbb{R}$ into itself. We note f^n, the n-th iterate of f. We will say that f is an *asymptotic contraction* of I if f has a single fixed point c, towards which converge all the sequences $(f^n(x))_{n\in\mathbb{N}}$, $x \in I$. Let us write $G(f)$, the graph of f. We note: $G_g(f) = G(f|_{[a,c]})$ and $G_d(f) = G(f|_{[c,b]})$. If $A \subset I^2 = I \times I$, we will write

$$A^{-1} = \{(y;\, x) \mid (x;\, y) \in A\}$$

$$\Delta = \{(x,\, x) \mid x \in I\}.$$

If $A \subset I$ or I^2, we will write $\mathring{A}$ the interior of A, and $\overline{A}$ its closure.

It is known that the asymptotic contractions of I of which c is the fixed point, are exactly the continuous mappings $f : I \to I$ such that:

$$G(f) \cap [G(f)]^{-1} = \{(c;\, c)\} \qquad (1) \quad \text{(cf. [1] and [3])}.$$

Our purpose here will be to study the numerical convergence of the sequences $(f^n(x))_{n\in\mathbb{N}}$.

The result above states that if f satisfies condition (1),

S. P. Singh et al. (eds.), Approximation Theory and Spline Functions, 327–338.

the sequences $(f^n(x))_{n\in\mathbb{N}}$ will lead to the fixed point c provided the two following assumptions hold:

- an infinite number of iterates have been computed;
- each iterate has been computed exactly, i.e., with infinite precision.

In practice, both assumptions are false and in the following we will give two models the purpose of which is to represent what really happens in practice.

1. RAMBLES

To "modelize" the error made at the n-th step of the iteration, the idea of the rambles is to say that, instead of f, the function used is an f_{λ_n}.

More precisely: Let $(f_\lambda)_{\lambda\in\Lambda}$ be a family of asymptotic contractions $I \to I$, having c as common fixed point. A *ramble* [associated with the family $(f_\lambda)_{\lambda\in\Lambda}$] is a mapping $B: \mathbb{N} \times I \to I$, $x \mapsto B(n, x) = H_n(x)$ such that there exists a sequence $(\lambda_n)_{n\in\mathbb{N}} \in \Lambda^{\mathbb{N}}$ verifying:

$$H_0 = Id_I,\ \forall n \in \mathbb{N}^*,\ H_n = f_{\lambda_n} \circ H_{n-1}.$$

In [4] was published a necessary and sufficient condition for the family $(f_\lambda)_{\lambda\in\Lambda}$ to have the property that, for any $x \in I$, all the rambles converge. In order to be able to estimate the speed of convergence, we need a result stating uniform convergence instead of pointwise convergence. The following theorem suits our purpose . The notations are the ones introduced above, and d designates the usual euclidean distance.

<u>Theorem 1</u>. Assume there exist two asymptotic contractions $\overline{f}$ and $\underline{f}$, of which c is the fixed point, and such that:

- For each compact set $K \subset (I \setminus \{c\})$, there exists $\varepsilon_K \in \mathbb{R}_+^*$ such that:

$$\begin{cases} d(\Delta;\ G(\overline{f}_{|K}) \cup G(\underline{f})_{|K})) \geq \varepsilon_K \\ \text{and} \\ d(G(\overline{f})^{-1};\ G(\underline{f}_{\ K})) \geq \varepsilon_K \end{cases}$$

- For each $\lambda \in \Lambda$ and each $x \in I$, $\overline{f}(x) \geq f_\lambda(x)$ and $f(x) \leq f_\lambda(x)$.

Then

$\forall K$ compact $\subset I \setminus \{c\}$, $\exists n \in \mathbf{N}$, $\forall m \in \mathbf{N}$, $m \geq n$, $\forall x \in I$,

$\forall (h_1, \ldots, h_m) \in \{f_\lambda \mid \lambda \in \Lambda\}^m$,

$H_m(x) = h_m o \ldots o h_1(x) \notin K$.

In other words, the "heart" of the ramble is equal to $\{c\}$ and there is convergence towards c, uniformly with regards to x and H.

It follows from theorem 2 of [4] that all the rambles associated with $\overline{f}$ and $\underline{f}$ converge towards c.

Proof. Let us write $G(I)$ the set of subintervals of I. Let L be the multi-application $[a; b] \mapsto G(I)$, $x \to [\underline{f}(x); \overline{f}(x)]$. Consider the relation:

$$\forall\ \ \in \mathbf{R}_+^*,\ \exists n \in \mathbf{N},\ \forall m \in \mathbf{N},\ \forall x \in [a; b],\ \forall y \in [a; b],$$

$$[m \geq n \text{ and } y \in L^m(x)] \Rightarrow [d(y; c) \leq \varepsilon].$$

This relation implies the convergence of the rambles towards c, uniformly with regards to x and the realizations of rambles. Therefore, that is what we will prove.

Step 1. For each interval $[\alpha; \beta] \subset [a; b]$, $L([\alpha; \beta])$ is a compact interval $\subset [a; b]$.

This follows from the equality:

$$L([\alpha; \beta]) = [\inf\{\underline{f}(x) \mid x \in [\alpha; \beta]\};\ \sup\{\overline{f}(x) \mid x \in [\alpha; \beta]\}].$$

Step 2. The heart of L, $\mathcal{C}(L) = \bigcap_{n\in\mathbb{N}} L^n([a;b])$, is a compact interval which contains c, which is stable $(L(\mathcal{C}(L)) \subset \mathcal{C}(L))$ and which is stationary (i.e. $L(\mathcal{C}(L)) = \mathcal{C}(L)$).

Indeed the inclusion $L([a;b]) \subset [a;b]$ leads to: $L(L([a;b])) \subset L([a;b])$ etc.

Furthermore, c is a fixed point of L, and step 1 shows that for each $n \in \mathbb{N}$, $L^n([a;b])$ is a compact interval. Therefore, $\mathcal{C}(L)$ is a compact interval which contains c.

Let us show the stationarity. Let $x \in \mathcal{C}(L)$. For each $n \in \mathbb{N}$, there exists $x_n \in L^n(I)$ such that: $x \in L(x_n)$. Let $(x_{j_p})_{p\in\mathbb{N}}$ be a convergent sub-sequence of the sequence (x_n), and let y be the limit of the sequence $(x_{j_p})_{p\in\mathbb{N}}$. We have:

$$\begin{cases} L(y) = [\underline{f}(y); \overline{f}(y)] & \text{(from the definition of } L) \\ L(x_{j_p}) = [\underline{f}(x_{j_p}); \overline{f}(x_{j_p}) & \text{(again from the definition of L)} \\ x \in L(x_{j_p}). \end{cases}$$

From this, we derive:

$$x \geq \underline{f}(x_{j_p}) \quad \text{and} \quad x \leq \overline{f}(x_{j_p}).$$

Taking the limits, since $\underline{f}$ and $\overline{f}$ are continuous, we get:

$$x \geq \underline{f}(y) \quad \text{and} \quad x \leq \overline{f}(y)$$

and, at last: $x \in L(y)$.

Together with the relation $y \in \mathcal{C}(L)$, this last inclusion establishes the stationarity of $\mathcal{C}(L)$.

Step 3. $\{c\}$ is the only compact interval which contains c and which is stable and stationary for L.

Indeed, assume for example that $[\alpha; \beta]$ is such an interval. There exist z and t such that:

$\alpha \in L(z)$ and $\beta \in L(t)$,

which implies

- on the one hand: $t \in [\alpha; c[$ and $z \in]c; \beta]$

- and on the other hand: $\alpha = \underline{f}(z)$ and $\beta = \overline{f}(t)$.

Therefore in the square $[\alpha, \beta]^2$, one has the inclusions:

$$\begin{cases} (\alpha, z) \in (G(\underline{f}) \cap [(\alpha, \alpha); (\alpha, \beta)]) \\ (\beta, t) \in (G^{-1}(\overline{f}) \cap [(\beta, \alpha); (\beta, \beta)]). \end{cases}$$

The arc $G_d(f)$ divides the triangle whose vertices are $((\alpha, \alpha); (\beta, \alpha); (\beta, \beta))$ in disjoint open sets in such a way that:

- the segments $[(\alpha, \alpha); (z, \alpha)]$ and $[(\beta, \underline{f}(\beta)); (\beta, \beta)]$ belong to disjoint connected componenets;

  the point (c, c) and the path $[(z, \alpha); (\beta, \alpha)] \cup [(\beta, \alpha); (\beta, \underline{f}(\beta))]$ belong to disjoint connex components.

Hence, the path $G_g^{-1}(\overline{f})$ cannot contain the three points (c, c), $(\overline{f}(\alpha), \alpha)$ and (β, t) without intersecting $G_d(\underline{f})$. This contradicts the theorem 2 of [4] which implies that all the rambles with $\overline{f}$ and $\underline{f}$ converge towards c.

Step 4 - Conclusion of proof. Step 2 and 3 imply that $C(L) = \{c\}$. Let $\varepsilon \in \mathbb{R}_+^*$. The sequence

$$(([a; c - \varepsilon] \cup [c + \varepsilon; b]) \cap L^n(I))_{n \in \mathbb{N}}$$

is a decreasing sequence of compact sets whose intersection is empty. So, there exists $n \in \mathbb{N}$ such that:

$$L^n(I) \cap (I \backslash]c - \varepsilon; c + \varepsilon[) = \emptyset.$$

As the sequence $(L^n(I))_{n \in \mathbb{N}}$ is decreasing, we deduce:

$$\forall m \in \mathbb{N}, \forall x \in I, \forall y \in I,$$

$$[m \geq n \text{ and } y \in L^m(x)] \Rightarrow [|y - c| \leq \varepsilon]$$

which concludes the proof. □

We will not go further in the study of this first model, but we will use the theorem in the second one.

2. THE FIRST ORDER GENERAL AUTOREGRESSIVE PROCESS

This process is defined by:

$$X_o \in I \quad \text{and} \quad \forall n \in \mathbb{N},\ X_{n+1} = f(X_n) + \tau_n,$$

where $(\tau_n)_{n\in\mathbb{N}}$ is a sequence of independent equidistributed random variables. We call F the law of the τ_ns, $S = [-\gamma, \delta]$ $(\gamma > 0,\ \delta > 0)$ the support of F, and we assume: $F(\{-\gamma\}) = F(\{\delta\}) = 0$. We extend f on $\mathbb{R}$ by: $f(x) = f(a)$ if $x \leq a$ and $f(x) = f(b)$ if $x \geq b$. Let $L =]\alpha; \beta[$ be the smallest interval K containing c and such that:

$$f(K) + \overset{\circ}{S} = K.$$

In [4], we proved that L is an absorbing set for the Markov process (X_n). Here, we go further by stating the ergodicity of the process restricted to L and giving a statistical estimator. Starting from corollary 3 in [4], the sketch of the proof is the following:

Lemma 2. Let ν be a finite measure on S with respect to which the Lebesgue measure λ is absolutely continuous: $\lambda \ll \nu$. Let C be a Borel set of S satisfying $\lambda(C) > 0$.

Then, there exists $\rho > 0$ and $\sigma > 0$ such that:

$$\forall y \in [-\rho; \rho],\quad \nu(C + y) > \sigma.$$

Lemma 3. Let ν be a finite measure on S with respect to which the Lebesgue measure λ is absolutely continuous. Let J be an interval of S, and C be a Borel set of S satisfying: $\lambda(J \cap C) > 0$.

Then, there exists $\rho > 0$ and $\sigma > 0$ such that:

$$\forall y \in [-\rho; \rho],\quad \nu(J \cap (C + y)) > \sigma > 0.$$

Let us define a sequence $(K_n)_{n\in\mathbb{N}}$ of subsets of $I + S$ by:

$$K_o = C + \overset{\circ}{S}$$

$$\forall n \in \mathbb{N},\ K_{n+1} = f(K_n) + \overset{\circ}{S}.$$

Lemma 4. The sequence $(K_n)_{n\in\mathbb{N}}$ is an increasing sequence of open intervals.

Lemma 5. $L = \bigcup_{n\in\mathbb{N}} K_n$.

Lemma 6. For each $x \in L$ and each Borel set $C_o \subset K_o$ verifying: $\lambda(C_o) > 0$, there exists a positive integer n such that:

$$\mathbb{P}_n(x, C_o) > 0$$

where $\mathbb{P}_n(x, C_o)$ is the probability that $X_n \in C_o$ if $X_o = x$ (i.e. $\mathbb{P}_n$ is the n-th transition probability of the process).

The proof of Lemma 6 uses theorem 1.

Lemma 7. For each $n \in \mathbb{N}$, and each Borel set $C \subset K_{n+1}$ such that $\lambda(C) > 0$, there exists $\rho > 0$ and an interval $M \subset K_n$ of positive length satisfying

$$\forall x \in M,\ \mathbb{P}(x, C) > \rho.$$

Theorem 8. For each $x \in L$ and each Borel set $C \subset L$, with $\lambda(C) > 0$, there exists an integer $n \in \mathbf{N}$ such that: $\mathbb{P}_n(x, C) > 0$. In other words, the restriction of the Markov process (X_n) on L is λ-irreducible.

Theorem 8 allows to apply Tweedie's theorems [5] which lead to:

Corollary 9. The process restricted to L is λ-irreducible, aperiodic, λ-recurrent and ergodic.

Furthermore, the unique sub-invariant probability is invariant. At last, if the law F is equivalent to λ, so is this invariant probability.

This corollary ends the probabilistic study. We now move on the statistical viewpoint.

Assume K is a kernel $\mathbb{R} \to \mathbb{R}_+$. The function K is measurable, has an integral equal to 1, is symmetric, and bounded.

We now assume that, in addition to the properties mentioned above, f belongs to the class C^i, $i = 1$ or 2, i.e. f is i times derivable, with a continuous i-th derivative.

We also assume that τ_n has a density ϕ which is strictly positive on $\mathring{S}$ and belongs to the class C^i on $\mathbb{R}$.

Let $(h_n)_{n \in \mathbb{N}}$ be a sequence of elements of $\mathbb{R}_+^*$. Given a partial realization $X_1, \ldots, X_n$ of the process (an observation), and $x \in I$, we set:

$$D_n(x) = \frac{1}{n \cdot h_n} \sum_{s=0}^{n-1} K\left(\frac{X_s - x}{h_n}\right)$$

$$f_n(x) = \frac{\sum_{s=0}^{n-1} X_{s+1} \cdot K\left(\frac{X_s - x}{h_n}\right)}{\sum_{s=0}^{n-1} K\left(\frac{X_s - x}{h_n}\right)} .$$

Lemma 10. Let K be a compact interval $\subset \mathring{L}$. Then, the density D of the invariant probability Δ satisfies:

$$\mathrm{Inf}\{D(y) \mid y \in K\} > 0 .$$

Theorem 11. Under the former assumptions, the four following assertions hold, for each integer $p > 0$:

1) $$\sup_{x \in X} \mathbb{E}(|f_n(x) - f(x)|^{2p}) = 0\left(\frac{1}{(n \cdot h_n)^p} + h_n^{2ip}\right)$$

2) $$\sup_{x \in \mathbb{R}} E(|D_n(x) - D(x)|^{2p}) = 0\left(\frac{1}{(n \cdot h_n)^p} + h_n^{2ip}\right)$$

3) These risks are optimized by $0(n^{-2ip/(1+2i)})$ when $h_n \sim n^{-1/(1+2i)}$.

4) f_n (resp. D_n) converges a.s. towards f (resp. D).

3. CONCLUSION AND USER'S MANUAL

Let us now conclude with the practical meaning of the former theorems in order to improve the numerical convergence of a successive approximations scheme.

Example of Interpretation of the Results

The numerical values given hereafter are fictitions and only intend to illustrate the ideas.

Consider a computer attempting to evaluate the fixed point, c, of an asymptotic contraction $f : I \to I$, by using the process:

$$x_o \in I, \quad \forall n \in \mathbb{N}, \quad x_{n+1} = f(x_n) .$$

Each elementary operation is performed with a precision of $\varepsilon_1 = 10^{-10}$. If the function f is complicated enough (otherwise our study is of no use), the absolute value of the error in computing $f(x_n)$ is bounded by 10^{-7}. Here, we have $\gamma = \delta = 10^{-7}$. Corollary 3 in [4] states that, ultimately, the sequence (x_n) is in L. The construction of L given by Lemmas 4 and 5 asserts that the length of the interval L can be much greater than the error on the computation of $f(x_n)$ (cf. fig. 1). Let for example $10^{-5} = \lambda(L)$ be that length.

Theorem 8 and corollary 9 state that the sequence (x_n) is almost surely dense on L: that means that in our example, individual operations performed with a precision of 10^{-10} lead to an estimation of c whose precision is of the order of 10^{-5}. And if the iterative process is not modified, there is no way to improve the convergence of the x_ns towards c. More precisely, if $10^{-5} = \lambda(L)$, for any $\varepsilon \in \mathbb{R}_+^*$ and any $N \in \mathbb{N}$, there almost surely exists an $n \in \mathbb{N}$, $n > N$ such that:

$$|x_n - c| > \frac{1}{2} (10^{-5} - \varepsilon) .$$

However, the erratic behavior of the individual x_n is counterbalanced by the global behavior of the sequence (x_n) which the corollary 9 states to be ergodic. Using the estimators given by theorem 11, one is able from a sample $(x_1, \ldots, x_n)$ of the process to get an estimation of f which has or high probability of being more accurate than $\lambda(L)$, for example $10^{-8} = \varepsilon_2$. From this estimation of f, it is easy to get an estimation of c of the same order of accuracy (e.g. using Newton's method).

Let us note the inequalities:

$\varepsilon_1 < \varepsilon_2 < \gamma = \delta < \lambda(L)$.

It seems impossible to get $\varepsilon_2 < \varepsilon_1$. However, theorem 11 gives the convergence of the estimator towards f. This means that, for any $\rho \in]0; 1[$ and any $\varepsilon_2 > \varepsilon_1$, provided n is large enough, with a probability $>1 - \rho$, one has: $|f_n - f| < \varepsilon_2$.

Here is the procedure to follow in order to improve the convergence.

1) Assure that the computer truncates each individual operation by choosing either the next higher or next lower admissible value, and that this choice between these two values is made stochastically with probability $\frac{1}{2}$ for each one, and that all the choices are mutually independent. This can be obtained from the programme, or by using some technical devices which have been patented recently.

2) Begin the computation of the terms of the sequence. Stop when the process enters L (this can be detected by the oscillation of the process).

3) Start from the last computed term. Let x_1 be this term. Compute n terms $x_1, x_2, \ldots, x_n$, where n is deduced from theorem 11. Store all of them.

4) Compute the fixed point c_n of f_n given by the formula before Lemma 10.

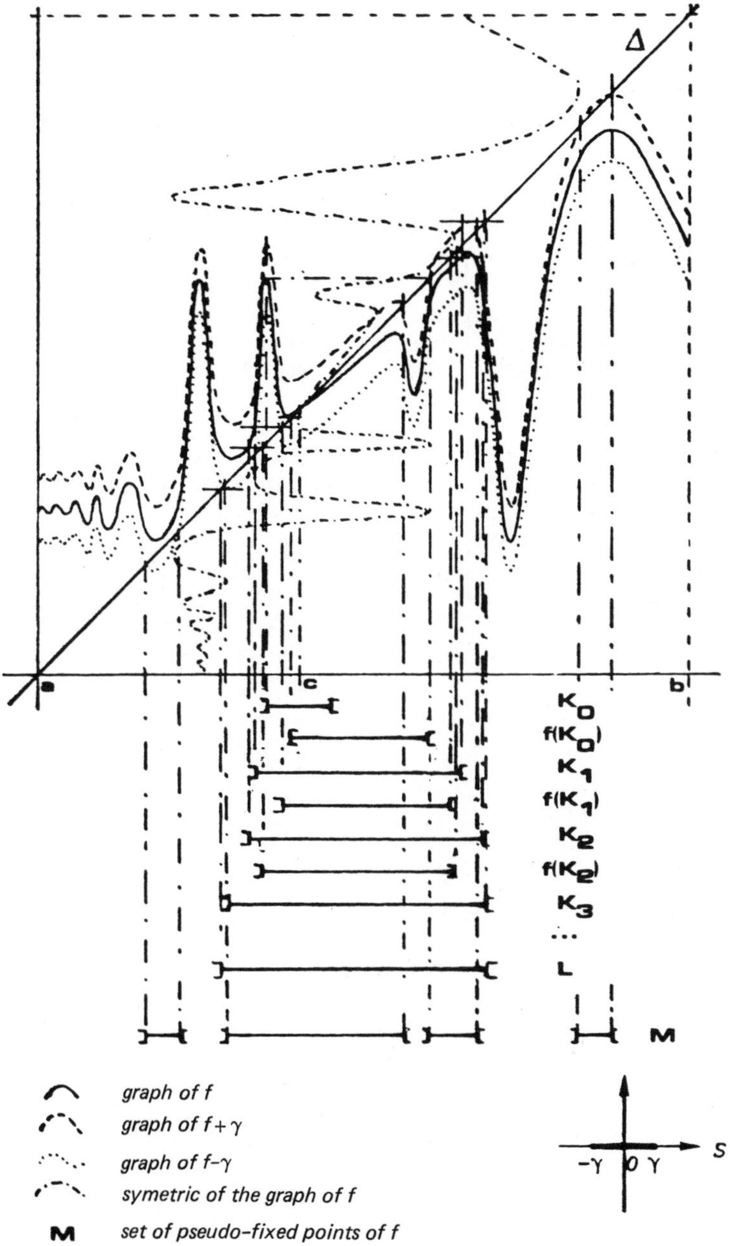

Δ
a
c
b
K_0
$f(K_0)$
K_1
$f(K_1)$
K_2
$f(K_2)$
K_3
...
L
M
graph of f
graph of f + γ
graph of f−γ
symetric of the graph of f
M set of pseudo-fixed points of f
−γ 0 γ
S

REFERENCES

1. Bashurov, V. V., and Ogibin, V. N., "Conditions for the convergence of iterative processes on the real axis", U.S.S.R. Comp. Math. and Math. Phys. 6, 5, 1966, pp. 178-184.

2. Doukhan, P., and Portal, F., "Moments de variables aléatoires mélangeantes", C. R. Acad. Sc. Paris, Série I, 297, 1983, pp. 129-132.

3. Guénard, F., "Caractérisations des contractions asymptotiques et des applications itérativement convergentes des intervalles", C. R. Acad. Sc. Paris, Série I, 292, 1981, pp. 55-58.

4. Guénard, F., "Itérations déterministes et stochastiques sur les intervalles", in Nonlinear Analysis and Applications, Singh and Burry ed., Lect. Notes, vol. 80, Dekker pub., New York, 1982.

5. Orey, S., "Limit Theorems for Markov Chain Transition Probabilities", Van Nostrand Reinhold Mathematical Studies, 34, London, 1971.

LEBESGUE CONSTANTS DETERMINED BY EXTREMAL SETS

Myron S. Henry

Central Michigan University

1. INTRODUCTION

Let $-1 \leq x_0^n < x_1^n < \dots < x_n^n < x_{n+1}^n \leq 1$ be $n + 2$ points in the interval $I = [-1, 1]$. Then

$$X_n = \{x_i^n\}_{i=0}^{n+1} \tag{1.1}$$

defines a set of nodes in I, and

$$X = \{X_n\}_{n=0}^{\infty} \tag{1.2}$$

is an infinite triangular array of nodes [9, p. 88]. Let

$$\{\ell_i^{(n+1)}(x)\}_{i=0}^{n+1} \tag{1.3}$$

be the fundamental Lagrange polynomials determined by X. The Lebesgue function determined by the n-th row of X (here we are using the convention that the n-th now contains $n + 2$ elements) is then

$$\lambda_n(X, x) = \sum_{i=0}^{n+1} |\ell_i^{(n+1)}(x)|, \tag{1.4}$$

and the Lebesgue constant determined by the n-th row of X is

S. P. Singh et al. (eds.), Approximation Theory and Spline Functions, 339–348.

defined to be

$$\Lambda_{n+1}(X) = \max_{I} \lambda_{n+1}(X, x). \tag{1.5}$$

Let $C(I)$ denote the space of real-valued, continuous functions on I, and designate the uniform norm by $\|\cdot\|$. Let $\pi_n \subseteq C(I)$ be the space of real polynomials of degree at most n.

Given $f \in C(I)$, the set of nodes X_n (the n-th row of X) determines a Lagrange interpolant to f; that is, a polynomial $Pf \in \pi_{n+1}$ that interpolates f at each node contained in X_n. The Lagrange interpolant to f on I is given by

$$(Pf)(x) = \sum_{i=0}^{n+1} f(x_i)\ell_i^{(n+1)}(x). \tag{1.6}$$

Let the distance between the function f and the subspace π_{n+1} be denoted by $\mathrm{dist}(f, \pi_{n+1})$. Then it can be shown [2, 7] that

$$\|f - Pf\| \leq [1 + \Lambda_{n+1}(X)]\mathrm{dist}(f, \pi_{n+1}); \tag{1.7}$$

thus infinite triangular arrays of nodes whose Lebesgue constants are of optimally small asymptotic order (hereafter referred to as optimal asymptotic order) have long been of interest. A principal objective of the current paper is to estimate $\Lambda_{n+1}(X)$ as a function of n and X for certain infinite triangular arrays of nodes.

Perhaps the most familiar infinite triangular array of nodes is the array T whose n-th row (again using the convention that the n-th row of T contains $n + 2$ elements) consists of the $n + 2$ zeros of C_{n+2}, the $(n + 2)$nd degree Chebyshev polynomial. The array T yields a Lebesgue constant whose asymptotic order is optimal ($\log(n+1)$, [9, p. 91]). Although $\Lambda_{n+1}(T)$ is not equal to

$$\min_{X} \Lambda_{n+1}(X), \tag{1.8}$$

Brutman [1, 2] has shown that if $\hat{T}$ is the infinite triangular array of nodes whose n-th row consists of the zeros of C_{n+2}, adjusted to I so that the first and last zero coincide with the end points of I, then $\Lambda_{n+1}(\hat{T})$ is very close to (1.8).

Although the infinite triangular array of nodes that yields (1.8) has yet to be discovered, DeBoor and Pinkus [2] and Kilgore [7] have succeeded in proving the long standing Bernstein conjecture dealing with the characterization of the optimal infinite triangular array of nodes (that is, the array that yields (1.8)). For further details on optimal nodes, interested readers are referred to [2, 7].

Neither the n-th row of T, nor the n-th row of $\hat{T}$ are extremal sets resulting from identifiable best approximation problems.

In what follows, we will focus on infinite triangular arrays of nodes whose n-th rows are extremal sets resulting from certain best approximation problems.

2. LEBESGUE CONSTANTS FROM BEST APPROXIMATIONS TO POLYNOMIALS

Let C_{n+1} be the Chebyshev polynomials of degree $n + 1$. If $g_n(x) = x^{n+1}$ and $B_n(g_n) \in \pi_n$ is the best uniform approximation to g_n on I, then it is well known that the error function

$$e_n(g_n)(x) = x^{n+1} - B_n(g_n)(x), \ x \in I, \tag{2.1}$$

satisfies

$$e_n(g_n)(x) = (1/2^n)C_{n+1}(x), \ x \in I. \tag{2.2}$$

The set of extreme points $E_n(g_n)$ of the error function is defined by $E_n(g_n) = \{x \in I: |e_n(g_n)(x)| = ||e_n(g_n)||\}$, thus

$$G = \{E_n(g_n)\}_{n=0}^{\infty} \tag{2.3}$$

is the infinite triangular array of nodes whose n-th row consists of the $n + 2$ extreme points of C_{n+1}. Ehlich and Zeller [3] have shown that the Lebesgue constant determined by the n-th row of G satisfies $\Lambda_{n+1}(G) = O(\log(n + 1))$. Thus $\Lambda_{n+1}(G)$ is of optimal asymptotic order.

A natural extension to the best approximation problem (2.1) results in an error function that defines the $(n + 2)$nd degree classical Zolotareff polynomial. More specifically, let

$$h_n(x) = x^{n+2} - \gamma_n x^{n+1}, \quad x \in I, \tag{2.4}$$

where

$$0 \le \gamma_n \le (n + 2)\tan^2(\pi/2(n + 2)). \tag{2.5}$$

If $B_n(h_n) \in \pi_n$ is the best uniform approximation to h_n, and if $e_n(h_n)(x) = h_n(x) - B_n(h_n)(x)$, $x \in I$, then it can be shown [8, p. 41] that the error function satisfies

$$e_n(h_n)(x) = (1/2^{n+1})(1+\gamma_n/(n+2))^{n+2} C_{n+2}\left(\frac{x-\gamma_n/(n+2)}{1+\gamma_n/(n+2)}\right), \tag{2.6}$$

$x \in I$.

The right side of (2.6), designated by Z_{n+2}, defines the $(n + 2)$nd degree Zolotareff polynomial [5, 8]. If $\gamma_n > 0$, then there are precisely $n + 2$ extreme points

$$-1 = x_0^n < x_1^n < \dots < x_n^n < x_{n+1}^n \le 1 \tag{2.7}$$

of (2.6) in the interval I, [5, 8]. Furthermore, if $-1 = t_0^{n+2} < t_1^{n+2} < \dots < t_{n+1}^{n+2} < t_{n+2}^{n+2} = 1$ are the $n + 3$ extreme points of C_{n+2}, then for $i = 0, 1, \dots, n + 1$,

$$t_i^{n+2} = \frac{x_i^n - \gamma_n/(n + 2)}{1 + \gamma_n/(n + 2)}. \tag{2.8}$$

For $\gamma_n > 0$ in (2.5), let

$$E_n(h_n) = \{x_i^n\}_{i=0}^{n+1} , \tag{2.9}$$

and let

$$H = \{E_n(h_n)\}_{n=0}^{\infty} .$$

Then H is an infinite triangular array of nodes of the type given by (1.2), and consequently the n-th row of H determines a Lebesgue constant $\Lambda_{n+1}(H)$. Considering the similarities between the best approximation problems that result in the error functions (2.2) and (2.6), and taking into account that both error functions involve Chebyshev polynomials, one might be tempted to conjecture that $\Lambda_{n+1}(G)$ and $\Lambda_{n+1}(H)$ have the same asymptotic order. However, the conclusions of Theorem 1 below assert that the asymptotic orders of $\Lambda_{n+1}(G)$ and $\Lambda_{n+1}(H)$ are actually very different. The proof of Theorem 1 is fairly complex and appears in [5].

Theorem 1. Assume γ_n in (2.5) is positive. Then there exists positive constants α and β not depending on n such that

$$\alpha < \Lambda_{n+1}(H)/n < \beta, \; n = 1, 2, \ldots . \tag{2.10}$$

Theorem 1 states that the Lebesgue constant $\Lambda_{n+1}(H)$ is of precise asymptotic order n. Hence, the asymptotic order of $\Lambda_{n+1}(H)$ is not optimal.

In the next section, we continue our examination of best approximation problems and infinite triangular arrays of nodes whose n-th rows are extremal sets of the error functions of these best approximation problems. The Lebesgue constants determined by the n-th rows of the infinite triangular arrays analyzed in the next section will be of optimal asymptotic order.

3. LEBESGUE CONSTANTS FROM BEST APPROXIMATIONS TO NON-POLYNOMIAL FUNCTIONS

For $f \in C(I)$ with best approximation $B_n(f) \in \pi_n$, let

$$e_n(f)(x) = f(x) - B_n(f)(x), \; x \in I. \tag{3.1}$$

The set of extreme points $E_n(f)$ of $e_n(f)$ is denoted by

$$E_n(f) = \{x \in I: |e_n(f)(x)| = \|e_n(f)\|\}. \tag{3.2}$$

Now let

$$W = \{w_n\}_{n=0}^{\infty} \subseteq C(I). \tag{3.3}$$

Suppose for each n that $E_n(w_n)$ contains precisely $n + 2$ points. Let

$$\hat{W} = \{E_n(w_n)\}_{n=0}^{\infty}. \tag{3.4}$$

Then $\hat{W}$ is an infinite triangular array of nodes of the type given in (1.2). Therefore, the n-th row of $\hat{W}$ determines a Lebesgue constant

$$\Lambda_{n+1}(\hat{W}). \tag{3.5}$$

In the remainder of this paper, we will focus on Lebesgue constants generated (in the sense described by (3.4) and (3.5)) by a certain class of rational functions and on Lebesgue constants generated by a certain class of smooth, non-rational functions.

Suppose that

$$\{a_n\}_{n=0}^{\infty} \tag{3.6}$$

is a (possibly unbounded) sequence of numbers satisfying

$$a_n \geq 2,\ n = 0, 1, \ldots . \tag{3.7}$$

Let

$$r_n(x) = 1/(a_n - x),\ x \in I. \tag{3.8}$$

Then $r_n^{(n+1)}(x) > 0$ for $x \in I$, and consequently, $E_n(r_n)$ contains precisely $n + 2$ points. If $R = \{r_n\}_{n=0}^{\infty}$, and if $\hat{R} = \{E_n(r_n)\}_{n=0}^{\infty}$, then the n-th row of the infinite triangular array $\hat{R}$ determines a Lebesgue constant $\Lambda_{n+1}(\hat{R})$.

Theorem 2. Let $\{a_n\}_{n=0}^{\infty}$ satisfy (3.7), and let r_n be defined as in (3.8). If $\hat{R} = \{E_n(r_n)\}_{n=0}^{\infty}$, then the Lebesgue constant determined by the n-th row of $\hat{R}$ is of optimal asymptotoc order $\log(n + 1)$. That is, there exist positive constants α and β not depending on n such that

$$\alpha < \Lambda_{n+1}(\hat{R})/\log(n + 1) < \beta, \ n \geq 2. \tag{3.9}$$

The proof of this theorem, which depends on a series of lemmas, will appear elsewhere [6].

Corollary 1. Let the n-th row of the infinite triangular array of nodes A be given by

$$A_n = \{t_0^n, t_1^n, \ldots, t_{n+1}^n\},$$

where $t_0^n = -1$, $t_{n+1}^n = 1$, and $\{t_i^n\}_{i=1}^n$ are the zeros of

$$n(a_n^2 - 1)^{1/2} C_n(x) + (a_n x - 1)C_n'(x) = 0. \tag{3.10}$$

Then $\Lambda_{n+1}(A)$ is of optimal asymptotic order $\log (n + 1)$.

Proof. From [6], $A_n = E_n(r_n)$. Therefore $A = \hat{R}$, and the conclusion follows from Theorem 2. □

Example 1. In (3.6), let $a_n = a$, $n = 0, 1, \ldots,$ where $a \geq 2$. Then (3.8) becomes $r(x) = 1/(a - x)$, and from Corollary 1, the extreme points $E_n(r)$ of the error function $e_n(r)$ satisfy $(1 - x^2)[n(a^2 - 1)^{1/2} C_n(x) - (ax - 1)C_n'(x)] = 0$. The Lebesgue constant determined by the n-th row of $\hat{R} = \{E_n(r)\}_{n=0}^{\infty}$ is of optimal asymptotic order $\log (n + 1)$.

Example 2. Let $\overline{\alpha} \geq \overline{\beta} > 0$ be positive constants not depending on n, and let

$$\begin{aligned} U_n(x) &= 1/(\overline{\alpha}(n + 2) + 2 - x), \ x \in I, \text{ and} \\ V_n(x) &= 1/(\overline{\beta}(n + 2) - 2 - x), \ x \in I. \end{aligned} \tag{3.11}$$

where n is large enough $(n \geq N)$ to insure that V_n does not vanish on I. Let $E_n(U_n)$ and $E_n(V_n)$ be the extreme points of the error functions $e_n(u_n)$ and $e_n(v_n)$, respectively. If $\hat{U} = \{E_n(U_n)\}_{n=N}^{\infty}$ and $\hat{V} = \{E_n(V_n)\}_{n=N}^{\infty}$, then the Lebesgue constants $\Lambda_{n+1}(\hat{U})$ and $\Lambda_{n+1}(\hat{V})$ determined by the n-th rows of $\hat{U}$ and $\hat{V}$ are of optimal asymptotic order $\log(n+1)$.

The rational functions U_n and V_n play key roles in the proof of Theorem 3 below. The final objective of the present paper is to present a class of non-rational functions with error functions (3.1) whose extreme sets (3.2) generate Lebesgue constants with optimal asymptotic orders.

Defintion 1. Let F be the set of all functions $f \in C^{\infty}(I)$ satisfying:

(a) $f^{(n+1)}(x) \neq 0$ on I,

and

$$\text{(b)} \quad 1/\alpha \leq \frac{f^{(n+2)}(x)}{f^{(n+1)}(x)} \leq 1/\beta \tag{3.12}$$

for all n sufficiently large, where $\alpha \geq \beta > 0$ are constants possibly depending on f, but not on n.

We observe that $f_\delta(x) = e^{\delta x}$, $\delta \neq 0$, is an element of F. Strong unicity constants for functions $f \in F$ are analyzed in [4], and a number of properties of F are noted in that reference. We observe (a) of Definition 1 implies that the cardinality of $E_n(f)$ is $n + 2$. The final theorem of this paper now follows.

Theorem 3. Let F be the class of functions given by Definition 1, and let E_f be the infinite triangular array of nodes whose n-th row is $E_n(f)$. Then the Lebesgue constant $\Lambda_{n+1}(E_f)$ generated by the n-th row of E_f is of optimal asymptotic order $\log(n+1)$.

Although the proof of this theorem will also appear elsewhere [6], it is worth repeating that properties of the rational functions U_n and V_n from (3.11) are instrumental to the proof of Theorem 3. More specifically, if $E_n(U_n) = \{u_0^n, u_1^n, \ldots, u_{n+1}^n\}$, and if $E_n(V_n) = \{v_0^n, v_1^n, \ldots, v_{n+1}^n\}$, then

$$\max_{1 \le k \le n} |u_k^n - v_k^n| = 0(\frac{1}{n^2}). \qquad (3.13)$$

This distance is to be contrasted with the maximum distance between corresponding extreme points of C_{n+1} and C_n, which is $0(\frac{1}{n})$. The extra degree of closeness demonstrated in (3.13) is essential to the proof of Theorem 3, [4, 6].

4. OBSERVATIONS AND CONCLUSIONS

In the preceding sections, Lebesgue constants generated by the rows of certain infinite triangular arrays of nodes are examined. The rows of these infinite triangular arrays are extreme points of error functions arising from best approximation problems. Except in the Zolotareff polynomial case, all of the Lebesgue constants considered turn out to be of optimal asymptotic order. For more details, the interested reader is referred to references [5, 6].

REFERENCES

1. Brutman, L, "On the Lebesgue function for polynomial interpolation", SIAM J. Numer. Anal. 15, 1978, pp. 694-704.

2. DeBoor, C., and Pinkus, A., "Proof of conjectures of Bernstein and Erdos concerning optimal nodes for polynomial interpolation", J. Approx. Theory 24, 1978, pp. 289-303.

3. Ehlich, H., and Zeller, K., "Auswertung der Normen von Interpolations-operatoren", Math. Ann, 164, 1966, pp. 105-112.

4. Henry, M. S., Swetits, J. J., and Weinstein, S. E., "On extremal sets and strong unicity constants for certain C^∞ functions", J. Approx. Theory, 37, 1983, pp. 155-174.

5. Henry, M. S., and Swetits, J. J., "Lebesgue and strong uniticy constants for Zolotareff polynomials", Rocky Mountain Journal of Mathematics, 12, 1982, pp. 547-556.

6. Henry, M. S., and Swetits, J. J., "Lebesgue constants for certain classes of nodes", J. Approx. Theory, 39, 1983, pp. 211-227.

7. Kilgore, T., "A characterization of the Lagrange interpolating projection with minimal Tchebycheff norm", J. Approx. Theory 24, 1978, pp. 273-288.

8. Meinardus, G., "Approximation of functions, theory and numerical methods", Springer-Verlag, New York-Berlin, 1967.

9. Rivlin, T. J., "An introduction to the approximation of functions", Ginn (Blaisdell), Boston, 1969.

10. Rivlin, T. J., "The Chebyshev Polynomials", Wiley-Interscience New York, 1974.

ERROR BOUNDS FOR INTERPOLATION BY FOURTH ORDER TRIGONOMETRIC SPLINES

P. E. Koch

ABSTRACT

The algorithm for solving interpolation problems with cubic splines is extended to fourth order generalized trigonometric splines. The attained tridiagonal linear system will be strictly diagonally dominant if the partition is sufficiently fine. If the function to be interpolated is in C^2 then the order of the error between this interpolating generalized spline and the cubic spline will be 4. Hence the interpolation error is of fourth order when the given function is in C^4. An upper bound for this error is found for a subclass of the generalized trigonometric splines.

1. INTRODUCTION

We will consider L-splines where the pieces are lying in the following space.

$$S_4 = \text{span}\{\cos\alpha x,\ \sin\alpha x,\ \cos\beta x,\ \sin\beta x\} \tag{1.1}$$

where α, β are parameters with $0 < \alpha < \beta$. Given any $n \in N$ and any partition

$$\Delta:\ a = x_0 < x_1 < \ldots < x_n = b$$

of $I = [a, b]$, set $I_i = [x_i, x_{i+1}]$, $h_i = x_{i+1} - x_i$, $i = 0,$

S. P. Singh et al. (eds.), Approximation Theory and Spline Functions, 349–360.

..., n - 1, and define the space of fourth order generalized trigonometric splines by

$$S_4(\Delta) = \{s \in C^2(I) \mid s|_{I_i} \in S_4,\ i = 0, \ldots, n-1\}.$$

We want to construct interpolants from $S_4(\Delta)$. Given function values $f_i = f(x_i)$, $i = 0, \ldots, n$, and endpoint slopes $f_i' = f'(x_i)$, $i = 0, n$, we seek an $s \in S_4(\Delta)$ satisfying

$$\begin{aligned} s(x_i) &= f_i,\ i = 0, .., n \\ s'(x_i) &= f_i',\ i = 0, n. \end{aligned} \tag{1.3}$$

We will use the technique from the theory of cubic splines. So let us briefly review some of that theory. Denote by P_4 the set of cubic polynomials and by $P_4(\Delta)$ the space of piecewise cubic polynomials which are in $C^2(I)$. The cubic spline $p \in P_4(\Delta)$ which solves (1.3) is given by (see e.g. [1, p. 10, 11])

$$\begin{aligned} p(x) = {} & \frac{x_{i+1} - x}{h_i} f_i + \frac{x - x_i}{h_i} f_{i+1} - \frac{h_i^2}{6}\left(\frac{x_{i+1} - x}{h_i} - \right. \\ & \left.\frac{(x_{i+1} - x)^3}{h_i^3}\right)p_i'' - \frac{h_i^2}{6}\left(\frac{x - x_i}{h_i} - \frac{(x - x_i)^3}{h_i^3}\right)p_{i+1}'', \end{aligned} \tag{1.4}$$

$$x \in I_i,\ i < n$$

where $p_0'', .., p_n''$ are given by

$$\begin{gathered} a_i^0 p_{i-1}'' + b_i^0 p_i'' + c_i^0 p_{i+1}'' = d_i^0 \\ a_i^0 = \frac{h_{i-1}}{h_{i-1} + h_i},\ b_i^0 = 2,\ c_i^0 = 1 - a_i^0,\ d_i^0 = 6[x_{i-1}, x_i, x_{i+1}]f. \end{gathered} \tag{1.5}$$

Here $x_{-1} = x_0$, $x_{n+1} = x_n$.

If $f \in C^4(I)$, then by [2, p. 209]

$$\|f - p\|_\infty \le \frac{5}{384} \|f^{(4)}\|_\infty h^4 \tag{1.6}$$

where $h = \max h_i$ and $\| \ \|_\infty$ is the sup-norm.

In the next section we will show that (1.3) has a unique solution if $h < \pi/\beta$ and that

$$\|s - p\|_\infty = O(h^4) \tag{1.7}$$

when $f \in C^2(I)$. If $f \in C^4(I)$ then by (1.6)

$$\|f - s\|_\infty = O(h^4). \tag{1.8}$$

In section 3 we analyze the interpolation error more thoroughly in the case that $\beta = 3\alpha$. We find that

$$\|f - s\|_\infty \le \frac{2}{3} K\left(\frac{\alpha h}{4}\right)\left(\frac{1}{\alpha}\sin\left(\frac{\alpha h}{4}\right)\right)^4 \|L_\alpha f\|_\infty \tag{1.9}$$

where

$$L_\alpha = (D^2 + \alpha^2)(D^2 + (3\alpha)^2), \quad D = d/dx$$

$$K(\theta) = \frac{1}{\cos^3(2\theta)}\left(1 - \frac{4}{3}\sin^2\theta + \frac{4\cos^4\theta}{2\cos(4\theta) - 1}\right). \tag{1.10}$$

Hence

$$\|f - s\|_\infty \le \left(\frac{5}{384} + (\alpha h)^2\right)h^4\|f^{(4)} + 10\alpha^2 f'' + 9\alpha^4 f\|_\infty . \tag{1.11}$$

When $\alpha \to 0$ the upper bound in (1.11) converges to that of (1.6). In [3] it was shown that the constant 5/384 in (1.6) is best possible. We think that the upper bound in (1.9) is as small as possible for general $f \in C^4$.

2. INTERPOLATION AT KNOTS

We want to solve (1.3). First we need a representation analog to (1.4) (see also [7] for the treatment of splines in tension.) Set

$$\phi(x) = \sin\alpha x, \ \psi(x) = \sin\beta x, \ \gamma^2 = \beta^2 - \alpha^2$$

and consider

$$\begin{aligned}\gamma^2\tau_i(x) = {} & (\beta^2 \frac{\phi(x_{i+1} - x)}{\phi(h_i)} - \alpha^2 \frac{\psi(x_{i+1} - x)}{\psi(h_i)}) f_i \\ & + (\beta^2 \frac{\phi(x - x_i)}{\phi(h_i)} - \alpha^2 \frac{\psi(x - x_i)}{\psi(h_i)}) f_{i+1} \\ & + (\frac{\phi(x_{i+1} - x)}{\phi(h_i)} - \frac{\psi(x_{i+1} - x)}{\psi(h_i)}) s''_i \\ & + (\frac{\phi(x - x_i)}{\phi(h_i)} - \frac{\psi(x - x_i)}{\psi(h_i)}) s''_{i+1}, \qquad x \in I_i.\end{aligned} \tag{2.1}$$

Clearly, $\tau_i \in S_4$ and $\tau_i(x_j) = f_j$, $j = i, i + 1$. It is also easy to see that $\tau''_i(x_j) = s''_j$, $j = i, i + 1$.

If we set $s_{|I_i} = \tau_i$, $i = 0, .., n - 1$, then $s \in S_4(\Delta)$ and s solves (1.3) iff $\tau'_{i-1}(x_i) = \tau'_i(x_i)$, $i = 1, .., n - 1$ and $\tau'_0(x_0) = f'_0$, $\tau'_{n-1}(x_n) = f'_n$, which is equivalent to

$$a'_i s''_{i-1} + b'_i s''_i + c'_i s''_{i+1} = d'_i, \ i = 0, .., n, \quad \text{where}$$

$$\begin{aligned}\gamma^2 a'_i &= -\frac{\phi'(0)}{\phi(h_{i-1})} + \frac{\psi'(0)}{\psi(h_{i-1})}, \quad c'_i = a'_{i+1} \\ \gamma^2 b'_i &= \frac{\phi'(h_{i-1})}{\phi(h_{i-1})} - \frac{\psi'(h_{i-1})}{\psi(h_{i-1})} + \frac{\phi'(h_i)}{\phi(h_i)} - \frac{\psi'(h_i)}{\psi(h_i)}\end{aligned} \tag{2.2}$$

$$\gamma^2 d_i' = (\beta^2 \frac{\phi'(0)}{\phi(h_{i-1})} - \alpha^2 \frac{\psi'(0)}{\psi(h_{i-1})}) f_{i-1} + (\beta^2 \frac{\phi'(0)}{\phi(h_i)} -$$

$$- \alpha^2 \frac{\psi'(0)}{\psi(h_i)}) f_{i+1} \tag{2.2}$$

$$- (\beta^2 \frac{\phi'(h_{i-1})}{\phi(_{i-1})} - \alpha^2 \frac{\psi'(h_{i-1})}{\psi(h_{i-1})} + \beta^2 \frac{\phi'(h_i)}{\phi(h_i)} - \alpha^2 \frac{\psi'(h_i)}{\psi(h_i)}) f_i \ .$$

Here, for $i = 0$ those terms involving h_{-1} is set equal to zero except for d_0' where they are replaced by $-f_0'$. Similarly, for $i = n$, where terms involving h_n is set equal to 0 except for d_n' where they are replaced by f_n'.

To insure that none of the denominators in (2.2) become 0, we suppose that

$$h < \frac{\pi}{\beta} \ . \tag{2.3}$$

This will be sufficient for (2.2) to have a unique solution. Because $x/\sin x$ is a positive, increasing function on $(0, \pi)$ and $x/\tan x$ is positive and decreasing on $(0, \pi/2)$ it is easy to see that

$$\frac{\phi'(t)}{\phi(t)} - \frac{\psi'(t)}{\psi(t)} > - \frac{\phi'(0)}{\phi\ (t)} + \frac{\psi'(t)}{\psi(t)} \quad > 0, \ 0 < t < \pi/\beta$$

and this implies that (2.2) is strictly diagonally dominant when (2.3) holds.

Let us now compare s with p. Using $\operatorname{cotan} t = 1/t - t/3 + 0(t^3)$ and $1/\sin t = 1/t + t/6 + 0(t^3)$ we get

$$a_i' = \frac{1}{6} h_{i-1} + 0(h_{i-1}^3), \ c_i' = \frac{1}{6} h_i + 0(h_i^3)$$

$$b_i' = \frac{1}{3}(h_{i-1} + h_i + 0(h_{i-1}^3 + h_i^3)) \tag{2.4}$$

$$d_i' = (\frac{1}{h_{i-1}} + 0(h_{i-1}^3)) f_{i-1} + (\frac{1}{h_i} + 0(h_i^3)) f_{i+1} -$$

$$- (\frac{1}{h_{i-1}} + \frac{1}{h_i} + 0(h_{i-1}^3 + h_i^3)) f_i \, .$$

To compare with (1.5) we divide by $(h_{i-1} + h_i)/6$ in (2.2). Call the results $a_i, .., d_i$. By (2.4)

$$a_i = a_i^0 + o(h^2),\ b_i = b_i^0 + 0(h^2),\ c_i = c_i^0 + 0(h^2)$$
$$d_i = d_i^0 + 0(h^2), \quad i = 0, .., n. \tag{2.5}$$

Put for a moment (1.5) and (2.2) in the forms $A^0p'' = d^0$, $As'' = d$. By (2.5), $\|A - A^0\|_\infty = 0(h^2)$ and $\|d - d^0\|_\infty = 0(h^2)$. If $f \in C^2$ then all the divided differences $[x_{i-1}, x_i , x_{i+1}]f$ and hence $\|d^0\|_\infty$ will be bounded independently of Δ. Since $\|A^0\|_\infty \leq 3$, $\|(A^0)^{-1}\|_\infty \leq 1$, we get

$$|s_i'' - p_i''| = 0(h^2),\ i = 0, .., n . \tag{2.6}$$

We may now compare the τ_i of (2.1) with (1.4). The formula

$$\frac{\sin\lambda t}{\sin\lambda h} = \frac{t}{h} + \frac{1}{6}\lambda^2 h^2 \left(\frac{t}{h} - \frac{t^3}{h^3}\right) + 0(h^4)$$

gives

$$\tau_i(x) = \frac{x_{i+1} - x}{h_i} f_i + \frac{x - x_i}{h_i} f_{i+1} - \frac{h_i^2}{6}\left(\frac{x_{i+1} - x}{h_i} - \frac{(x_{i+1} - x)^3}{h_i^3}\right)s_i'' - \frac{h_i^2}{6}\left(\frac{x - x_i}{h_i} - \frac{(x - x_i)^3}{h_i^3}\right)s_{i+1}'' + 0(h_i^4) \tag{2.7}$$

which, in view of (2.6) and (1.4), implies (1.7).

By (1.6) we obtain

$$\|f - s\|_\infty = 0(h^4), \quad f \in C^4(I),\ h < \pi/\beta . \tag{2.8}$$

We may proceed similarly when instead of S_4 we use span$\{1, x, \cos\beta x, \sin\beta x\}$. We get the strict diagonal dominance when (2.3) holds. (2.8) is still valid. In fact, we get these results by letting $\alpha\to 0+$. Similary when S_4 is replaced by

$$H_4 = \text{span}\{\cosh(\alpha x), \sinh(\alpha x), \cosh(\beta x), \sinh(\beta x)\} .$$

The strict diagonal dominance holds for all $0 < \alpha < \beta < \infty$. (2.8) holds. Letting $\alpha\to 0+$ we get the same conclusion for span $\{1, x, \cosh\beta x, \sinh\beta x\}$, i.e. for splines in tension (the strict diagonal dominance is shown in [7]).

3. ERROR BOUNDS IN THE CASE $\beta = 3\alpha$

In this section we let $\beta = 3\alpha$. In that case we have explicit expressions and bounds for the error in Hermite interpolation by S_4 (see [4, p. 232], [6, p. 193]). We may proceed similar to [2] to prove

Theorem. Let $\beta = 3\alpha$. Given $f \in C^4(I)$ and a partition Δ with $h < \pi/\beta$ let $s \in S_4(\Delta)$ solve (1.3). Then (1.9) holds, where K and L_α are given by (1.10).

The rest of this section will be a proof of this theorem. Set $e = f - s$, $\varepsilon_j' = e'(x_j)$, $j = 0, .., n$. We first want to bound the (ε_j')'s.

Fix i. Set $e_i = e|_{I_i}$. It is easy to see that

$$\mu_i(x) = \frac{\sin\alpha(x - x_i)\sin\alpha(x_{i+1} - x)}{\alpha\sin^2\alpha h_i} \cdot \qquad (3.1)$$

$$\cdot\ (\varepsilon_i'\sin\alpha(x_{i+1} - x) - \varepsilon_{i+1}' \sin\alpha(x - x_i))$$

solves $\mu_i(x_j) = 0$, $\mu_i'(x_j) = \varepsilon_j'$, $j = i, i + 1$, and that $\mu_i \in S_4$.

By [6, p. 192] the interpolation error $e_i - \mu_i$ is given by

$$r_i(x) = e_i(x) - \mu_i(x) = \sin^2\alpha(x - x_i)\sin^2\alpha(x - x_{i+1})u_i(x) \quad (3.2)$$

where, with $\ell(y) = 2\alpha(y - x_{i+1/2}) + x_{i+1/2}$,

$$u_i(x) = [\ell(x_i), \ell(x_i), \ell(x_{i+1}), \ell(x_{i+1}), \ell(x)]_t(f \circ \ell^{-1}) \quad (3.3)$$

and where $[\ ]_t$ is the trigonometric divided differenced introduced in [5] and $f \circ \ell^{-1}$ is the composite function.

We will bound $|e_i|$ by bounding both $|\mu_i|$ and $|r_i|$. Let us consider μ_i first. Since $e \in C^2(I)$,

$$\mu''_{i-1}(x_i) - \mu''_i(x_i) = r''_i(x_i) - r''_{i-1}(x_i),\ i = 1, .., n - 1 .$$

Differentiating (3.1) and (3.2) twice gives

$$\frac{2\alpha}{\sin\alpha h_{i-1}}\varepsilon'_{i-1} + 4\alpha\left(\frac{\cos\alpha h_{i-1}}{\sin\alpha h_{i-1}} + \frac{\cos\alpha h_i}{\sin\alpha h_i}\right)\varepsilon'_i + \frac{2\alpha}{\sin\alpha h_i}\varepsilon'_{i+1} =$$

$$= 2\alpha^2\sin^2\alpha h_i u_i(x_i) - 2\alpha^2\sin^2\alpha h_{i-1}u_{i-1}(x_i)$$

or, normalized

$$\frac{\sin\alpha h_i}{\sin\alpha(h_{i-1} + h_i)}\varepsilon'_{i-1} + 2\varepsilon'_i + \frac{\sin\alpha h_{i-1}}{\sin\alpha(h_{i-1} + h_i)}\varepsilon'_{i+1} =$$

$$= \alpha\sin^3(\alpha h_i)\frac{\sin\alpha h_{i-1}}{\sin\alpha(h_{i-1} + h_i)}u_i(x_i) - \quad (3.4)$$

$$- \alpha\sin^3\alpha h_{i-1}\frac{\sin\alpha h_i}{\sin\alpha(h_{i-1} + h_i)}u_{i-1}(x_i)$$

and

$$\varepsilon'_0 = 0,\ \varepsilon'_n = 0.$$

Let for the moment $M = \max_i \max(|u_i(x_i)|, |u_{i-1}(x_i)|)$. Using the formulas $\sin 2A = 2\sin A\cos A$ and $\sin A + \sin B = 2\sin((A + B)/2 \cdot \cos((A - B)/2)$ we can bound the right side of (3.4) by

$\alpha\sin^3(\alpha h)M/\cos(\alpha h)$, where $h = \max h_i$. Likewise,

$$2 - \frac{\sin\alpha h_i}{\sin(h_{i-1} + h_i)} - \frac{\sin\alpha h_{i-1}}{\sin\alpha(h_{i-1} + h_i)} \geq \frac{2\cos\alpha h - 1}{\cos\alpha h}.$$

By a standard diagonal dominance argument

$$|\varepsilon_i'| \leq \frac{\alpha\sin^3\alpha h}{2\cos\alpha h-1} M, \; i = 0, \ldots, n, \; \alpha h < \pi/3. \tag{3.5}$$

Let us now find an upper bound for M. We use the integral-representation in [5, p. 276]

$$[y_0, \ldots, y_n]_t g = \frac{2^{n-1}}{(n-1)!} \int T_{0,n} L_n g(y)\,dy \tag{3.6}$$

where $T_{0,n}$ is the trigonometric B-spline with knots $y_0, \ldots, y_n$ and

$$L_n = \begin{cases} D(D^2 + 1^2)\ldots(D^2 + k^2) & , n = 2k + 1 \\ (D^2 + (\frac{1}{2})^2)\ldots(D^2 + (k - \frac{1}{2})^2), & n = 2k. \end{cases}$$

This trigonometric divided difference possesses the following recurrence relation (see [5, p. 271]), $[y]_t g = g(y)$ and

$$[y_0,\ldots,y_n]_t g = \lambda[y_0,\ldots,y_{n-2}]_t g + \mu[y_1,\ldots,y_{n-1}]_t g + \nu[y_2,\ldots,y_n]_t g$$

$$1/\lambda = \sin\frac{y_n - y_0}{2} \sin\frac{y_{n-1} - y_0}{2}$$

$$1/\mu = -\sin\frac{y_n - y_0}{2} \sin\frac{y_{n-1} - y_0}{2} \sin\frac{y_n - y_1}{2} \Big/ \sin\frac{y_n + y_{n-1} - y_1 - y_0}{2} \tag{3.7}$$

$$1/\nu = \sin\frac{y_n - y_0}{2} \sin\frac{y_n - y_1}{2}.$$

If n is even, $n = 2k$, $L_n 1 = ((2k)!/4^k k!)^2$. So by (3.6)

$$\int T_{0,n} = \frac{2^{2k} k!(k-1)!}{(2k)!} [y_0, .., y_n]_t 1, \quad n = 2k. \tag{3.8}$$

By (3.3), (3.6),

$$|u_i(x)| \le \frac{4}{3} \int T_{0,4} \cdot (2\alpha)^{-4} \| L_\alpha f \|_\infty \tag{3.9}$$

where $T_{0,4}$ has the knots $\ell(x_i)$, $\ell(x_i)$, $\ell(x_{i+1})$, $\ell(x_{i+1})$ and $\ell(x)$. (3.7) plus some computation shows that

$$[y_0, y_0, y, y_1, y_1]_t 1 = \frac{2\cos\frac{y_1 - y_0}{4} + \cos\left(\frac{y - y_0}{2} - \frac{y_1 - y_0}{4}\right)}{8\cos^2\frac{y - y_0}{4} \cos^2\frac{y_1 - y}{4} \cos^3\frac{y_1 - y_0}{4}} . \tag{3.10}$$

By (3.8), (3.9) and (3.10)

$$|u_i(x)| \le \frac{2}{9} \cdot \frac{2\cos\frac{\alpha h_i}{2} + \cos\left(\alpha(x - x_i\gamma - \frac{\alpha h_i}{2}\right)}{\cos^2\frac{\alpha(x - x_i)}{2} \cos^2\frac{\alpha(x_{i+1} - x)}{2} \cos^3\frac{\alpha h_i}{2}} \cdot$$

$$\cdot (2\alpha)^{-4} \| L_\alpha f \|_\infty . \tag{3.11}$$

So $|u_i(x_i)|, |u_i(x_{i+1})| \le \dfrac{2}{3\cos^4\frac{\alpha h_i}{2}} (2\alpha)^{-4} \| L_\alpha f \|_\infty$. Since $1/\cos x$ is increasing in x we have

$$M \le \frac{2}{3\cos^4(\frac{\alpha h}{2})} (2\alpha)^{-4} \| L_\alpha f \|_\infty . \tag{3.12}$$

Inserting this in (3.5) and using $\sin 2A = 2\sin A\cos A$ gives

$$|\varepsilon_i'| \le \frac{\sin^3(\frac{\alpha h}{2})}{3\alpha^3 \cos\frac{\alpha h}{2}(2\cos\alpha h - 1)} \| L_\alpha f \|_\infty , \; i = 0, .., n . \tag{3.13}$$

Consider (3.1). The formula $\sin A + \sin B = 2\sin\frac{A+B}{2}\cos\frac{A-B}{2}$ yields

$$|\mu_i(x)| \le \frac{2\sin\frac{\alpha h_i}{2}}{\alpha\sin^2\alpha h_i}\sin\alpha(x - x_i)\sin\alpha(x_{i+1} - x)\cos\alpha(x - x_{i+1/2})\cdot$$

$$\cdot \max_i |\varepsilon_i'| .$$

Both $\cos\alpha(x - x_{i+1/2})$ and $\sin\alpha(x - x_i)\sin\alpha(x_{i+1} - x)$ take their maxima at $x = x_{i+1/2}$. So $|\mu_i(x)| \le \max_j|\varepsilon_j'|\sin\frac{\alpha h_i}{2}/(2\alpha\cos^2(\frac{\alpha h_i}{2}))$, which is monotone in h_i. (3.13) gives

$$|\mu_i(x)| \le \frac{\sin^4(\frac{\alpha h}{2})}{6\alpha^4\cos^3(\frac{\alpha h}{2})\ (2\cos(\alpha h) - 1)}\|L_\alpha f\|_\infty,\ x \in I_i,\ i < n. \tag{3.14}$$

Now switch to r_i. Set $t = x - x_i$. Using $\sin 2A = 2\sin A\cos A$, (3.11) and (3.2), we obtain

$$|r_i(x)| \le 2\sin^2(\frac{\alpha t}{2})\sin^2(\frac{\alpha(h_i - t)}{2})$$

$$\frac{2\cos\frac{\alpha h_i}{2} + \cos(\alpha(t - h_i/2))}{9\alpha^4\cos^3(\frac{\alpha h_i}{2})}\|L_\alpha f\|_\infty . \tag{3.15}$$

This upper bound takes its maximum for $t = h_i/2$. This implies

$$|r_i(x)| \le \frac{2}{9\alpha^4}\|L_\alpha f\|_\infty \sin^4(\frac{\alpha h_i}{4})(1 + 2\cos\frac{\alpha h_i}{2})/\cos^3(\frac{\alpha h_i}{2}) . \tag{3.16}$$

We have $1 + 2\cos\theta = 3 - 4\sin^2(\frac{\theta}{2})$. Hence the upper bound in

(3.16) increases in h_i. Therefore we may replace h_i by h in (3.16). (3.14) and (3.16) now imply (1.9) and the theorem is proved.

A similar result can be given for the hyperbolic case.

REFERENCES

1. Ahlberg, J. H., Nilson, E. N., and Walsh, J. L., "The theory of splines and their applications", Academic Press, New York, 1967.

2. Hall, C. A., "On error bounds for spline interpolation", J. Approximation Theory 1, 1968, pp. 209-218.

3. Hall, C. A., and Meyer, W. W., "Optimal error bounds for cubic spline interpolation", J. Approximation Theory 16, 1976, pp. 105-122.

4. Lyche, T., "A Newton form for trigonometric Hermite interpolation", BIT 19, 1979, pp. 229-235.

5. Lyche, T., and Winther, R., "A stable recurrence relation for trigonometric B-splines", J. Approximation Theory, 25, 1979, pp. 266-279.

6. Koch, P. E., and Lyche, T., "Bounds for the error in trigonometric Hermite interpolation", in Quantitative Approximation, eds. R. Devore and K. Scherer, 1980, pp. 185-196.

7. Rentrop, P., "An algorithm for the computation of the exponential spline", Num. Math. 35, 1980, pp. 81-93.

8. Schoenberg, I. J., "On trigonometric spline interpolation", J. Math. and Mech. 13, 1964, pp. 795-825.

9. Schweikert, D. G., "An interpolation curve using a spline in tension", J. Math. and Phys. 45, 1966, pp. 312-317.

APPROXIMATION OF DERIVATIVES IN $\mathbb{R}^n$ APPLICATION: CONSTRUCTION OF SURFACES IN $\mathbb{R}^2$

Alain Le Mehaute

Institut National des Sciences Appliquees de Rennes

A popular method of calculating derivatives of a function of one variable from its values on a given set of knots uses splines, or spline-on-spline technique. In this paper, we discuss a strategy to use such techniques for more than one variable. Computational examples show that our choice of strategy is not so bad, when applied to the construction of surfaces in $\mathbb{R}^2$, using finite elements method as developed in [9].

1. $X^{m,s}$-Splines in $\mathbb{R}^n$ [7]

Let $\tilde{H}^s$ the set of tempered distributions u whose Fourier transform $\hat{u}$ (or $\mathcal{F}u$) is such that

$$\int_{\mathbb{R}^n} |\hat{u}(\tau)|^2 \; |\tau|^{2s} \, d\tau < \infty .$$

It is an Hilbert space for the norm

$$\|u\|_s = \| \; |\tau|^s \; \hat{u}\|_{L^2} \qquad \text{if} \quad s < \frac{n}{2} .$$

Let $X^{m,s}$ (or $D^{-m}(\tilde{H}^s)$) the space of all distributions whose all derivatives of order m are in $\tilde{H}^s$, with the norm

$$\|u\|_{ms} = \left(\sum_{|\alpha|=m} \|\partial^\alpha u\|_s^2 \right)^{1/2}$$

S. P. Singh et al. (eds.), Approximation Theory and Spline Functions, 361–378.

and scalar product

$$((u,v))_{ms} = \sum_{|\alpha|=m} \int_{\mathbb{R}^n} F\partial^\alpha u(\tau)\ \overline{F\partial^\alpha v(\tau)}\ |\tau|^{2s}\, d\tau$$

$$\left(\text{where} \quad \partial^\alpha u = \frac{\partial^{\alpha_1 + \dots + \alpha_n} u}{\partial x_1^{\alpha_1}\, \partial x_2^{\alpha_2} \dots \partial x_n^{\alpha_n}} \right).$$

Duchon [7] shows that, if $-m + \frac{n}{2} < s < \frac{n}{2}$, $X^{m,s}$ is a semi-Hilbert space of continuous functions on $\mathbb{R}^n$, of null-space P_{m-1} (polynomials of degree $\leq m-1$), separating points of $\mathbb{R}^n$, and with a reproducing kernel $K(t,t') = A_{2m+2s-n}(t-t')$, where

$$A_\theta(t) = \begin{cases} |t|^\theta \operatorname{Log} |t| & \text{if } \theta \text{ is even} \\ |t|^\theta & \text{if } \theta \text{ is odd.} \end{cases}$$

Let $\Sigma = \{a_1, \dots, a_N\} \subset \mathbb{R}^n$ be a set of points which contains a P_{m-1}-unisolvent set (i.e.: if $p \in P_{m-1}$ is zero on Σ, then p is identically zero), and let $\{z_1, \dots, z_n\} \subset \mathbb{R}$. Then there exists one and only one $\sigma \in X^{m,s}$ such that

$$\begin{cases} \sigma(a_i) = z_i & i=1,\dots,N \\ \|\sigma\|_{ms} = \inf \{ \|v\|_{ms},\ v(a_i) = z_i,\ i=1,\dots,N\} \end{cases}$$

σ is the $X^{m,s}$ spline interpolating z on Σ.

If $K(t,t')$ is the reproducing kernel of $X^{m,s}$, if $(\nu_1, \dots, \nu_p)$ is a basis of P_{m-1}, then we have explicitly

$$\sigma(t) = \sum_{j=1}^{N} \lambda_j\, K(t,a_j) + \sum_{k=1}^{p} \alpha_k\, \nu_k(t)$$

where λ_i, $i=1,\dots,N$ and α_k, $k=1,\dots,p$ are real solutions of the linear system

$$\begin{cases} \sum_{j=1}^{N} \lambda_j K(a_i,a_j) + \sum_{k=1}^{p} \alpha_k \nu_k(a_i) = z_i & i=1,\ldots,N \\ \sum_{j=1}^{N} \lambda_j \nu_k(a_j) = 0 & k=1,\ldots,p \end{cases}$$

Remarks.

1.1. Thin plates splines are obtained for $n=2$, $m=2$, $s=0$ and pseudo-cubics splines for $m=2$, $s = \frac{n-1}{2}$ ($\theta = 3$ in $\mathbb{R}^2$).

1.2. If $s=0$, it is equivalent to consider $X^{m,o}$ splines as the solution (unique if Σ contains a P_{m-1}-unisolvent set) of the problem:

$$\text{"minimize } \|D^m\|_{L^2(\mathbb{R}^n)}, \ u \in D^{-m}(L^2(\mathbb{R}^n)), \ u|_\Sigma = z\text{"}$$

[6], [7], [11], [13].

1.3. $X^{m,s} \subset C^k$ if and only if $2m+2s-n > 2k$.

2. Approximation of Derivatives Using $X^{m,s}$-Splines

Let Ω be a bounded open set in $\mathbb{R}^n$, Σ a set of separated points of $\overline{\Omega}$, C a "triangulation" of $\overline{\Omega}$ based on Σ, and f a function such that we know only the values $f(A_i)$, $A_i \in \Sigma$.

(To obtain such a triangulation, see for instance [9].)

We want to obtain an approximation of some derivatives of f, on Σ. Let A be a point of Σ, and denote

$$C_A = \{T \in C, A \in \Sigma \cap T\}$$

$$\Sigma_A = \Sigma \cap C_A$$

$$\Omega_A = \bigcup_{T \in C_A} T .$$

1. *Splines Technique. On each Ω_A, we approach f with the unique $X^{m,s}$-spline interpolating f on Σ_A, denoted by*

f^A, *and we approach the wanted derivative of* f *by the corresponding derivative of* f^A.

It is possible to obtain error bounds (see below).

2. *Spline on spline technique. On each* $_A$, *we approach* f *with the* $X^{m,s}$*-spline* f^A. *To approach first derivatives of* f, *we use first derivatives of the* $X^{m,s}$*-spline interpolating the first derivatives of* f^{A_i} *for* $A_i \in \Sigma_A$, *and so one to calculate upper order derivatives.*

This second approach gives nice results for computing.

3. Error Bounds When Using $X^{m,o}$-splines (Or θ is Even)

Let s be an integer $s > \frac{n}{2}$, θ an open bounded polyedrical domain in $\mathbb{R}^n$, $L(\theta)$ a set of separated points in $\overline{\theta}$, $C(\theta)$ a triangulation of $\overline{\theta}$ based on $L(\theta)$, regular in the sense that there exists α such that, for each triangle $T \in C(\theta)$, $\frac{h_T}{\rho_T} < \alpha$, where h_T is the diameter of T and ρ_T the diameter of the inscribed sphere in T.

Let $h = \max\{h_T, T \in C(\theta)\}$ and $\rho = \inf\{\rho_T, T \in C(\theta)\}$.

Assume that for each triangle $T \in C(\theta)$, $L(\theta) \cap \overline{T}$ contains a r_T-unisolvent set, with $0 \le r_T \le s-1$, and let $r = \inf\{r_T, T \in C(\theta)\}$.

Using the method used in [2], we obtain:

Theorem. For all $k \in \mathbb{N}$, $0 \le k \le r+1$, there exists a constant $C(n,h,r,\frac{h}{\rho})$ such that for all triangle $T \in C(\theta)$, for all $v \in H^s(\theta)$ such that $v|_{L(\theta)} = 0$, we have

$$|v|_{k,T} \le C\left(n,h,r,\frac{h}{\rho}\right) h_T^{r+1-h} |v|_{r+1,T} .$$

It follows that, for each $\theta = \Omega_A$, with the above assumptions:

$$|v|_{k,\Omega_A} \leq C\left(n, h_A, r, \frac{h_A}{\rho_A}\right) h_A^{r+1-k} |v|_{r+1,\Omega_A} .$$

We must now compare the norms in $H^s(\Omega)$ instead of $H^s(\Omega_A)$.

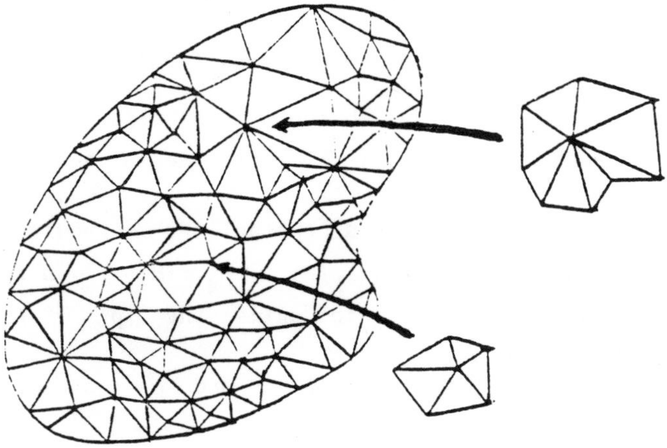

Figure 1.1. Spline technique

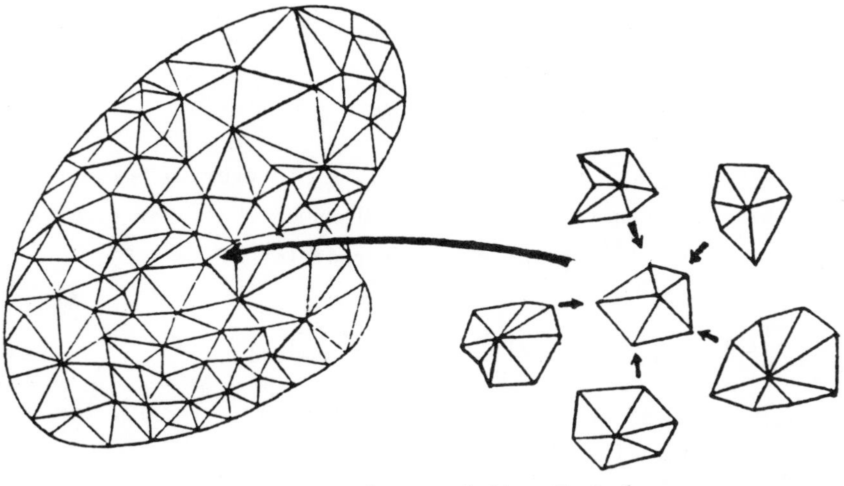

Figure 1.2. Spline on Spline Technique

Figure 1

Definition. Let p and q positive integers, Ω an open bounded domain in $\mathbb{R}^n$, $VL(p,q)\ (\Omega)$ is the set of all open sets δ such that $\delta \subset \Omega$ and $\forall\ x \in \partial\delta$, $\exists\ V_x$, open set in $\mathbb{R}^n$, and $T_x : B(x,p) \to V_x$, bilipschitz of constant q, and such that $T_x(B(x,p) \cap \delta) = V_x \cap \mathbb{R}^{n+1}$.

In other words, $VL(p,q)\ (\Omega)$ is the set of all open sets of $\mathbb{R}^n$, included in Ω and "uniformly locally mapped of an half-space by a bilipschitz one to one transform".

In our case, each Ω_A has obviously the uniform cone property, and thus each $\Omega_A \in VL(p,q)\ (\Omega)$ for some p and q.

Theorem. [4] There exists a constant K, depending only on (p,q) such that $\forall \delta \in VL(p,q)\ (\Omega)$, $\exists\ p_\delta : H^m(\delta) \to H^m(R^n)$, linear continuous extension operator satisfying $\|p_\delta\| \leq K$.

In our case, K depends only on h and ρ (not h_A and ρ_A). Therefore we obtain

Lemma. There exists a constant K, depending on $\mathcal{C}$ (by h and ρ) such that

$$\forall\ A \in \Sigma,\ \forall\ f \in H^m(\Omega_A),\ |f^A|_{m,\Omega_A} \leq K|f|_{m,\Omega}.$$

It follows easily that:

Proposition. Let $f \in H^{r+1}(\Omega)$, k integer, $0 \leq k \leq r+1$. There exists a constant $C(n,k,r,\frac{h}{\rho},\mathcal{C})$ such that, for each $A \in \Sigma$, if f^A is the D^{r+1}-spline that interpolates f on Σ_A,

$$|f-f^A|_{k,\Omega_A} \leq C(n,k,r,\frac{h}{\rho},\mathcal{C})\ h^{r+1-h}\ |f|_{r+1,\Omega}.$$

Remark. If we cannot assume more than $m-1 \geq 1$, then $r = 1$ and, for $f \in H^2(\Omega)$,

$$|f-f^A|_{0,\Omega_A} \le C_o\, h^2 |f|_{2,\Omega}$$

$$|f-f^A|_{1,\Omega_A} \le C_1\, h\, |f|_{2,\Omega}\ ,$$

and we cannot use second derivatives because of (1.3).

4. Error Bounds Where Using $X^{m,\frac{n-1}{2}}$-splines (Or θ is Odd)

Let $\theta = 2m-1$, if $s = \frac{n-1}{2}$ (or $s = \frac{\theta-2m+n}{2}$ if θ, odd, is known). We assume that each Ω_A contains a(m-1)-unisolvent subset of Σ_A. Let f^A be the unique $X^{m,s}$ spline that interpolates f on Σ_A and let us write

$$f^A(t) = \sum_{A_j\in\Sigma_A} \lambda_j\, K(t,A_j) + \sum_{|\alpha|=0}^{m-1} \mu_\alpha\, t^\alpha$$

the λ_j and μ_α are solutions of a linear system of $\nu_A = N_A + \begin{pmatrix} m-1+n \\ n \end{pmatrix}$ equations, where $N_A = \text{card } I_A$.

Let $\nu_A(\theta) = N_A\theta + \begin{pmatrix} m-1+n \\ n \end{pmatrix}$, thus $\nu_A = \nu_A(1)$.

Using "change of scale" method and Cramer's rule to calculate the solution of this system, we obtain

$$|\lambda_j| \le (\nu_A)!\ 2^{(N_A-1)\theta}\ (\tfrac{h}{d})^{\nu_A(\theta)}\ \frac{1}{|\hat{\Delta}|}\ \frac{1}{h^\theta} \left\{ \max_{A_i\in\Sigma_A} |f(A_i)| \right\}$$

$$|\mu_\alpha| \le (\nu_A)!\ 2^{(N_A-1)\theta}\ (\tfrac{h}{d})^{\nu_A(\theta)}\ \frac{1}{|\hat{\Delta}|}\ \frac{1}{h^{|\alpha|}} \left\{ \max_{A_i\in\Sigma_A} |f(A_i)| \right\}$$

where $d = \inf\,\{d(A_i,A_j),\ A_i \in \Sigma_A,\ A_j \in \Sigma_A\} > 0$, and $\hat{\Delta}$ a constant that depends only on the "geometry" of Σ_A, not on its scale.

Using Faa de Bruno's formula [10], we are able to calculate explicitly the derivatives of f^A, and to obtain:

Proposition. For all $x \in \bigcup_{A_i \in \Sigma_A} B(A_i, h)$

$$|f^A(x)| \le \left\{ (\nu_A)! \left(\frac{h}{d}\right)^{\nu_A(\theta)} \frac{1}{|\hat{\Delta}|} 2^{(N_A - 1)\theta} \right\} \left\{ \max_{A_i \in \Sigma_A} |f(A_i)| \right\}$$

$$|\partial^\beta f^A(x)| \le \left\{ C(n,\beta)\ (\nu_A)! \left(\frac{h}{d}\right)^{\nu_A(\theta)} \frac{1}{|\hat{\Delta}|} 2^{(N_A - 1)\theta} \right\} \frac{1}{h^{|\beta|}} \left\{ \max_{A_i \in \Sigma_A} |f(A_i)| \right\}$$

where $C(n,\beta)$ is a constant depending only on n and β.

Let $\Sigma_{A,u}$ a (m-1)-unisolvent set of Σ_A.

If Ω_A is star-shaped relatively to $\Sigma_{A,u}$, we can use the Peano approach and the Taylor-Sobolev formula ([12]).

Let $f = p_{m-1}f + R_{m-1}f$, where $p_{m-1}f$ is a polynomial of degree $\le m-1$.

As $(p_{m-1}f)^A = p_{m-1}f$, we can write $(f-f^A)(x) = R_{m-1}f(x) + (R_{m-1}f)^A(x)$.

Using the fact that

$$|R_{m-1}f(x)| \le \frac{\sqrt{m+n}\ 2^m}{\sqrt{\mathrm{mes}\,\Omega}(m-1)!\ \sqrt{4m^2-n^2}}\ h^m \|D^m f\|_{L^2(\Omega_A)}$$

and

$$|(R_{m-1}f)^A(x)| \le \frac{\sqrt{m+n}\ 2^{m+(N_A-1)\theta}}{\sqrt{\mathrm{mes}\,\Omega}\ (m-1)!\ \sqrt{4m^2-n^2}} \left(\frac{h}{d}\right)^{\nu_A(\theta)} (\nu_A)!$$

$$\frac{1}{|\hat{\Delta}|} \ h^m \|D^m f\|_{L^2(\Omega_A)}$$

we obtain:

Proposition. For all $f \in C^m(\Omega)$, $|\alpha| \le m - \frac{n}{2}$, $x \in \Omega_A$, Ω_A star-shaped relatively to a (m-1)-unisolvent subset of Σ_A, these exists a constant C (that we can calculate explicitly) such that

$$|\partial^\alpha f(x) - \partial^\alpha f^A(x)| \le C \ h^{m-|\alpha|} \|D^m f\|_{L^2(\Omega)} .$$

5. Application: Construction of Surfaces in $\mathbf{R}^2$

We use the method introduced in [9] to construct surfaces of C^k, in $\mathbf{R}^2$, using finite elements. This method uses triangular elements, namely Bell's element for C^1-surfaces and its generalization for surfaces of class C^k, $k \ge 2$.

The interpolating function is piecewise polynomial of degree 4k+1. Each polynomial is calculated independently on each triangle, without the use of a reference element, unlike in the usual way for finite elements. For C^1-interpolation, we need to use, on each triangle, all the derivatives of the function up to order 2 on each vertex, to calculate a polynomial of degree 5. For C^2-interpolation, the order must be 4, and the polynomial is of degree 9.

Now, we replace the derivatives by the estimates given by spline-method or by spline-on-spline method (see figures: 3, 4, 5).

Remark. When we cannot assume more than m-1 = 1 (that is, there is only 3 points not on the same line, for each Ω_A), the error estimate of the whole method (estimation of derivatives and calculus of the surface with finite elements method) is of the same order that the approximation given by finite elements of class C^0. (Compare the two in figures 4,5), namely, $0(h^2)$.

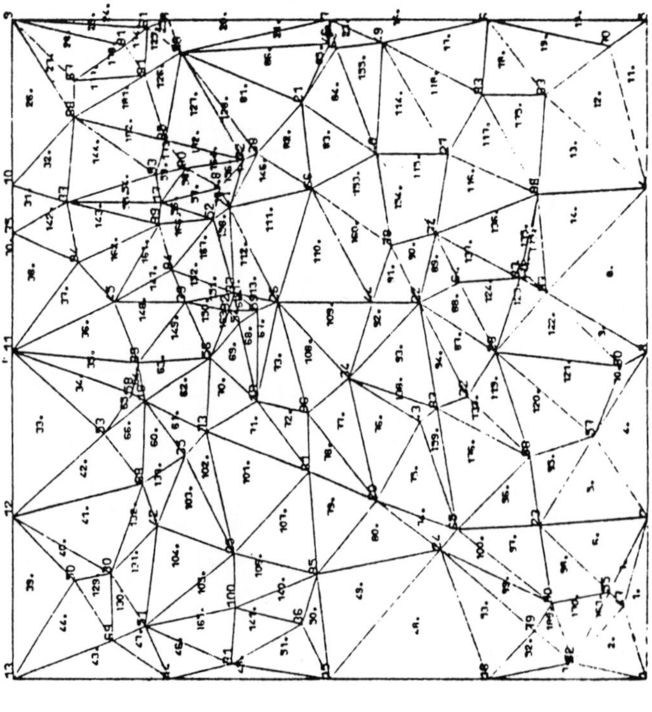

Figure 2.2.
Σ_2 consists of 100 pts

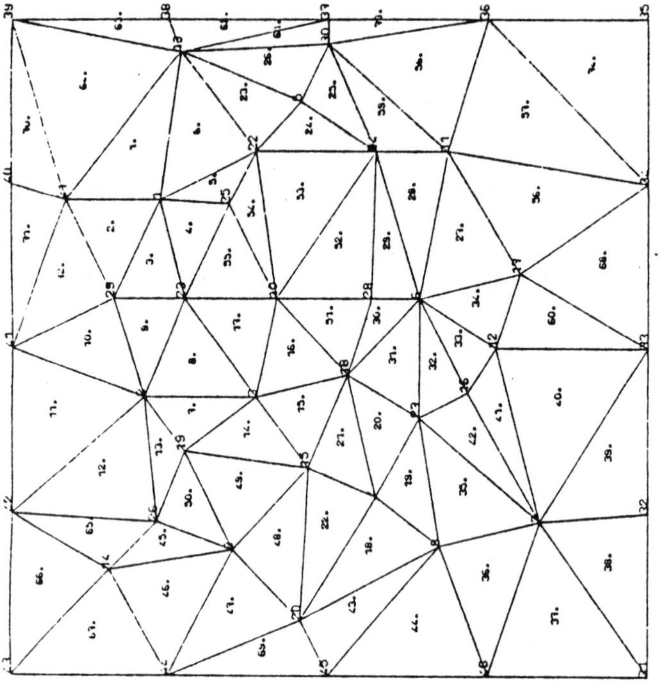

Figure 2.1.
Σ_1 consists of 46 pts

Figure 2

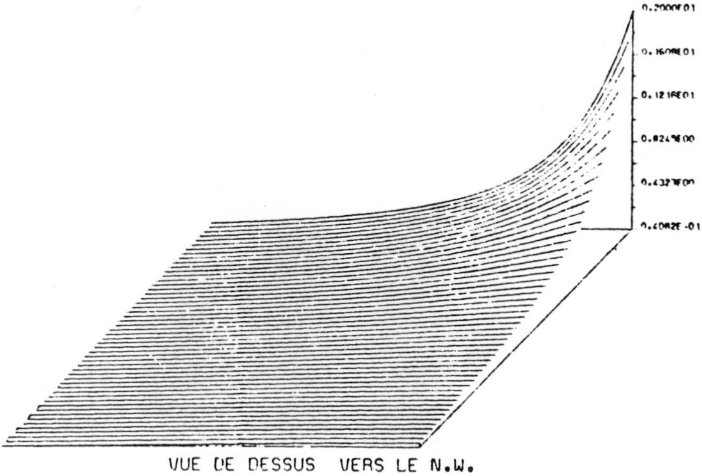

Figure 3.1.

The test function

$$f(x,y) = (x-3.5)^2 + (y-3.5)^2$$

Figure 3

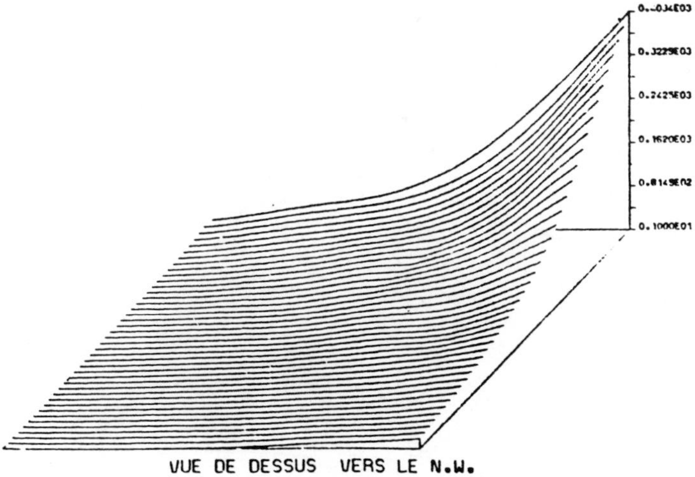

Figure 3.2.

C_1 interpolation on a set of 17 pts, using Spline Technique for calculating the derivatives up to order 2 a the vertex

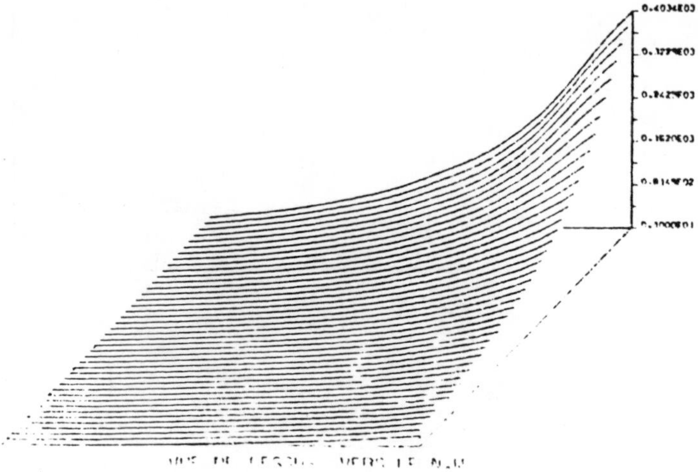

Figure 3.3.

C_1 interpolation on the same set of points, using SOS technique

Figure 3

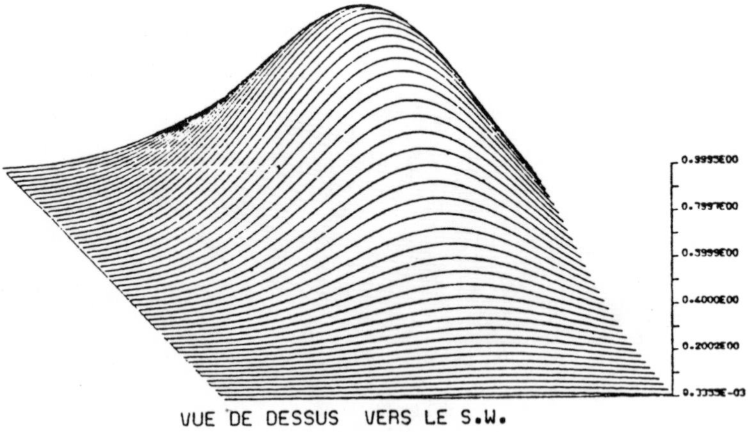

Figure 4.1.

The test function

$f(x,y) = \exp((x-1)^2+(y-1)^2)$

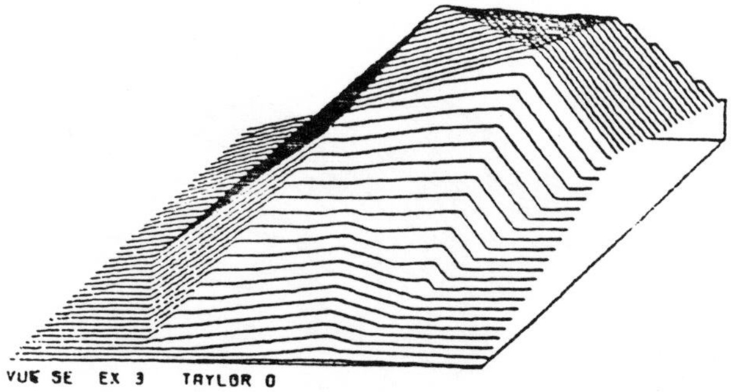

Figure 4.2.

C_0 interpolation using f.e.m. method, with only the values of the function on Σ, 17 pts.

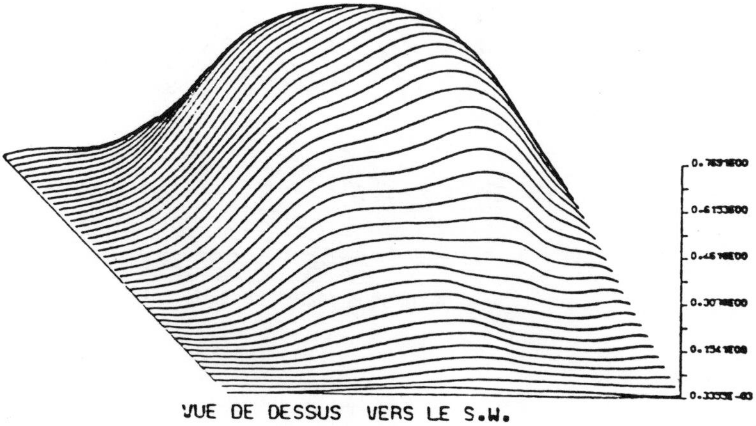

Figure 4.3.

C_1 interpolation, spline technique, using the same values as in (4.2.)

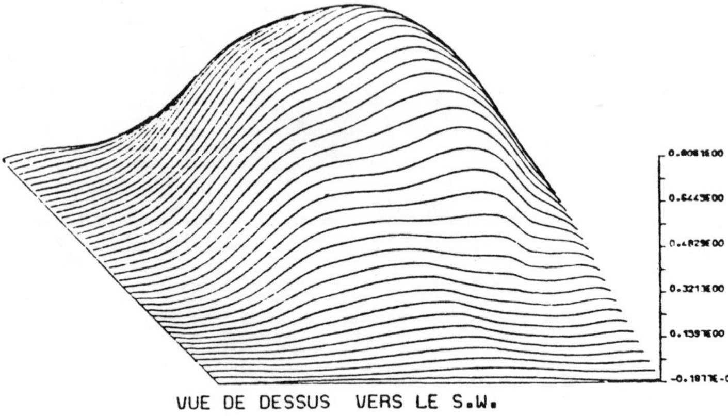

Figure 4.4.

C_1 interpolation,
SOS technique
Same values as in (4.2.)

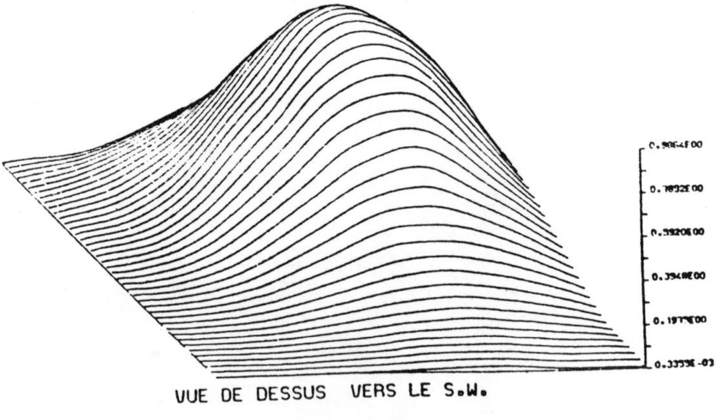

Figure 4.5.

C_1 interpolation
Spline technique
Σ consists of 46 pts

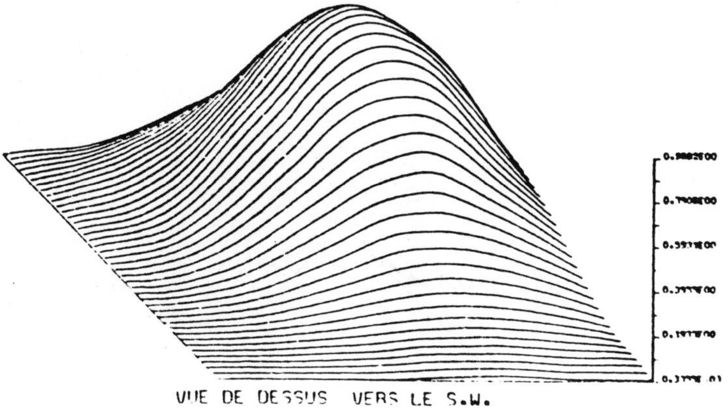

Figure 4.6.

C_1 interpolation
SOS technique with 46 pts

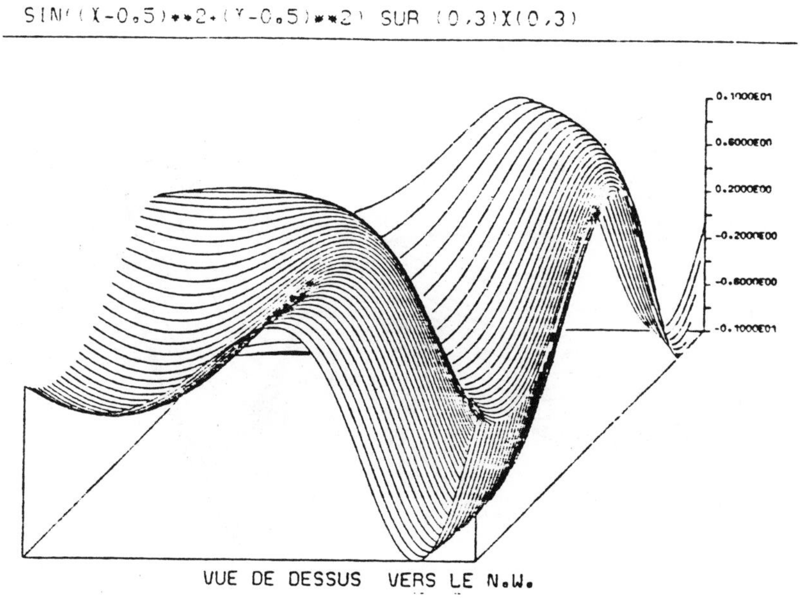

Figure 5.1.

$$f(x,y) = \sin\left[(x - \frac{1}{2})^2 + (y - \frac{1}{2})^2\right]$$

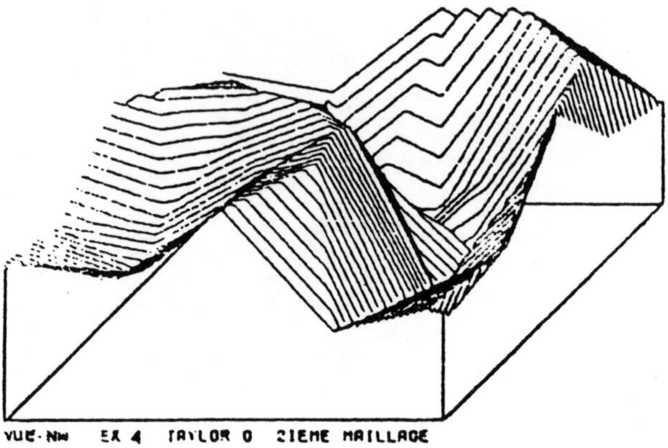

Figure 5.2.

C_o interpolation

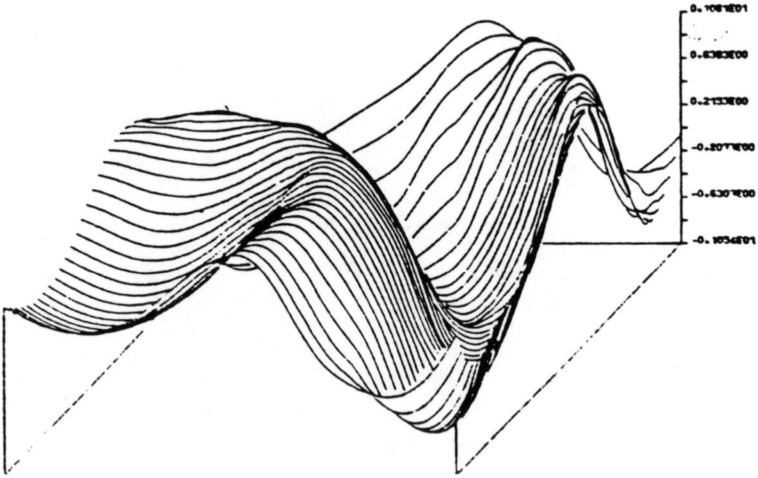

Figure 5.3.

C_1 interpolation with the same data

REFERENCES

1. Akima, H., "A method of bivariate interpolation and smooth surface fitting for values given arbitrary distributed points", Report COM 75 - 11285, 1975.

2. Arcangeli, R., "Etude de problèmes de type elliptique ou parabolique avec conditions ponctuelles", Thèse. Université de Toulouse, 1974.

3. Arcangeli, R., and Gout, J. L., "Sur l'evaluation de l'erreur d'interpolation de Lagrange dans un ouvert de R^n", RAIRO, Anal. Num. Vol. 10, N. 3, 1976, pp. 5-27.

4. Chenais, D., "Sur une famille de variétés à bord Lipschitzien", Annales de l'Institut Fourier, 1977.

5. Dolezal, V., and Tewarson, R., "Error bounds for spline on spline interpolation", J. of Approx. Theory, N. 36, 1982.

6. Duchon, J., "Fonctions splines à énergie invariante par rotation", Rapport N. 27, Université de Grenoble, 1976.

7. Duchon, J., "Interpolation des fonctions de deux variables suivant le principe de la flexion des plaques minces", RAIRO, Anal. Num. Vol. 10, N. 12, 1976, pp. 5-12.

8. Franke, R., "Scattered data interpolation. Test of some methods", Math of computation. Vol. 38, Jan. 1982.

9. Le Mehaute, A., "Construction of surfaces of class C^k on a domain $\Omega \in \mathbb{R}^2$ after triangulation", Oberwolfach 1982, Multivariate Approximation Theory II, ISNM 61, Birkhäuser Verlag, 1982.

10. Le Mehaute, A., "Quelques méthodes explicites de prolongements de fonctions numériques de plusieurs variables", Thèse 3e cycle, Rennes., 1976.

11. Meinguet, J., "Surface spline interpolation. Basic theory and computational aspects", These proceeding.

12. Meinguet, J., "A practical method for obtaining a priori error bounds in pointwise and mean square approximation problems", These proceeding.

13. Michelli, C. A., "Positive definite function and interpolation of scattered data", These proceeding.

14. Schumaker, L. L., "Fitting surfaces to scattered data", In Approximation Theory II, Academic Press, 1976.

MEROMORPHIC FUNCTIONS, MAPS AND THEIR RATIONAL APPROXIMANTS IN $\mathbb{C}^n$

C. H. Lutterodt

Howard University

ABSTRACT

We discuss Montessus and non-Montessus type of convergence for (μ,ν) rational approximants to meromorphic functions and their generalizations to meromorphic maps in $\mathbb{C}^n$.

§0. INTRODUCTION

The extension of Padé Approximant theory to several variables began a little over a decade ago from sources in Birmingham, England. In fact, a preliminary talk, which was a first on the subject, was presented by the author at Abingdon near Oxford, England, in August 1972. Since that time, considerable progress has been made in a number of specific directions; nevertheless, the overall development of the theory continues to be sluggish.

There are two main approaches used in the construction of rational approximants from power series in several variables. These approaches are inherently rooted in the kind of power series expansion and polynomials employed in the construction. The two main representations of power series about the origin in $\mathbb{C}^n$ are:

1) <u>non homogeneous/normal expansion</u>: $\sum_{\lambda \varepsilon \mathrm{IN}^n} c_\lambda z^\lambda$

S. P. Singh et al. (eds.), Approximation Theory and Spline Functions, 379–396.

2) homogeneous expansions: $\sum_{k=0}^{\infty} \sum_{|\lambda|=k} c_\lambda z^\zeta$.

Taking either approach, Gonĉar [1974], Karlsson and Wallin [1977] and Lutterodt [1974, 1976] have all indicated that in several variables, there are as many types of rational approximants as there are index interpolation sets (IIS). The class of index interpolation sets (IIS), for each fixed (μ,ν), even though it is finite, is a large class. But it is a restricted subclass of this large class of IIS which gives rise to rational approximants that possess simple algebraic properties and interesting convergence behaviour. From practical and application points of view, see Cuyt (1981) for the ε-algorithm method.

As suggested by the title, our main focus is on meromorphic functions, maps and their rational approximants. One of the main uniform convergence results in relation to meromorphic functions is the theorem of Montessus de Ballore. This result is not generally well known and so we have provided a simple scheme to illustrate the main point of the theorem and at the same time indicate how Montessus' convergence is a generalization of Taylor's convergence.

$g \nearrow$ g meromorphic in D with exactly m poles

$g \searrow$ $f=g|U$ holomorphic in $U \ni 0$, $U \subsetneq D$

$\downarrow$

$f(z)=\sum_{\lambda\in\mathbb{N}} c_\lambda z^\lambda$ in U, domain of convergence of f.

$\swarrow \quad \searrow$

Taylor partial sequence	$\{p_\mu(z)\}_\mu$	$\{R_{\mu m}(z)\}_\mu$	Padé rational sequence with m poles counting multiplicities
	as $\mu \to \infty$		
convergence uniform on any compact subset of U.	$p_\mu(z)=R_{\mu 0}(z)$	$R_{\mu m}(z)$	(i) poles of $R_{\mu m}$ tend to poles of g in D.
	$\downarrow$	$\downarrow$	
	$f=g\|U$	$g(z)$	(ii) convergence is uniform on any $K \subset D \setminus \{\text{poles of } g\}$.

It turns out that within U, $R_{\mu m}(z)$ converges much faster than $p_\mu(z)$ to $g\,|\,U = f$ and outside U in $D \setminus U$ $R_{\mu m}(z)$ converges to g uniformly on compact subsets of D except at the poles of g. Thus $R_{\mu m}(z)$ over-converges in U and may be used in analytically continuing f to its extension g on D.

The extension of the Montessus' result to several variables was successfully carried out by Karlsson and Wallin [1977] in two variables using homogeneous expansions. In Lutterodt [1980] we extended the result to $\mathbb{C}^n$ using non-homogeneous expansions. Our [1983] paper provides the proof of the 1980 result and our paper [1983] has a further generalization of the Montessus result for meromorphic maps in $\mathbb{C}^n$.

§2. NOTATION

We shall write $z = (z_1, \ldots, z_n)$ for $z \in \mathbb{C}^n$ and

$\hat{z}: = (z_1, \ldots, z_{j-1}, z_{j+1}, \ldots, z_n)$, for $\hat{z} \in \mathbb{C}^{n-1}$

$\lambda: = (\lambda_1, \ldots, \lambda_n)$ for $\lambda \in \mathbb{N}^n$ and

$\hat{\lambda}: = (\lambda_1, \ldots, \lambda_{j-1}, \lambda_{j+1}, \ldots, \lambda_n)$ for $\hat{\lambda} \in \mathbb{N}^{n-1}$

Let $\rho \geq 0$ and let $\Delta^n_\rho: = \{z \in \mathbb{C}^n: |z_j| < \rho,\ j = 1, \ldots, n\}$ be a polydisk in $\mathbb{C}^n$ centered at 0. We shall write

$$\frac{1}{\lambda!}\frac{\partial^{|\lambda|}}{\partial z^\lambda} \equiv \frac{1}{\lambda_1!\ldots\lambda_n!}\frac{\partial^{\lambda_1+\ldots+\lambda_n}}{\partial z_1^{\lambda_1}\ldots z_n^{\lambda_n}}, \quad dz \equiv dz_1\ldots dz_n$$

$$\sum_{\alpha\in\mathbb{N}^n} \equiv \sum_{\alpha_1=0}^{\infty} \cdots \sum_{\alpha_n=0}^{\infty} .$$

We shall introduce a partial ordering on $\mathbb{N}^n$ as follows:
For $\lambda, \tau \in \mathbb{N}^n$ $0 \preceq \lambda \preceq \tau \Longleftrightarrow 0 \leq \lambda_j \leq \tau_j$, $j = 1\ldots, n$. We let

$E_\tau: = \{\lambda \in \mathbb{N}^n: 0 \preceq \lambda \preceq \tau\}$, so that

$$\sum_{\alpha \in E_\tau} \equiv \sum_{\alpha_1=0}^{\tau_1} \cdots \sum_{\alpha_n=0}^{\tau_n} .$$

Let Ω be a domain in $\mathbb{C}^n$ such that $0 \in \Omega$, then $H(\Omega)$ is the ring of all functions holomorphic on Ω and continuous on $\overline{\Omega}$; $M(\Omega)$ is the set of all functions analytic at the origin but meromorphic in Ω with finitely many polar sections in Ω.

A convenient notation for a polynomial $P_\lambda(z)$ in $\mathbb{C}^n$ of multiple degree or simply 'degree' at most $\lambda \in \mathbb{N}^n$ is

$$P_\lambda(z) = \sum_{\gamma \in E_\lambda} g_\gamma z^\gamma$$

where $g_\gamma \equiv g_{\gamma_1 \ldots \gamma_n}$ and $z^\gamma \equiv z_1^{\gamma_1} \ldots z_n^{\gamma_n}$.

Let $R_{\mu\nu}$ be the class of rational functions of the form $R_{\mu\nu}(z) = P_\mu(z)/Q_\nu(z)$ where $Q_\nu(0) \neq 0$ and $P_\mu(z)$ and $Q_\nu(z)$ are polynomials of 'degree' at most μ and ν respectively; moreover, $(P_\mu(z), Q_\nu(z)) = 1$ for all $z \in \Delta_\rho^n$ for some $\rho > 0$ for all (μ,ν), except on a subvariety of codimension ≥ 2.

§2. RATIONAL APPROXIMANT; ITS DEFINITION AND UNIQUENESS

Let U be an 0-neighbourhood in $\mathbb{C}^n$.

Definition 2.1. Suppose $f \in H(U)$ with $f(0) \neq 0$. A rational function $R_{\mu\nu}(z) \in R_{\mu\nu}$ is called a (μ,ν)-rational approximant to f if

$$\frac{\partial^{|\lambda|}}{\partial z^\lambda} (Q_\nu(z)f(z) - P_\mu(z))\Big|_{z=0} = 0 \tag{2.1}$$

for all $\lambda \in E^{\mu\nu} \subset \mathbb{N}^n$, an index interpolation set with the following properties:

(i) $0 \in E^{\mu\nu}$

(ii) $\lambda \in E^{\mu\nu} \Rightarrow \gamma \in E^{\mu\nu} \ \forall \gamma \in E_\lambda$

(iii) $E_\mu \subset E^{\mu\nu}$

(iv) Each projected variable has the Padé indexing set

(v) $|E^{\mu\nu}| \le \prod_{j=1}^{n} (\mu_j + 1) + \prod_{j=1}^{n} (\nu_j + 1) - 1$ where

$|E^{\mu\nu}|$ is the cardinality of $E^{\mu\nu}$.

Remark. If $f(0) = 0$, i.e. f is regular up to some finite order in some variable z_j, then Weierstrass Preparation theorem may be used to write $f = W \cdot g$ where $g(0) \neq 0$ and $W = W(\hat{z}, z_j)$ is a pseudo polynomial in z_j satisfying $W(\hat{0},0) = 0$. Definition 2.1 then applies to g.

Remark. For the rest of this paper we shall assume that $\forall\ (\mu,\nu)\ \nu \preceq \mu$.

The rational approximants defined above are in general, not unique. In order to obtain the kind of uniqueness known in the Padé case, the index interpolation set $E^{\mu\nu}$ must be maximal.

Definition 2.2. The index interpolation set $E^{\mu\nu}$ is said to be maximal if $|E^{\mu\nu}| \ge \prod_{j=1}^{n} (\mu_j + 1) + \prod_{j=1}^{n} (\nu_j + 1) - 1$.

There are many maximal $E^{\mu\nu}$ associated with a pair (μ,ν). Each maximal $E^{\mu\nu}$ determines its corresponding Padé Table: this feature of rational approximants in several variables has no Padé analogue.

Theorem 2.1. Let the pair $\langle P_\mu(z), Q_\nu(z)\rangle$ define a (μ,ν) approximant to $f \in \mathcal{H}(u)$ w.r.t. $E^{\mu\nu}$ maximal. Then $\langle P_\mu(z), Q_\nu(z)\rangle$ is uniquely determined w.r.t. $E^{\mu\nu}$ maximal, except on some subvariety of codimension ≥ 2.

From equation (2.1) we separate out the following system of equations

$$\frac{\partial^{|\lambda|}}{\partial z^\lambda} (Q_\nu(z) f(z) - P_\mu(z)) \Big|_{z=0} = 0, \ \lambda \in E_\mu \qquad (2.7)$$

$$\frac{\partial^{|\lambda|}}{\partial z^{\lambda}} (Q_{\nu}(z)f(z)) \Big|_{z=0} = 0, \qquad \lambda \in E^{\mu\nu} \setminus E_{\mu} \tag{2.8}$$

where $E^{\mu\nu}$ is maximal. The solution of (2.8) provided a certain rank or determinantal condition is satisfied, gives rise to a rational approximant we shall call unisolvent. For the rest of this paper, we shall assume that all (μ,ν)-rational approximants considered are (μ,ν) unisolvent rational approximants (URA) w.r.t. some maximal $E^{\mu\nu}$. We shall denote (μ,ν)-URA's by $\pi_{\mu\nu}(z) = P_{\mu\nu}(z)/Q_{\mu\nu}(z)$, where $Q_{\mu\nu}$ is assumed to be normalized, i.e. its largest coefficient is unity.

§3. CONVERGENCE

Under this heading we state the Montessus de Ballore's theorem and then provide an elementary example† to show the denominator of a URA with an appropriate 'degree' tends to that of some meromorphic function with similar 'degree'. We then discuss a convergence in measure result.

Theorem 3.1. Let $\rho > 0$ and $\nu = (\nu_1, \ldots, \nu_n)$ be fixed. Suppose $f \in M(\Delta_{\rho}^{n})$ with a finitely sectioned polar set defined $G_{\nu} := \{z \in \mathbb{C}^n : q_{\nu}(z) = 0\}$ and $q_{\nu} f \in C(\bar{\Delta}_{\rho}^{n})$, where $q_{\nu}(z)$ is a polynomial with exact minimal 'degree' ν and $\Delta_{\rho}^{n} \cap G_{\nu} \neq \phi$.

Suppose $\pi_{\mu\nu}(z)$ is a (μ,ν) URA to $f(z)$ with its polar set given by $Q_{\mu\nu}^{-1}(0) = \{z \in \mathbb{C}^n : Q_{\mu\nu}(z) = 0\}$ where 'degree' of $Q_{\mu\nu}(z)$ is exactly ν and for sufficiently large μ, $\Delta_{\rho}^{n} \cap Q^{-1}(0) \neq \phi$. Then as $\mu' = \min_{1\le i\le n} (\mu_i) \to \infty$, we obtain

(i) $\Delta_{\rho}^{n} \cap Q_{\mu\nu}^{-1}(0) \to \Delta_{\rho}^{n} \cap G_{\nu}$

(ii) $\pi_{\mu\nu}(z) \to f(z)$ uniformly on compact subsets of $\Delta_{\rho}^{n} \setminus G_{\nu}$.

The proof of this theorem and that of a partial converse of the theorem may be found in Lutterodt [1984].

<u>Example</u>. Let $f(z_1,z_2) = \frac{F(z_1,z_2)}{q_{11}(z_1,z_2)}$ where $f(z_1,z_2)$ is entire in $\mathbb{C}^2$ and it is given by $F(z_1,z_2) = \frac{z_1 e^{z_1} - z_2 e^{z_2}}{z_1 - z_2}$ and $q_{11}(z_1,z_2) = 1 - \frac{1}{3}z_1 - \frac{1}{2}z_2 + \frac{1}{6}z_1 z_2$. Then $f(z_1,z_2)$ is meromorphic in any polydisk $\Delta_\rho^2 = \{(z_1,z_2) : |z_1| < \rho, |z_2| < \rho, \rho \geq 4\}$ in $\mathbb{C}^2$ but $f(z_1,z_2)$ is holomorphic in the polydisk $U = \{(z_1,z_2) : |z_1| < 3, |z_2| < 2\} \subset \Delta_\rho^2$, $\rho \geq 4$ and has a congent Taylor development in U given by

$$f(z_1,z_2) = \sum_{m=0}^{\infty} \sum_{n=0}^{\infty} c_{mn} z_1^m z_2^n$$

where

$$c_{mn} = \frac{1}{3^m 2^n} \sum_{r=0}^{m} \sum_{s=0}^{n} \frac{3^r 2^s}{(r+s)!} \; ; \; m, n = 0, 1, \ldots$$

Let $\pi_{\mu 1} = P_{\mu 1}(z_1,z_2)/Q_{\mu 1}(z_1,z_2)$ be a $(\mu_1\mu_2 11)$ - URA to $f(z_1,z_2)$ w.r.t. a maximal $E^{\mu 1}$ with $|E^{\mu 1}| = \prod_{j=1}^{2} (\mu_j + 1) + \prod_{j=1}^{2} (\nu_j + 1) - 1$, with $\nu_1 = \nu_2 = 1$. Let $E_\mu = \{\gamma_1,\gamma_2)$: $0 \leq \gamma_1 \leq \mu_1$, $0 \leq \gamma_2 \leq \mu_2\}$ and we select $E^{\mu 1} \setminus E_\mu = \{\gamma_1 = \mu_1 + 1, 0 \leq \gamma_2 \leq 1\} \cup \{\gamma_1 = 0, \gamma_2 = \mu_2 + 1\}$. Recall that we want to show that $Q_{\mu 1}(z_1,z_2) \to q_{11}(z_1,z_2)$ as $\mu \to (\infty, \infty)$. We write

$Q_{\mu 1}(z_1,z_2) = 1 + b_{10}^{(\mu)} z_1 + b_{01}^{(\mu)} z_2 + b_{11}^{(\mu)} z_1 z_2$ and from equation (2.8) using the above $E^{\mu 1} \setminus E_\mu$ we get

$$\begin{aligned} 0 &= c_{\mu_1+1,0} + b_{10}^{(\mu)} c_{\mu_1,0} \\ 0 &= c_{\mu_1+1,1} + b_{10}^{(\mu)} c_{\mu_1,1} + b_{01}^{(\mu)} c_{\mu_1+1,0} + b_{11}^{(\mu)} c_{\mu_1,0} \\ 0 &= c_{0,\mu_2+1} + b_{01}^{(\mu)} c_{0,\mu_2}. \end{aligned} \tag{3.1}$$

Now using the form of c_{mn} computed above, we get

$$b_{10}^{(\mu)} = -\frac{c_{\mu_1+1,0}}{c_{\mu,0}} = -\frac{1}{3}\,\frac{\sum_{r=0}^{\mu_1+1} \frac{3^r}{r!}}{\sum_{r=0}^{\mu_1} \frac{3^r}{r!}} \to -\frac{1}{3} \quad \text{as} \quad \mu \to (\infty, \infty)$$

$$b_{01}^{(\mu)} = -\frac{c_{0,\mu_2+1}}{c_{0,\mu_2}} = -\frac{1}{2}\,\frac{\sum_{s=0}^{\mu_2+1} \frac{2^s}{s!}}{\sum_{s=0}^{\mu_2} \frac{2^s}{s!}} \to -\frac{1}{2} \quad \text{as} \quad \mu \to (\infty, \infty)$$

$$b_{11}^{(\mu)} = -\frac{c_{\mu_1+1,1}}{c_{\mu_1,0}} + \frac{c_{\mu_1,1}}{c_{\mu_1,0}} \cdot \frac{c_{\mu_1+1,0}}{c_{\mu_1,0}} + \frac{c_{\mu_1+1,0}}{c_{\mu_1,0}} \cdot \frac{c_{0,\mu_2+1}}{c_{0,\mu_2}}$$

$$\to -\frac{1}{3}\left(\frac{5}{6}\right) + \frac{1}{3}\left(\frac{5}{6}\right) + \frac{1}{6} = \frac{1}{6} \quad \text{as} \quad \mu \to (\infty, \infty).$$

This shows that $Q_{\mu 1}(z_1, z_2) \to q_{11}(z_1, z_2)$ as $\mu \to (\infty, \infty)$.

Next we state the following Lemmas which are used in the proof of the next theorem.

Lemma 3.2. For $0 < \rho' < \rho$ and $\mu' = \min_{1 \le j \le n} (\mu_j)$

$$\sum_{\lambda \in \mathbb{N}^n \setminus E_\mu} \left(\frac{\rho'}{\rho}\right)^{|\lambda|} \le \frac{n\left(\frac{\rho'}{\rho}\right)^{\mu'+1}}{\left(1 - \frac{\rho'}{\rho}\right)n} .$$

Lemma 3.3. (Bishop.) Let $F_{\underline{d}}$ be a normalized polynomial of 'degree' $\underline{d} = (d, \ldots, d)$ in $\mathbb{C}^n$. Let $\rho > 0$ and $0 < \eta < 1$. Then $\exists\ c = c(n,\rho)$ such that

$$Z_\eta = \{z \in \bar{\Delta}^n_\rho : \ |F_{\underline{d}}(z)| < \eta^d\}$$

has $\mathbb{C}^n$-Lebeague measure satisfying $m(Z_\eta) \le c\eta^{2/n}$.

Theorem 3.4. Let $\varepsilon > 0$ and $0 < \eta < 1$ be given. Let $\omega \in \mathbb{N}$ be fixed. Suppose $f \in M(\Delta^n_\rho)$ and its polar set given by $G_{\underline{\omega}} := \{z \in \mathbb{C}^n : q_{\underline{\omega}}(z) = 0\}$ with $q_{\underline{\omega}} f \in C(\bar{\Delta}^n_\rho)$, is defined by the polynomial $q_{\underline{\omega}}(z)$ with minimal 'degree' $\underline{\omega} = (\omega, \ldots, \omega)$. Suppose $\mu := (\mu_1, \ldots, \mu_n)$ and $\underline{\nu} := (\sigma, \ldots, \sigma)$ are such that $0 < \omega < \sigma < \mu'$ and $\sigma = \sigma(\mu') \to \infty$ as $\mu' \to \infty$ but $(\sigma + \omega)/\mu' \to 0$.

Suppose $\pi_{\mu\nu}(z)$ is a (μ, ν) - URA to $f(z)$ at 0. Then for any $K \subset \Delta^n_\rho$ compact, $\exists\ c > 0,\ 0 < \delta < 1$ and μ'_0 such that $\mu' > \mu'_0 \Rightarrow |f(z) - \pi_{\mu\nu}(z)| < \delta^{\mu'}\ \forall\ z \in K \setminus Z^{\mu\underline{\nu}}_\eta$ where $m(Z^{\mu\underline{\nu}}_\eta) < \varepsilon$.

Proof 1°. Without loss of generality we shall assume that $q_{\underline{\omega}}(z)$ and $Q_{\mu\underline{\nu}}(z)$ are normalized in a similar manner with largest coefficient unity. Now $q_{\underline{\omega}} f \in H(\Delta^n_\rho)$ and since $Q_{\mu\underline{\nu}}$ and $P_{\mu\underline{\nu}}$ are polynomials, therefore $H_{\mu\underline{\nu}} = Q_{\mu\underline{\nu}} q_{\underline{\omega}} f - q_{\underline{\omega}} P_{\mu\underline{\nu}} \in H(\Delta^n_\rho)$. Hence by Cauchy's integral formula

$$H_{\mu\underline{\nu}}(z) = \frac{1}{(2\pi i)^n} \int_T \frac{H_{\mu\underline{\nu}}(t)}{\prod_{j=1}^{n} (t_j - z_j)} dt_1 \ldots dt_n \qquad (3.2)$$

where T is the distinguished boundary of Δ_ρ^n. Now observe

$(\prod_{j=1}^{n} (t_j - z_j))^{-1} = \sum_{\lambda \in \mathbb{N}^n} \frac{z^\lambda}{t^{\lambda+1}}$ is convergent absolutely in

Δ_ρ^n and uniformly on compact subsets of Δ_ρ^n. Thus substituting in (3.2) and interchanging $\int$ and Σ

$$H_{\mu\underline{\nu}}(z) = \sum_{\lambda \in \mathbb{N}^n} h_{\mu\underline{\nu}\lambda} z^\lambda \qquad (3.3)$$

where

$$h_{\mu\underline{\nu}\lambda} = \frac{1}{(2\pi i)^n} \int_T \frac{H_{\mu\underline{\nu}}(t)}{t^{\lambda+1}} dt_1 \ldots dt_n. \qquad (3.4)$$

We claim that

$$H_{\mu\underline{\nu}}(z) = \sum_{\lambda \in \mathbb{N}^n \setminus E^{\mu\nu}} h_{\mu\underline{\nu}\lambda} z^\lambda \qquad (3.5)$$

with

$$h_{\mu\underline{\nu}\lambda} = \frac{1}{(2\pi i)^n} \int_T \frac{Q_{\mu\underline{\nu}}(t) q_{\underline{\omega}}(t) f(t)}{t^{\lambda+1}} dt_1 \ldots dt_n \qquad (3.6)$$

To see the claim recall that $H_{\mu\underline{\nu}} = q_{\underline{\omega}}(Q_{\mu\underline{\nu}} f - P_{\mu\underline{\nu}})$. Then

$$h_{\mu\underline{\nu}\lambda} = \frac{1}{(2\pi i)^n} \int_T \frac{H_{\mu\underline{\nu}}(t)}{t^{\lambda+1}} dt_1 \ldots dt_n = \qquad (3.7)$$

$$\frac{1}{\lambda_1! \ldots \lambda_n!} \frac{\partial^{|\lambda|}}{\partial z^\lambda} [H_{\mu\underline{\nu}}(z)] \Big|_{z=0}$$

Now apply (2.7) and (2.8) together with Leibnitz's rule and we immediately obtain

$$\frac{\partial^{|\lambda|}}{\partial z^{\lambda}}[q_{\underline{\omega}}(z)(Q_{\mu\underline{\nu}}(z)f(z)-P_{\mu\underline{\nu}}(z))]\Big|_{z=0}=0,\ \lambda\in E_{\mu}$$

$$\frac{\partial^{|\lambda|}}{\partial z^{\lambda}}[q_{\underline{\omega}}(z)Q_{\mu\underline{\nu}}(z)f(z)]\Big|_{z=0}=0,\quad \lambda\in E^{\mu\underline{\nu}}\setminus E_{\mu}$$

from which the claim follows.

Since $Q_{\mu\underline{\nu}}(z)$ is normalized and locally bounded in $\mathbb{C}^n$ and $q_{\underline{\omega}}(z)f(z)\in C(\bar{\Delta}^n_{\rho})$ by hypothesis, maximum principle provides $M = M(\rho)$ independent of μ on T such that

$$|h_{\mu\underline{\nu}\lambda}|\le\frac{M}{\rho^{|\lambda|}}\qquad \lambda\in \mathrm{IN}^n\setminus E^{\mu\underline{\nu}} \tag{3.8}$$

from (3.6). Combining (3.5) and (3.8) yield

$$|H_{\mu\underline{\nu}}(z)|\le M\sum_{\lambda\in \mathrm{IN}^n\setminus E^{\mu\underline{\nu}}}\frac{|z^{\lambda}|}{\rho^{|\lambda|}}\le M\sum_{\lambda\in \mathrm{IN}^n\setminus E_{\mu}}\frac{|z^{\lambda}|}{\rho^{|\lambda|}} \tag{3.9}$$

where $E^{\mu\underline{\nu}}\supset E_{\mu}$. From this inequality we get for $z\in\Delta^n_{\rho}$

$$|f(z)-\pi_{\mu\underline{\nu}}(z)|\le\frac{M\kappa}{|Q_{\mu\underline{\nu}}(z)q_{\underline{\omega}}(z)|}\Big(\sum_{\lambda\in \mathrm{IN}^n\setminus E_{\mu}}\frac{|z^{\lambda}|}{\rho^{|\lambda|}}\Big) \tag{3.10}$$

Now let K be any compact subset of Δ^n_{ρ} and choose $\rho' > 0$ so that $0<\rho'<\rho \Rightarrow \Delta^n_{\rho'}\subset\Delta^n_{\rho}$ and $K\subset\Delta^n_{\rho'}$. Then by Lemma 3.2, (3.10) is sharpen to give for $z\in K$

$$|f(z)-\pi_{\mu\underline{\nu}}(z)|\le\frac{M}{|Q_{\mu\underline{\nu}}(z)q_{\underline{\omega}}(z)|}\Big(\frac{\rho'}{\rho}\Big)^{\mu'+1} \tag{3.11}$$

where $\kappa=\kappa(\rho',\rho,n)$ is a constant depending on n, ρ, ρ'.

2°. Given $0<\eta<1$, let $\underline{d}=\underline{\mu}+\underline{\omega}=(\sigma+\omega,\dots,\sigma+\omega)$

and $F_{\underline{d}}(z) = Q_{\mu\underline{\nu}}(z)q_{\underline{\omega}}(z)$. Then $F_{\underline{d}}$ is a normalized polynomial of 'degree' $\underline{d}$. Now let

$$Z_{\eta}^{\mu\underline{\nu}} = \{z \in K: \ |F_{\underline{d}}(z)| \le \eta^{d}\} \quad \text{with} \quad d = \sigma + \omega.$$

then by Lemma 3.3 we have the $\mathbf{C}^n$-Lebesgue measure of $Z_{\eta}^{\mu\underline{\nu}}$ satisfying $m(Z_{\eta}^{\mu\underline{\nu}}) \le c\ \eta^{2/n}$; which is independent of d.

For $z \in K \setminus Z_{\eta}^{\mu\underline{\nu}}$ we must have $|F_{\underline{d}}(z)| \ge \eta^{d} > 0$ so that from (3.11)

$$|f(z) - \pi_{\mu\underline{\nu}}(z)| \le \frac{M_K}{\eta^{\sigma+\omega}} \left(\frac{\rho'}{\rho}\right)^{\mu'+1} \tag{3.12}$$

Since $0 < \frac{\rho'}{\rho} < 1$ and $\mu' \to \infty$, we can find $\delta \in \,]0, 1[$ and μ_0' such that $\mu' > \mu_0'$ with $\sigma = o(\mu') \Longrightarrow$

$$|f(z) - \pi_{\mu\underline{\nu}}(z)| < \delta^{\mu'}$$

for $z \in K$ except for $z \in Z_{\eta}^{\mu\underline{\nu}}$ with $m(Z_{\eta}^{\mu\underline{\nu}}) < \varepsilon$ where $c\eta^{2/n} < \varepsilon$.

The above result is an n-dimensional generalization of Gonĉar's [1975] result; see Wallin [1979, 1980].

§4. MEROMORPHIC MAPS AND RATIONAL APPROXIMANTS

A map $\underset{\sim}{f}: U \to \mathbb{C}^m$ where $U \subset \mathbb{C}^n$, is called holomorphic if each component f_j, $j = 1,\dots, m$ is holomorphic on U. $\underset{\sim}{f}(z) := (f_1(z),\dots, f_m(z))$.

We let $\mu_j := (\mu_{1_j},\dots,\mu_{n_j})$, $\nu_j := (\nu_{1_j}, \dots, \nu_{n_j}) \in \mathrm{IN}^n$, IN^n, $j = 1,\dots, m$. To each pair (μ_j,ν_j) we introduce $R_{\mu_j\nu_j}$

the class of rational functions in $\mathbb{C}^n$ of the form $R_{\mu_j\nu_j}(z) = P_{\mu_j\nu_j}(z)/Q_{\mu_j\nu_j}(z)$ with $Q_{\nu_j}(0) \neq 0$ and $(P_{\mu_j}(z), Q_{\nu_j}(z)) = 1$, $j = 1,\ldots, m$, except on a set of codimension ≥ 2. $R_{\underset{\sim}{\mu\nu}}(z) := (R_{\mu_1\nu_1}(z)\ldots, R_{\mu_m\nu_m}(z)) \in \mathcal{R}_{\mu_1\nu_1} \times \ldots \times \mathcal{R}_{\mu_m\nu_m}$ is a rational map provided it is defined away from an analytic subvariety of codimension ≥ 2 of the common polar locus.

Definition 4.1. Suppose $\underset{\sim}{f}(z)$ is a holomorphic map on an 0-neighbourhood U so that $f_j \in H(U)$. $f_j(0) \neq 0$, $j = 1,\ldots, m$. A rational map $R_{\underset{\sim}{\mu\nu}}(z) \in \mathcal{R}_{\mu_1\nu_1} \times \ldots \times \mathcal{R}_{\mu_m\nu_m}$ is called a multiple $(\underset{\sim}{\mu},\underset{\sim}{\nu})$ - rational approximant to $\underset{\sim}{f}(z)$ at 0 if to each j-component

$$\frac{\partial^{|\lambda_j|}}{\partial z^{\lambda_j}} (Q_{\nu_j}(z)f_j(z) - P_{\mu_j}(z)) \Big|_{z=0} = 0, \ \lambda_j \in E^{\mu_j\nu_j} \qquad (4.1)$$

where $E^{\mu_j\nu_j}$ satisfies the conditions laid down in definition 2.1 for each j.

Definition 4.2. Suppose $E^{\mu_j\nu_j}$ is maximal for $j = 1,\ldots, m$ in $\mathbb{N}^n$. Then a multiple $(\underset{\sim}{\mu},\underset{\sim}{\nu})$ rational approximant $\pi_{\underset{\sim}{\mu\nu}}(z)$ is called a URA to $\underset{\sim}{f}(z)$ at 0 if each j-component $\pi_{\mu_j\nu_j}(z)$ is URA to f_j at 0.

Let $D \subset \mathbb{C}^n$. A map $\underset{\sim}{f}: D \to \bar{\mathbb{C}}^m$ is called moromorphic if each component be a domain $f_j \in M(D)$, $j = 1,\ldots, m$ and $\underset{\sim}{f}$ is defined away from the common polar locus of codim ≥ 2. We shall assume for the rest of this section that meromorphic maps and their multiple rational approximants, all defined away from proper analytic sub-varieties of codimension ≤ 2 are replaced by their meromorphic extensions in accordance with Levi's theorem, see Griffith and Harris (1978).

The following theorem which represents a further generalization of Montessus de Ballore's theorem was first stated in Lutterodt [1983].

Theorem 4.1. Let ν_j and $\rho > 0$ be fixed $j = 1,\ldots,m$. Suppose $f_j \in M(\Delta^n_\rho)$ with its polar set $G_{\nu_j} := \{z \in \mathbb{C}^n : q_{\nu_j}(z) = 0\}$ and $q_{\nu_j} f_j \in C(\bar{\Delta}^n_\rho)$ where $q_{\nu_j}(z)$ is a polynomial of minimal 'degree' ν_j and $G_{\nu_j} \cup \Delta^n_\rho \neq \phi$, $j = 1,\ldots,m$ so that the polar set of $\underset{\sim}{f}(z)$ given by $\bigcup_{j=1}^m G_{\nu_j}$ satisfies $\bigcup_{j=1}^m G_{\nu_j} \cap \Delta^n_\rho \neq \phi$.

Let $\pi_{\underset{\sim}{\mu}\underset{\sim}{\nu}}(z)$ be a (μ,ν)-mulptiple URA to $\underset{\sim}{f}(z)$ at 0 with its polar set given by $\bigcup_{j=1}^m Q^{-1}_{\mu_j\nu_j}(0)$. Suppose for μ_j sufficiently large, $j = 1,\ldots,m$

$$\bigcup_{j=1}^m Q^{-1}_{\mu_j\nu_j}(0) \cap \Delta^n_\rho \neq \phi.$$

Then as $\mu' = \min_{1\le i\le n} \min_{1\le j\le m} (\mu_{ij}) \to \infty$

(i) $\bigcup_{j=1}^m Q^{-1}_{\mu_j\nu_j}(0) \cap \Delta^n_\rho \to \bigcup_{j=1}^m G_{\nu_j} \cap \Delta^n_\rho$

(ii) $\pi_{\underset{\sim}{\mu}\underset{\sim}{\nu}}(z) \to \underset{\sim}{f}(z)$ uniformly on compact subsets of

$$\Delta^n_\rho \setminus \bigcup_{j=1}^m G_{\nu_j}.$$

Proof: (i) follows from (i) of theorem 3.1, taking unions over $j = 1,\ldots,m$ and (ii) uniform convergence of each $\pi_{\mu_j\nu_j}(z)$ to $f_j(z)$ on compact subsets of $\Delta^n_\rho \setminus G_{\nu_j}$ is equiva- to, given $K \subset \Delta^n_\rho \setminus G_{\nu_j}$ compact, $j = 1,\ldots,m$,

$$\lim_{\mu'\to\infty} \sup \| f_j(z) - \pi_{\mu_j\nu_j}(z)\|_K^{1/\mu'} \le \frac{\rho'}{\rho} < 1$$

for $z \in K$. Thus by introducing a norm on $\mathbb{C}^m$ in terms of f_j's, $\pi_{\mu_j\nu_j}$'s w.r.t. K compact in $\Delta_\rho^n \setminus \bigcup_{j=1}^m G_{\nu_j}$ as

$$|||\underset{\sim}{f}(z) - \pi_{\underset{\sim\sim}{\mu\nu}}(z)||| = \max_{1\le j\le m} (\| f_j(z) - \pi_{\mu_j\nu_j}(z)\|_K)$$

the desired result follows. The rate of convergence in the case of the maps is no different from that of the functions. Both are geometric controlled by the appropriate μ', as the latter tends to infinity.

Next we move on to the generalization of the convergence in measure discussed in theorem 3.4.

Theorem 4.2. Let $\varepsilon > 0$ and $0 < \eta < 1$ be given. Let $\omega \in \mathbb{N}$ be fixed. Suppose $f_j \in M(\Delta_\rho^n)$ and its finitely sectioned polar set given by $G_{j\underline{\omega}} := \{z \in \mathbb{C}^n : q_{j\underline{\omega}}(z) = 0\}$ with $q_{j\underline{\omega}} f_j \in C(\bar{\Delta}_\rho^n)$, is specified by a polynomial with minimal 'degree' $\underline{\omega}$ for each j. Suppose $\mu_j := (\mu_{1j},\ldots,\mu_{nj})$ and $\underline{\nu}_j = (\sigma,\ldots,\sigma)$ are such that $0 < \omega < \sigma < \mu'$, where $\sigma = \sigma(\mu')$ and $\mu' = \min_{1\le i\le n} \min_{1\le j\le m} (\mu_{ij})$ but $\mu' \to \infty \Rightarrow \sigma = o(\mu')$.

Suppose $\pi_{\underset{\sim\sim}{\mu\nu}}(z)$ is a multiple $(\underset{\sim}{\mu}, \underset{\sim}{\underline{\nu}})$ - URA to $\underset{\sim}{f}(z)$ at 0. Then for $K \subset \Delta_\rho^n$ compact, $\exists\, c > 0$, $0 < \delta < 1$ and μ' such that $\mu' > \mu_0'$

$$\Rightarrow \|\underset{\sim}{f}(z) - \pi_{\underset{\sim\sim}{\mu\nu}}(z)\| = \max_{1\le j\le m} |f_j(z) - \pi_{\mu_j\underline{\nu}_j}(z)| < \delta^{\mu'}$$

$$\forall\, z \in K \setminus \bigcup_{j=1}^m Z_\eta^{\mu_j\underline{\nu}_j} \quad \text{where} \quad m(\bigcup_{j=1}^m Z_\eta^{\mu_j\underline{\nu}_j}) < \varepsilon$$

Proof. This follows easily from the result of theorem 3.4 applied to each component of $\underset{\sim}{f} - \pi_{\underset{\sim}{\mu\nu}}$ and taking suitable unions of $Z_\eta^{\mu_j \nu_j}$.

Appendix. In the example discussed in §3 to illustrate how the Montessud de Ballore's theorem works in several variables we merely showed the (i) part of the conclusion of the theorem, i.e. as $(\mu_1,\mu_2) \to (\infty, \infty)$, $Q_{\mu_1\mu_2 11}(z_1,z_2) \to q_{11}(z_1,$ in Δ_ρ^2. In this appendix we want to establish (ii) part of the theorem for that example.

Recall that $f(z_1,z_2) = F(z_1,z_2) / q_{11}(z_1,z_2)$ where

$F(z_1, z_2) = \dfrac{z_1 e^{z_1} - z_2 e^{z_2}}{z_1 - z_2}$ is entire in $\mathbb{C}^2$ and $q_{11}(z_1,z_2) = 1 - \frac{1}{3} z_1 - \frac{1}{2} z_2 + \frac{1}{6} z_1 z_2$. To show that $\pi_{\mu_1\mu_2 11}(z_1,z_2) = \dfrac{P_{\mu_1\mu_2 11}(z_1,z_2)}{Q_{\mu_1\mu_2 11}(z_1,z_2)}$ tends to $f(z_1,z_2)$ as $(\mu_1,\mu_2) \to (\infty, \infty)$ we need to check that $P_{\mu_1\mu_2 11}(z_1,z_2) \to F(z_1,z_2)$ since we have we have already established that $Q_{\mu_1\mu_2 11}(z_1,z_2) \to q_{11}(z_1,z_2)$.

$$P_{\mu_1\mu_2 11}(z_1,z_2) = \sum_{(\lambda_1,\lambda_2)\varepsilon E_{\mu_1\mu_2}} a^{(\mu)}_{\lambda_1\lambda_2} z_1^{\lambda_1} z_2^{\lambda_2}$$

$$a^{(\mu)}_{\lambda_1\lambda_2} = \sum_{\gamma_1=0}^{\min(\lambda_1,1)} \sum_{\gamma_2=0}^{\min(\lambda_2,1)} b^{(\mu)}_{\gamma_1\gamma_2} c_{\lambda_1-\gamma_1,\lambda_2-\gamma_2}$$

which reduces to

$$a^{(\mu)}_{\lambda_1\lambda_2} = b^{(\mu)}_{00} c_{\lambda_1\lambda_2} + b^{(\mu)}_{10} c_{\lambda_1-1,\lambda_2} + b^{(\mu)}_{01} c_{\lambda_1,\lambda_2-1} +$$

$$+ b_{11}^{(\mu)} c_{\lambda_1-1,\lambda_2-1}$$

with $b_{00}^{(\mu)} = 1$.

The desired result is proved if we can show that as $(\mu_1,\mu_2) \to (\infty, \infty)$, $a_{\lambda_1\lambda_2}^{(\mu)} \to \frac{1}{(\lambda_1+\lambda_2)!}$, $\forall\ \lambda_1, \lambda_2 \geq 0$, i.e. the coefficients in the expansion of $F(z_1,z_2)$. To see this, we compute the $\lim_{(\mu_1,\mu_2)\to(\infty,\infty)} a_{\lambda_1\lambda_2}^{(\mu)}$ directly. Note that we have already computed limists of $b_{10}^{(\mu)}$, $b_{01}^{(\mu)}$ and $b_{11}^{(\mu)}$ and thus using these we obtain

$$\lim_{(\mu_1,\mu_2)\to(\infty,\infty)} a_{\lambda_1\lambda_2}^{(\mu)} = c_{\lambda_1\lambda_2} - \frac{1}{3} c_{\lambda_1-1,\lambda_2} - \frac{1}{2} c_{\lambda_1,\,\lambda_2-1} +$$

$$\frac{1}{6} c_{\lambda_1-1,\lambda_2-1}$$

$$= \frac{1}{3^{\lambda_1}2^{\lambda_2}} \Big[\sum_{r=0}^{\lambda_1} \sum_{s=0}^{\lambda_2} \frac{3^r\,2^s}{(r+s)!} - \sum_{r=0}^{\lambda_1-1} \sum_{s=0}^{\lambda_2} \frac{3^r\,2^s}{(r+s)!}$$

$$- \sum_{r=0}^{\lambda_1} \sum_{s=0}^{\lambda_2-1} \frac{3^r\,2^s}{(r+s)!} + \sum_{r=0}^{\lambda_1-1} \sum_{s=0}^{\lambda_2-1} \frac{3^r\,2^s}{(r+s)!} \Big]$$

$$= \frac{1}{3^{\lambda_1}2^{\lambda_2}} \Big[\frac{3^{\lambda_1}2^{\lambda_2}}{(\lambda_1+\lambda_2)!}\Big] = \frac{1}{(\lambda_1+\lambda_2)!}$$

†See appendix for the other half of the example not discussed in §3.

REFERENCES

1. Bishop, E., "Holomorphic Completions, Analytic Continuations and interpolation of semi norms", Ann. Math 78, No. 3, 1963, pp. 468-500.

2. Goncar, A. A., "A local condition for single-valuedness of analytic functions of several variables", Math. USSR Sb. 22, No. 1, 1974, pp. 305-322.

3. __________, "On the convergence of generalized Padé approximants of meromorphic functions", Math. USSR Sb. 27, 1976.

4. Cuty, A., Preprint, 1981.

5. Griffith, P. and Harris, J., "Principles of Algebraic Geometry", 1978, John Wiley & Sons.

6. Karlsson, J. and Wallin, H., "Rational approximation by an Interpolation procedure in several variables", Padé & Rational Approximation - Eds. Saff & Varga, AP, 1977, pp. 83-100.

7. Lutterodt, C. H., "A two dimension analogue of Padé Approximant Theory", J. Phys. A. Math. Nucl. Gen. 7, 1974, pp. 1027-1037.

8. __________, "Rational Approximants to Holomorphic Functions in n-Dimensions", J. Math. Anal. & Applic. 53, No. 1, 1976, pp. 89-98.

9. __________, "On a theorem of Montessus de Ballore for (μ,ν)-type Rational Approximants in $\mathbb{C}^n$", Approximation Theory III, 1980, pp. 603-609.

10. __________, "On Uniform Convergence for (μ, ν)-type Rational Approximants in $\mathbb{C}^n$ - II", Internat. J. Math. & Math. Sci. 4, No. 4, 1981, pp. 655-660.

11. __________, "On a Partial Converse of Montessus de Ballore theorem in $\mathbb{C}^n$", to appear J. Approx., 1984.

12. __________, "Rational Approximants to meromorphic maps in $\mathbb{C}^n$", to appear Approximation Theory IV, 1983.

13. Wallin, H., "Potential Theory and Approximation of Analytic Functions by Rational Interpolations", Lecture Notes in Math. 747, Springer-Verlag, 1979, pp. 434-450.

14. __________, "Rational Interpolation to Meromorphic functions", Umea preprint, 1980.

SPLINES AND COLLOCATION FOR ORDINARY INITIAL VALUE PROBLEMS

Syvert P. Norsett

ABSTRACT

Normally, the methods for solving ordinary initial value problems are viewed as discrete algorithms. However, a subclass of these schemes can be constructed by aiming at a global continuous approximation to the unknown solution. In this paper we approximate the solution by a spline of degree m and continuity k, $0 \le k \le m - 1$. In each subinterval the $m - k$ free parameters are determined by collocation. The choice of collocation points is discussed with respect to zero-stability, superconvergence and A-stability.

1. INTRODUCTION

For solving numerically the problem

$$\begin{cases} y'(x) = f(y(x)),\ y(x) \in R^S, & a \le x \le b \\ y(a) = y_0 \end{cases} \tag{1}$$

one normally uses a discrete scheme, i.e., one computes approximations y_n to $y(x_n)$, $n \ge 0$ where $x_n \in G$ and G the set of distinct step points. Methods of this type are Runge-Kutta methods, linear multistep methods and extrapolation methods. (A treatment of these schemes can be found in Henrici (1962), Gear (1969) and Lambert (1973).)

S. P. Singh et al. (eds.), Approximation Theory and Spline Functions, 397–417.

In order to give approximations to $y(x)$ for x not a step point, some kind of interpolation is used. For linear multistep methods it is easy whereas for Runge-Kutta methods and in particular for extrapolation methods it is difficult.

One of the most popular linear multistep methods is the Adams family. Their construction is based on a local interpolation of f. This polynomial is then integrated and the scheme and its corresponding interpolation formula is obtained. A global continuous approximation to y is obtained by joining together from each interval these integrated local interpolants.

The general idea in this approach for solving (1) numerically is to choose in an interval (x_n, x_{n+1}) an "interpolating" function and tie it to the differential equation in one way or another. In this paper we will discuss the use of splines and collocation. In order to get a simple treatment, we will assume that we have constant stepsizes, i.e., $x_{n+1} = x_n + h$, $n = 0, 1, \ldots$, with $x_0 = a$. $h > 0$ is called the stepsize.

Let the space $S_m^{(k)}(h)$ of splines of degree m and continuity k be defined by:

$$S_m^{(k)}(h) = \{u \in C^{(k)}([a, b]),\ u|[x_n, x_{n+1}] \in \Pi_m,\ n \geq 0 \text{ and } u(a) = y_0\},\ 0 \leq k \leq m - 1. \tag{2}$$

On each interval $I_n = [x_n, x_{n+1}]$, $n \geq 1$ there are $m - k$ free parameters to be determined. (For the first interval I_0 we have m parameters.) Let $c_1, \ldots, c_{m-k}$ be $m - k$ real and distinct numbers. For determining these parameters on I_n, $n \geq 1$ we use the collocation condition

$$u'(x_n + c_i h) = f(u(x_n + c_i h)),\quad i = 1, \ldots, m - k. \tag{3}$$

The problems to be discussed in this connection are:

1) Existence of u
2) Zero-stability
3) Global error
4) Superconvergence
5) Special choices of the collocation points and

6) A-stability.

The existence of u for h small enough is trivial when f satisfies a Lipschitz condition in a neighbourhood of the exact solution y of (1).

We will end this presentation with some generalizations of this approach.

2. CONTINUOUS SPLINES, $k = 0$

The first serious treatment for $k = 0$ was given by Wright (1970). His main conclusion was that the discrete version was equivalent to certain implicit Runge-Kutta methods. The breakthrough regarding order and A-stability was the papers by Norsett & Wanner (1979) and Wanner, Hairer and Norsett (1978).

For the points $c_1, \ldots, c_m$ we define the Lagrange polynomials $\ell_i(x)$,

$$\ell_i(x) = \prod_{\substack{j=1 \\ j \neq i}}^{m} \frac{(x - c_j)}{(c_i - c_j)}, \qquad i = 1, \ldots, m$$

Let u be computed for $x \in [x_0, x_n]$. On the interval I_n, u is represented by

$$u(x) = y_n + h \sum_{j=1}^{m} k_j \int_0^{(x-x_n)/h} \ell_j(t)dt, \qquad k_j \in R^S \text{ for } \tag{4}$$

$j = 1, \ldots, m.$

The collocation condition (3) is then written as

$$k_i = f(y_n + h \sum_{j=1}^{m} a_{ij}k_j), \quad i = 1, \ldots, m \tag{5}$$

where

$$a_{ij} = \int_0^{c_i} \ell_j(t)dt.$$

The value of y_{n+1} is found from (4) to be

$$y_{n+1} = y_n + h \sum_{i=1}^{m} b_i k_i, \quad b_i = \int_0^1 \ell_i(t)dt, \quad i = 1, \ldots, m. \quad (6)$$

Hence the collocation method is in this case equivalent to special implicit m-stage Runge-Kutta methods. The methods are all zero-stable, independent on the choice of collocation points.

When viewed as RK-methods, we are interested in the order. For continuous approximation methods this is called superconvergence. The natural way to analyze this question is by using the Alekseev-Gröbner theorem, Norsett and Wanner (1979).

Theorem 1. The spline method with $k = 0$ possesses the same order $p \geq m$ as the corresponding quadrature formula based on $c_1, \ldots, c_m$. For the global error $e_n = y_n - y(x_n)$ we then have when f is sufficiently smooth $e_n = O(h^p)$. □

For the spline error this gives,

Theorem 2. When f is sufficiently smooth and $x \in G$

$$u(x) - y(x) = O\,(h^{\min(m,p-1)+1})$$

$$u^{(j)}(x) - y^{(j)}(x) = O\,(h^{m+1-j}), \quad j = 1, \ldots, m.$$

If one c_i is equal 1, $u_n^{(1)} - y^{(1)}(x_n) = O\,(h^p)$. □

The A-stability is studied by applying the method to the test equation $y' = \lambda y$, $\lambda \in C$ with $x \geq 0$ and $y(0) = y_0$. From Norsett and Wanner (1979),

Theorem 3. When the spline method with $k = 0$ is applied to $y' = \lambda y$, $x \geq 0$, $y(0) = y_0$ we get for $x \in I_n$

$$u(x) = \left[\sum_{j=0}^{m} N^{(j)}((x - x_n)/h) z^{m-j} \Big/ \sum_{j=0}^{m} N^{(j)}(0) z^{m-j} \right] y_n, \quad z = \lambda h$$

where $N(t)$ is the N-polynomial,

$$N(t) = \prod_{i=1}^{m} (t - c_i) \frac{1}{m!}. \quad □$$

Let us remark that this way of writing $u(x)$ was for $x = x_{n+1}$ given in Norsett (1975). He called a slightly related version of N the C-polynomial. For $x_n < x < x_{n+1}$ this formula was also proved by Iserles (1981).

The stability function is from Theorem 3:

$$R(z) = \sum_{j=0}^{m} N^{(j)}(1) z^{m-j} / \sum_{j=0}^{m} N^{(j)}(0) z^{m-j}. \tag{7}$$

With $N(t)$ as the scaled Legendre polynomials,

$$N(t) = \frac{1}{(n+m)!} \frac{d^n}{dt^n} [t^n (t-1)^m], \quad n \le m$$

different entries $R_{n/m}(z)$ of the Padé-approximations to $\exp(z)$ appear as the stability function.

The first treatment of A-acceptability[1] of $R_{n/m}(z)$ was given by Varga (1961) and Birkhoff and Varga (1965). They only considered $R_{m/m}(z)$. Later Ehle (1971) showed that indeed $R_{n/m}(z)$ is A-acceptable for $m - 2 \le n \le m$. Further he conjected that $R_{n/m}(z)$ is not A-acceptable for $n < m - 2$. This was an open problem until Wanner, Hairer and Norsett (1978) introduced the new and powerful technique, the order star. The result is,

Theorem 4. $R_{n/m}(z)$ is A-acceptable iff $m - 2 \le n \le m$. □

In order to compute u at each interval we need to solve (5) for $k_1, \ldots, k_m$. For simplicity let us assume $s = 1$ and set

$$K = h[k_1, \ldots, k_m]^T$$

$$f(ey_n + AK) = \{f(y_n + h \sum_{j=1}^{m} a_{ij} k_j)\}_{i=1,\ldots,m}$$

$$A = \{a_{ij}\}_{i,j=1}^{m}, \quad e = [1, \ldots, 1]^T \in R^m.$$

Then (5) is equivalent to

$$R(K) := K - hf(ey_n + AK) = 0. \tag{7}$$

Due to numerical efficiency (7) is solved by a Quasi-Newton method,

$$\begin{aligned} &N(J)\,(K^{(\ell+1)} - K^{(\ell)}) = -R(K^{(\ell)}),\ \ell \geq 0 \\ &N(J) = I - hJA,\quad J \approx (\partial f/\partial y)(y_n). \end{aligned} \tag{8}$$

(8) is for each ℓ a system of m linear equations. One of the nice ways of solving this is to assume

$$A = SDS^{-1}$$

where D is the Jordan matrix for A, Butcher (1976). Instead of (8) we then solve

$$N(D)\,(L^{(\ell+1)} - L^{(\ell)}) = -S^{-1}R(SL^{(\ell)}),\quad \ell \geq 0;\ K^{(\ell)} = SL^{(\ell)}.$$

Since diag $D = [\mu_1, \ldots, \mu_m]$ (μ_i eigenvalues of A), one should construct the method such that $\mu_1 = \ldots = \mu_m = \gamma \in R$. This is possible when $c_1, \ldots, c_m$ are chosen in the following way.

$$c_j = \gamma\,\gamma_{m,j},\qquad j = 1, \ldots, m$$

where $\gamma_{m,j}$ are the zeros of the Laguerre polynomial $L_m(x)$, Norsett (1976). The discrete order is m for all $\gamma \in R$ and $m + 1$ when $L'_{m+1}(1/\gamma) = 0$.

With these special collocation points the stability function turns out to be

$$R(z) = P_m(z)/(1 - \gamma z)^m;\quad P_m \in \Pi_m.$$

When the order is $m + 1$ we have restricted diagonal Padé-approximations and with $L_m(1/\gamma) = 0$ first restricted subdiagonal Padé-approximations are obtained, Norsett (1978). The A-acceptability of the restricted approximations has been studied by Norsett (1974a), Norsett (1974b), Norsett (1975), Wolfbrandt (1977) and Wanner, Hairer and Norsett (1978). For example, does there not exist restricted Padé-approximations of order greater than 6.

A program STRIDE Burrage, Butcher and Chipman (1980) has been written based on allied methods.

Let us end this section with some comments on the relation between collocation methods and Runge-Kutta methods. The former constitute a subclass of Runge-Kutta methods. In order to cover all Runge-Kutta methods we have to modify the collocation condition. One way of doing that is to perturb the value of u in f to Pu where P is a linear map from Π_m to Π_m. The perturbed condition is then

$$u'(x_n + c_i h) = f((Pu)(x_n + c_i h)), \quad i = 1, \ldots, m.$$

If we in addition perturb y_{n+1} to

$$y_{n+1} = (Qu)(x_{n+1}).$$

Where Q also is a linear map from Π_m to Π_m, we are able to get an equivalence between these two classes of methods, Norsett and Wanner (1981).

3. THE C'-CASE

The step from $k = 0$ to $k = 1$ was recently taken up by Butcher (1981) in an attempt to construct general linear methods, in particular, singly implicit type methods. No final theory has so far been found.

We proceed as for $k = 0$ and define

$$c_0 = 0, \quad k_0^n = y_n^{(1)}.$$

Now the k_i-values are denoted by k_i^n in order to indicate their dependence on the interval I_n. $u(x)$ will in I_n be represented by

$$u(x) = y_n + h \sum_{j=0}^{m} k_j^n \int_0^{(x-x_n)/h} \ell_j(t)\, dt \quad \text{where now}$$

$$\ell_j(t) = \prod_{i=0,\, i \neq j}^{m} [(t - c_i)/(c_j - c_i)].$$

The collocation condition becomes

$$k_i^n = f(y_n + a_{i0}hy_n^{(1)} + h\sum_{j=1}^{m} a_{ij}k_j^n) \quad i = 1, \ldots, m.$$

The discrete approximations to $y(x_{n+1})$ and $y'(x_{n+1})$ are

$$y_{n+1} = y_n + b_0hy_n^{(1)} + h\sum_{i=1}^{m} b_ik_i^n$$

$$y_{n+1}^{(1)} = \bar{b}_0y_n^{(1)} + \sum_{i=1}^{m} \bar{b}_ik_i^n .$$

The coefficients are defined by

$$a_{ij} = \int_0^{c_i} \ell_j(t)dt; \qquad i = 1, \ldots, m, \qquad j = 0, 1, \ldots, m$$

$$\left.\begin{aligned} b_i &= \int_0^1 \ell_i(t)dt \\ \bar{b}_i &= \ell_i(1). \end{aligned}\right\} i = 0, 1, \ldots, m$$

For $k > 0$ the schemes give rise to multivalue methods. From step to step we carry along more information than the differential equations require to define a unique solution. We should therefore expect problems with zero-stability. The extra information for $k = 1$ is $y_n^{(1)}$. We eliminate this and obtain the formulas

$$k_i^{n+1} = f((1 - \alpha_i)y_{n+1} + \alpha_iy_n + h\sum_{j=1}^{m} a_{ij}k_j^{n+1} +$$

$$h\sum_{j=1}^{m} (a_{i0}\bar{b}_i + b_i\alpha_i)k_j^n \quad i = 1, \ldots, m$$

$$y_{n+2} - (1 - \alpha_0)y_{n+1} - \alpha_0y_n = h\sum_{i=1}^{m} [b_ik_i^{n+1} +$$

$$(b_0\bar{b}_i - \bar{b}_0b_i)k_i^n]$$

with

$$\alpha_i = -\bar{b}_o\, a_{io}/b_o, \quad i = 0, 1, \ldots, \quad (\alpha_o = -\bar{b}_o).$$

The zero-stability of this general linear method is given by,

Theorem 5. The spline method with $k = 1$ is zero-stable iff

$$-1 \le \frac{M(1)}{M(0)} < 1$$

where

$$M(t) = \prod_{i=j}^{m} (t - c_i). \qquad \square$$

We will assume $M(0) \neq 0$. If $M(0) = 0$, we have $y_n^{(1)} = f(y_n)$ and a normal collocation method with $k = 0$ is obtained or a higher derivative method. Zero-stability is always present. Similar methods have been studied by Mülthei (1980a, 1980b).

Example 6. For $m = 1$ and $m = 2$ we find:

Zero-stability for

$$c_1 \ge \frac{1}{2} \quad \text{when} \quad m = 1$$

$$c_1 + c_2 > 1 \quad \text{and} \quad 1 - (c_1 + c_2) + 2c_1c_2 \ge 0$$

or

$$c_1 + c_2 > 1 \quad \text{for} \quad c_1, c_2 > 0 \quad \text{when} \quad m = 2. \qquad \square$$

The difficult question is superconvergence. Order $\ge m$ is trivial by way of construction.

From step to step both y_n and $y_n^{(1)}$ are approximated and carried along. y_n is natural, $y_n^{(1)}$ is extra information and unnatural in connection with first order differential equations. The question of order of approximation of y_n is therefore also connected to how well we can approximate $y_n^{(1)}$. Our methods are based on polynomials of degree m. Hence, order m for y_n and $m - 1$ for $y_n^{(1)}$ is directly produced. The condition for order $m + 1$ is, Norsett and Wanner (198x).

Theorem 6. The spline-collocation method with $k = 1$ has order $m + 1$ if,

$$\int_0^1 M(\tau)[M(0) - M(1)\tau + M(1)]d\tau = 0. \quad \square$$

The question of order $> m + 1$ is not solved. Obviously, one can use the B-series approach[2], but it is difficult to solve the resulting set of non-linear equations.

Example 8. For $m = 2$ the condition on c_1 for order 3 is

$$c_1 - c_1 + \frac{1}{6} = 0$$

or

$c_1 = (3 \pm \sqrt{3})/6$. Only $c_1 = (3 + \sqrt{3})/6$ will give a zero-stable method. $\square$

For A-stability we apply the method at $y' = \lambda y_1, \lambda \in C$ and find

$$\begin{bmatrix} y_{n+1} \\ hy_{n+1}^{(1)} \end{bmatrix} = R(z) \begin{bmatrix} y_n \\ hy_n^{(1)} \end{bmatrix}, \quad z = \lambda h$$

where $R(z)$ is a 2×2-matrix with polynomial coefficients in z. The condition for A-stability is that $\rho(R(z)) < 1$ for $z \in C^-$.

Example 9. For $m = 2$ $R(z)$ is computed to be

$$R(z) = \begin{bmatrix} 1 & 1 - \frac{1}{2c_1} \\ 0 & 1 - \frac{1}{c_1} \end{bmatrix} + \frac{z}{1 - \frac{c_1 z}{2}} \begin{bmatrix} \frac{1}{2c_1} & \frac{1}{4} \\ \frac{1}{c_1} & \frac{1}{2} \end{bmatrix}$$

Hence $\rho(R(z) < 1$ for $z \in R^-$ iff $c_1 \geq 1$. The method of order 3 cannot be A-stable. □

A general discussion of A-stability is not known. Butcher (1981) has given conditions for singly implicitness.

4. CLASSICAL SPLINES; $k = m - 1$

Let us set $\theta = c_1$. The case $\theta = 1$ was studied by Loscalzo and Talbot (1967). One of their conclusions was that the methods are not zero-stable for $m \geq 4$. However, one can introduce more freedom by going for $\theta \neq 1$. The clue in proving these results is to find equivalent multistep-type methods. The well-known B-spline $Q_{m+1}(x)$ is defined by

$$Q_{m+1}(x) = \frac{1}{m!} \sum_{i=0}^{m+1} (-1)^i \binom{m+1}{i} (x - 1)_+^m.$$

We have,

<u>Theorem 7</u>. For $u \in S_m^{(m-1)}(h)$ the following consistency relations are valid.

$$\sum_{k=0}^{m} Q'_{m+1} (k + \theta)\, u_{n-k} = h \sum_{k=0}^{m} Q_{m+1}(k)\, f_{n+\theta-k} \tag{9}$$

$$\sum_{k=0}^{m} Q'_{m+1} (k + \theta)\, u_{n+\theta-k} = h \sum_{k=0}^{m} Q_{m+1} (k + \theta)\, f_{n+\theta-k} \tag{10}$$

$$\sum_{k=0}^{m} Q_{m+1} (k + \theta)\, u_{n-k} = \sum_{k=0}^{m} Q_{m+1} (k)\, u_{n+\theta-k} \tag{11}$$

where $u_m = u(x_m)$ and $f_m = f(u_m)$. □

The first main conclusion to be drawn is then,

<u>Theorem 8</u>. The spline method with $k = m - 1$ is zero-stable iff the polynomials $\rho(z)$ are Schur-polynomials,

$$\rho(z) = \sum_{k=0}^{m} Q'_{m+1} (k + \theta) z^{m-k} =$$

$$(z-1)\sum_{k=0}^{m-1} Q_m(k+\theta)z^{m-1-k} =: (z-1)\,\hat{\rho}\,(z). \qquad \square$$

Let us remark that normally $\theta \in [0, 1]$. But as we will see also, $\theta > 1$ is of interest.

Example 9. For $m = 2, 3$ and 4, we get zero-stability for

$$\theta = 0 \quad \text{or} \quad \theta \geq \frac{1}{2} \quad \text{for} \quad m = 2$$

$$\theta = 0 \quad \text{or} \quad \theta \geq 1 \quad \text{for} \quad m = 3$$

$$\theta \geq \frac{1+\sqrt{3}}{2} = 1.366025404 \quad \text{for} \quad m = 4. \qquad \square$$

In the case of $m = 4$, θ needs to be greater than 1. Compare the results of Loscalzo and Talbot (1967) with $\theta = 1$.

In general, we can prove

Theorem 10. The spline methods with $k = m - 1$ are not zero-stable for

$m \geq 3$ and $0 < \theta < 1$

$m \geq 4$ and $\theta = 1$

$m \geq [4\theta^2 - 2\theta + 2]$ and $\theta > 1$. $\square$

The order of convergence is studied by using the relations of Theorem 7. By way of construction the order is at least m. The freedom in θ might give rise to methods of order $m + 1$ at the step points. For this purpose we use relation (10) and define the local truncation error T_{n-m} by

$$T_{n-m} = (m-1)! \sum_{k=0}^{m} \{Q'_{m+1}(k+\theta)y(x_{n+\theta-k}) -$$
$$hQ_{m+1}(k+\theta)y'(x_{n+\theta-k})\}.$$

Example 11. For $m = 2, 3$ and 4 we obtain

$$m = 2: \quad T_n = -\frac{h^3}{2}\left[\theta^2 - \theta + \frac{1}{6}\right] y^{(4)}(x_n) + 0(h^4)$$

$$= 0(h^4) \quad \text{for} \quad \theta = \frac{3 \pm \sqrt{3}}{6} \ .$$

Only $\theta = \dfrac{3 + \sqrt{3}}{6}$ is of interest from zero-stability.

$$m = 3: \quad T_n = -\frac{\theta}{12}(2\theta - 1)(\theta - 1)h^4 \, y^{(4)}(x_n) + 0(h^5)$$

$$= 0(h^5) \quad \text{for} \quad \theta = 0, \tfrac{1}{2}, 1.$$

Only $\theta = 0$ or $\theta = 1$ will give zero-stable methods.

$$m = 4: \quad T_n = \frac{h^5}{120}(1 - 30\,\theta^2(1 - \theta)^2)y^{(5)}(x_n) + 0(h^6)$$

$$= 0(h^6) \quad \text{for} \quad \theta = 0.5 \pm 1.272519702.$$

Due to zero-stability only $\theta = 1.772519702$ is of interest. □

By using the theory for linear multistep methods,

Theorem 12. For $n \geq 0$ and those θ which produce zero-stable methods

$$\left.\begin{array}{l} y_{n+\theta} - y(x_{n+\theta}) \\ \\ y^{(1)}_{n+\theta} - y^{(1)}(x_{n+\theta}) \end{array}\right\} = 0(h^m) \quad \text{or} \quad \left(0(h^{m+1}) \quad \text{when} \quad T_n = 0(h^{m+2})\right)$$

□

In order to give results for y_n and $y_n^{(1)}$ we use the interpolation formula (11) and its equivalent for $y_n^{(1)}$.

Theorem 13. For $\theta \neq 0, 1$ and those θ which give zero-stable methods

$$y_n - y(x_n) = \begin{cases} 0(h^m) \\ 0(h^{m+1}) & \text{when} \quad T_n = 0(h^{m+2}) \end{cases}$$

and

$$y_n^{(1)} - y^{(1)}(x_n) = O(h^m). \quad \square$$

When $m = 2, 3$ and 4 at least, results similar to the results of Lascalzo and Talbot (1967) can also now be given for the global error $u^{(i)}(x) - y^{(i)}(x)$, $i = 0, 1, 2, \ldots, m$.

The A-stability is studied by applying (10) at $y' = \lambda y$, $\lambda \in C$. The characteristic polynomial is

$$\phi(r, z) = \rho(r) - z\sigma(r), \qquad z = \lambda h$$

with

$$\rho(r) = \sum_{k=0}^{m} Q'_{m+1}(k + \theta)\, r^{m-k}$$

$$\sigma(r) = \sum_{k=0}^{m} Q_{m+1}(k + \theta)\, r^{m-k}.$$

By the Wanner-Hairer-Norsett result (Wanner and Co. (1978)), order 2 is maximum for A-stability. This is attained by $m = 1$ and $\theta = 1$, the trapezoidal rule.

5. MODIFICATIONS

For k between 1 and $m - 1$ no general results are available. A full knowledge of the properties of those methods would be interesting.

Stable methods of higher order are obtained in the case of $k = m - 1$ by moving θ from 1 to values greater than 1. But still the continuity is C^{m-1}. Another possibility is to fix θ at 1 and to loosen the continuity requirements from C^{m-1} to C^{m-2}. The extra freedom is used for other purposes. One possibility is to interpolate u at a nonstep point. Let u in $[x_{n-1}, x_n]$ be computed and let us denote it by $u_{n-1}(x)$. For the next interval we assume continuity C^{m-2} and in addition

$$u_n(x_n - \delta h) = u_{n-1}(x_n - \delta h)$$

and collocation at x_{n+1}. A basis function for the corresponding set of functions is for $m \geq 3$

$$B_{m+1}(x;\ \delta) = Q_{m+1}(x) - \frac{\delta}{m} Q'_{m+1}(x).$$

With the collocation parameter set to 1, the relevant multistep method turns out to be

$$\sum_{k=0}^{m-1} \dot{B}'_{m+1}(k+1;\ \delta) y_{n-k} = h \sum_{k=0}^{m-1} B_{m+1}(k+1);\ \delta) f_{n-k}.$$

The zero-stability is given by

$$\rho(r;\ \delta) = \sum_{k=0}^{m-1} B'_{m+1}(k+1;\ \delta) r^{m-1-k}.$$

Example 14. For $m = 3$ and 4 we find

$$m = 3: \quad \rho(r;\ \delta) = \frac{1}{2}(r^2 - 1) - \frac{\delta}{3}(r-1)^2$$

and zero-stability for $\delta \leq 0$.

$$m = 4: \quad \rho(r;\ \delta) = \frac{1}{6}(r-1)\{r^2 + 4r + 1 - \frac{3}{4}\delta(r^2 - 1)\}$$

and zero-stability for no $\delta \in R$!! □

The extreme modification in this direction is to lower the continuity from $C^{(m-1)}$ to $C^{(0)}$ and to use one collocation point plus interpolation. This discrete spline approach has for lower values of k been discussed by Aurdal (1983).

The relevant basis functions are

$$B_{m+1}(x) = \frac{1}{m!} \sum_{i=0}^{m+1} (-1)^i \binom{m+1}{i} (x-i)_+ (x-i-\delta_1)_+ \dots (x-i-\delta_{m-1})_+$$

where $\delta_1, \dots, \delta_{m-1}$ are the interpolation parameters. With the collocation parameter set to 1, the consistency relation is given by

$$\sum_{k=0}^{m} B'_{m+1}(k+1)y_{n-k} = h \sum_{k=0}^{m} B_{m+1}(k+1)\, f_{n-k} \,.$$

The zero-stability is therefore related to the zeros of

$$\rho(r) = \sum_{k=0}^{m} B'_{m+1}\,(k+1)r^{m-k}.$$

<u>Example 15</u>. For $m = 2, 3, 4$ we find

$$m = 2: \quad \rho(r) = \frac{1}{2}\,(r-1)[(2-\delta_1)r + \delta_1].$$

Zero-stability for $\delta_1 \le 1$.

$$m = 3: \quad \rho(r) = \frac{1}{6}\,(r-1)[\{\delta_1\delta_2 - 2(\delta_1+\delta_2) + 3\}r^2 +$$

$$\{-2\delta_1\delta_2 + 2(\delta_1+\delta_2) + 3\}r + \delta_1\delta_2].$$

Zero-stability for $\delta_1 < 1$ and $\delta_2 < \delta_1/(\delta_1 - 1)$.

$m = 4$: Set $A = \delta_1\delta_2\delta_3$, $B = \delta_1\delta_2 + \delta_2\delta_3 + \delta_1\delta_3$,

$C = \delta_1 + \delta_2 + \delta_3$. Then

$$\rho(r) = \frac{1}{24}\,(r-1)[(-A + 2B - 3C + 4)r^3 + (3A - 4B + 16)r^2 +$$

$$(-3A + 2B + 3C + 4)r + A].$$

With $\delta_2 = 2\delta_1$ and $\delta_3 = 3\delta_1$, we get zero-stability for $\delta_1 \le (3 - \sqrt{17})/4 = -0.28$. □

The consistency relation is a linear multistep method. The theory of Dahlquist, Henrici (1962) can therefore be used, implying that the order can at most be $2[(m+2)/2]$ with zero-stability preserved. By way of construction the starting error is of the correct order for order m and $m+1$ globally. Order $m+2$ is not analyzed. A short proof would be of interest.

<u>Example 16</u>. For $m = 2, 3$ and 4 we have,

$m = 2$: Order 3 for $\delta_1 = \frac{1}{3}$.

The corresponding method is zero-stable.

$m = 3$: The order is 4 for $\delta_2 = \delta_1/(3\delta_1 - 1)$.

The condition for zero-stability is then $\delta_1 < \frac{1}{3}$. When $\delta_2 = 2\delta_1$ only the classical case $\delta_1 = 0$ is zero-stable.

$m = 4$: For order 5 we need

$$\delta_1\delta_2 + \delta_2\delta_3 + \delta_1\delta_3 = \frac{1}{5} + 3\delta_1\delta_2\delta_3.$$

When $\delta_i = i\delta_1$, $i = 2, 3$, order 5 is obtained for

$$\delta_1 = -0.13253662$$

$$\delta_2 = 0.189022974$$

$$\delta_3 = 0.44351365.$$

However, no value gives zero-stability. □

As a verification of the discrete spline results, several tests have been done. In all cases the above statements on order and zero-stability were verified.

Continuous error-estimates can also be given. For $m = 2, 3, 4$, theorems similar to Loscalzo and Talbot (1967) are valid.

The discrete spline idea is local in nature and in use. The polynomial for $[x_n, x_{n+1}]$ is constructed from the knowledge of the approximation for $[x_{n-1}, x_n]$, $n \geq 1$. However, a more global approach can be given. For example, let y_{n+i}, f_{n+i}, $i = 0, 1, \ldots, k - 1$ be given. A polynomial u is computed such that

$$u(x_p) = y_p \quad \text{and} \quad u'(x_q) = f_q$$

for a proper set of indexes p and q between n and $n + k - 1$. The remaining free parameters in u are determined by collocation.

This idea is not new. The well-known BDF-methods are of this type. For the superconvergence discussion the theorem of Alekseev and Gröbner need to be extended. This has been done by Norsett and Wanner (198x) for a special choice of the set of p and q. u is constructed by

$$u(x_{n+i}) = y_{n+i}, \quad i = 0, 1, \ldots, k - 1$$

$$u'(x_n + c_j h) = f(u(x_n + c_j h)), \quad j = 1, \ldots, m$$

and then

$$y_{n+k} = u(x_{n+k}).$$

Instead of discussing the general case we give these results for $m = 1$ and $m = 2$.

m = 1: The methods obtained in this case are a subclass of the one-leg methods of Dahlquist. The order is $k + 1$ when

$$\Pi'(c_1) = 0$$

where

$$\Pi(t) = \prod_{i=0}^{k} (t - i).$$

There are k possible methods for each k. For $k = 1$ we have $c_1 = \frac{1}{2}$, the midpoint method and for $k = 2$, $c_1 = 1 \pm \sqrt{3}/3$.

m = 2: Let $\Delta = c_1 - c_2$. The result is

Theorem 17. The two-leg methods $(m = 2)$ are of order $k + 3$ when c_1 and c_2 are taken as one of the $k(k + 1)/2$ solutions in $(0, k)$ of

$$2\Pi(c_1) + \Delta\Pi'(c_1) = 0$$

$$2\Pi(c_2) - \Delta\Pi'(c_2) = 0. \qquad \square$$

For $k = 1, 2$ the 'k-step' Gauss-points are

$k = 1$: $c_1 = 0.2113246703$ $c_2 = 0.7886753297$

$k = 2$: $c_1 = 0.1880060962$ $c_2 = 0.7536913702$

$c_1 = 0.2928930738$ $c_2 = 1.707106495$

$c_1 = 1.246307547$ $c_2 = 1.811992988$

Let us finally remark that these new 'k-step' Gauss-points are nicely located in different subintervals of $[0, k]$.

REFERENCES

1. Aurdal, P. H. (1983), "Cand. Scient. Thesis", Department of Computer Science, University of Oslo, Norway (In Norwegian).

2. Birkhoff, G., and Varga, R. S. (1965), "Discretization errors for well set Cauchy problems", J. of Math and Phy. 44, 1965, pp. 1-23.

3. Burrage, K., Butcher, J. C., and Chipman, F. (1980), "An implementation of singly-implicit Runge-Kutta methods", BIT 20, 1980, pp. 326-340.

4. Butcher, J. C., (1976), "On the implementation of Runge-Kutta processes", BIT 16, 1976, pp. 237-240.

5. Butcher, J. C., (1981), "A generalization of singly-implicit methods", BIT 21, 1981, pp. 175-189.

6. Elhe, B. L., (1971), "A-stable methods and Padé-approximations to the exponential", SIAM J. Math. Anal. 4, 1971, pp. 384-388.

7. Gear, C. W., (1971), "Numerical initial-value problems in ordinary differential equations", Prentice-Hall.

8. Hairer, E., and Wanner, G., (1975), "On the Butcher group and general multi-value methods", Computing 13, 1975, pp. 1-15.

9. Henrici, P., (1962), "Discrete variable methods in ordinary differential equations", Wiley & Sons, New York.

10. Iserles, A., (1981), "On multi-valued exponential approximation", SIAM J. Num. Anal. 18, 1981, pp. 480-499.

11. Lambert, J. D., (1973), "Computational methods in ordinary differential equations", Wiley & Sons.

12. Loscalzo, F. R., and Talbot, T. D., (1967), "Spline function approximations for solutions of ordinary differential equations", SIAM J. Num. Anal. 4, 1967, pp. 433-445.

13. Mülthei, H. N., (1980a), "Numerische Lösung gewöhnlicher differential-gleichungen mit splinefunctionen", Computing 25, 1980, pp. 317-335.

14. Mülthei, H. N., (1980b), "Zur numerischen Lösung gewöhnlicher differential gleichungen mit splines in einem Sonderfall", Math Meth. in the Appl. Sci. 2, 1980, pp. 419-428.

15. Norsett, S. P., (1974a), "One-step methods of Hermite type for numerical integration of stiff systems", BIT 14, 1974, pp. 63-77.

16. Norsett, S. P., (1974b), "Multiple Padé-approximations to the exponential function", Dept. of Math, Report No. 4, 1974, NTH, Trondheim, Norway.

17. Norsett, S. P., (1975), "C-polynomials for rational approximation to the exponential function", Numer. Math 25, 1975, pp. 39-56.

18. Norsett, S. P., (1976), "Runge-Kutta methods with a multiple real eigenvalue only", BIT 16, 1976, pp. 388-393.

19. Norsett, S. P., (1978), "Restricted Padé-approximations to the exponential function", SIAM J. Num. Anal. 15, 1978, pp. 1008-1029.

20. Norsett, S. P., Wanner, G., (1979), "The real-pole sandwich for rational approximations and oscillation equations", BIT 19, 1979, pp. 79-94.

21. Norsett, S. P., and Wanner, G., (1981), "Perturbed collocation and Runge-Kutta methods", Numer. Math 38, 1981, pp. 193-208.

22. Norsett, S. P., and Wanner, G., (198x), "Spline-collocation and superconvergence", to appear.

23. Varga, R. S., (1961), "On higher order stable implicit methods for solving parabolic differential equations", J. of Math and Phys., 40, 1961, pp. 220-231.

24. Wanner, G., Hairer, E., and Norsett, S. P., (1978), "Order stars and stability theorems", BIT 18, 1978, pp. 475-489.

25. Wolfbrandt, A., (1977), "A study of Rosenbrock processes with respect to order conditions and stiff stability", Report 77.01R Chalmer Tech. Goteborg, Sweden.

26. Wright, K., (1970), "Some relationships between implicit Runge-Kutta, collocation and Lanczos τ-methods, and their stability properties", BIT 10, 1970, pp. 217-227.

---

1. A-acceptability of $R(z)$ means $|R(z)| < |$ for $z \in C^-$.

2. See for example Hairer and Wanner (1975).

DEGREE OF APPROXIMATION OF QUASI-HERMITE-FEJÉR INTERPOLATION BASED ON JACOBI ABSCISSAS $P_n^{(\alpha,\alpha)}(x)$

J. Prasad and A. K. Varma

1. Let $\alpha, \beta > -1$ and let us denote by $x_k = x_{kn}^{(\alpha,\beta)}$ $(k = 1, 2, \ldots, n)$ the roots of the Jacobi polynomial $P_n^{(\alpha,\beta)}(x)$ of degree n. It is well known that for a given function f on $[-1, 1]$ the Hermite-Fejér interpolation polynomial $H_n^{(\alpha,\beta)}(f, x)$ based on the zeros of Jacobi polynomial is uniquely determined by the following conditions:

$$H_n^{(\alpha,\beta)}(f, x) \text{ is a polynomial of degree } \leq 2n + 1 \tag{1.1}$$

and

$$H_n^{(\alpha,\beta)}(f, x_k) = f(x_k),\ H_n^{(\alpha,\beta)'}(f, x_k) = 0,\ k = 1, 2, \ldots, n. \tag{1.2}$$

Further, in the case $\alpha, \beta < 0$ $||\Delta_n^{(\alpha,\beta)}(f, x)||_{[-1,1]} = ||f(x) - H_n^{(\alpha,\beta)}(f, x)||_{[-1,1]} \to 0$ as $n \to \infty$ for all continuous continuous $f(x)$ in $[-1, 1]$ where $||\cdot||_{[-1,1]}$ denotes the maximum norm in $[-1, 1]$. As for the rate of uniform convergence is concerned the case $\alpha = \beta = -\frac{1}{2}$ has received at most attention. In this case R. Bojanic [4] and independently

S. P. Singh et al. (eds.), Approximation Theory and Spline Functions, 419–440.

P. Vertesi [21] have proved that

$$\|H_n^{(-\frac{1}{2},\ -\frac{1}{2})}(f,\ x) - f(x)\| = 0(\frac{1}{n} \sum_{k=1}^{n} \omega_f(\frac{1}{k}))$$

where $\omega_f(h)$ is the modulus of continuity of f. Later, J. Szabadös [17] in a more general setting proved that

$$\|H_n^{(\alpha,\beta)}(f,\ x) - f(x)\|_{[-1,1]} = 0(\omega_f(n^{\nu})),$$

$$= \max(\alpha,\ \beta,\ -\frac{1}{2}) < 0,$$

for all continuous functions.

Further from G. Szegö ([19], Theorem 14.6) it is known that in the case $\alpha \geq 0$ the sequence $\{H_n^{(\alpha,\beta)}(f,\ x)$ is in general divergent at $x = 1$ and in the case $\beta \geq 0$ the sequence $\{H_n^{(\alpha,\beta)}(f,\ x)\}$ is divergent at $x = -1$ $(f \in C[-1,\ 1])$. These results led P. Szász [18] to introduce the concept of quasi-Hermite-Fejér interpolation. For f defined on $[-1,\ 1]$ the quasi-Hermite-Fejér interpolation polynomial $S_n^{(\alpha,\beta)}(f,\ x)$ is uniquely determined by the following conditions:

$$S_n^{(\alpha,\beta)}(f,\ x) \text{ is a polynomial of degree } \leq 2n + 1, \tag{1.3}$$

$$S_n^{(\alpha,\beta)}(f,\ x_k) = f(x_k),\ S_n^{(\alpha,\beta)\,'}(f,\ x_k) = 0, \quad k = 1,\ 2,\ .-.,\ n \tag{1.4}$$

and

$$S_n^{(\alpha,\beta)}(f,\ 1) = f(1),\ S_n^{(\alpha,\beta)}(f,\ -1) = f(-1), \tag{1.5}$$

where x_k are the zeros of Jacobi polynomial of degree n. Regarding the convergence of $\{S_n^{(\alpha,\beta)}(x)\}$ Szasz proved that if $f(x) \in C[-1,\ 1]$, $\{S_n^{(\alpha,\beta)}(f,\ x)\}$ converges to $f(x)$ uniformly on $[-1, 1]$ provided $0 \le \alpha < 1$ and $0 \le \beta < 1$. Throughout this paper $Q_n(x) = P_n^{(\alpha,\alpha)}(x)$. Recently in the important case $\alpha = \beta = 0$ we have proved [12] that if $f \in C[-1, 1]$

$$|S_n^{(0,0)}(f, x) - f(x) \le C \sum_{i=1}^{n} \frac{1}{i^2} w_f \left(\frac{i \sin \theta}{n}\right) \tag{1.6}$$

where C is a positive constant independent of f, n, x. The object of this paper is to prove the following Theorems.

Theorem 1. Let $f \in C[-1, 1]$, $0 \le \alpha \le \frac{1}{2}$, $0 \le x \le 1$ (a similar statement is valid if $-1 \le x \le 0$) and let x_j be that zero $Q_n(x)$ which is nearest to x then

$$|S_n^{(\alpha,\alpha)}(f, x) - f(x)| \le C_1 \sum_{k=1}^{j-1} \frac{1}{(k-j)^2} w_f \left(\frac{j-k}{n} \sin \theta\right)$$

$$+ C_2 w_f \left(\frac{\sin}{n}\right) + C_4 (1 - x^2) Q_n^2(x) w_f(1) \tag{1.7}$$

$$C_3 n(1 - x^2)^{\frac{1+2\alpha}{2}} Q_n^2(x) \sum_{k=j+1}^{[\frac{3n}{4}]} \frac{1}{(k-j)^2} \left(\frac{\sin \theta}{\sin \theta_k}\right)^{1-2\alpha} w_f$$

$$\left(\frac{k-j}{n} \sin \theta_k\right)$$

where C_1, C_2, C_3, C_4 are positive constant independent of f, n, x.

Corollary 1. For $f \in C[-1, 1]$, $0 \le \alpha \le \frac{1}{2}$, $-1 \le x \le 1$,

$$|S_n^{(\alpha,\alpha)}(f, x) - f(x)| \le C_5 \sum_{i=1}^{n} \frac{1}{i^2} \left[w_f \left(\frac{i \sin \theta}{n}\right) + w_f \left(\frac{i^2}{n^2}\right)\right]. \tag{1.8}$$

It is interesting to point out that (1.8) was proved earlier for

$\alpha = 0$ by Prasad and Saxena [11] and for $\alpha = \frac{1}{2}$ it was done by Mather and Saxena [15]. It should be noted that we are able to obtain the same result for all α, $0 \le \alpha \le \frac{1}{2}$.

Corollary 2. Let $f(x) \in C[-1, 1]$, $0 \le \alpha \le \frac{1}{2}$, $-1 \le x \le 1$, then we have

$$|S_n^{(\alpha,\alpha)}(f, x) - f(x)| \le C_6 \sum_{i=1}^{n} \frac{1}{i^2} w_f\left(\frac{i}{n} \sin^{1-2\alpha}\theta\right) + C_7 \frac{\sin^{1-2\alpha}\theta}{n} w_f(1) \tag{1.9}$$

where C_6, C_7 are absolute constant independent of n, f, x.

Thus, for $\alpha = 0$ we obtain (1.6).

Corollary 3. Let $f(x) \in C[-1, 1]$, $0 \le \alpha \le \frac{1}{2}$, $-1 \le x \le 1$ and let $f(x) \in \text{Lip}\ \beta$, $\beta < 1 - \alpha$, then

$$|S_n^{(\alpha,\alpha)}(f, x) - f(x)| \le C_8 \left(\frac{\sqrt{1 - x^2}}{n}\right)^{\beta}. \tag{1.10}$$

Thus, we have obtained quantitative estimates of $S_n^{(\alpha,\alpha)}(f, x) - f(x)$ which reflect both the pointwise behavior at the end points as well as the dependence on the smoothness of the function. More precisely we may conclude from (1.10) that for $f \in \text{Lip}\ \beta$, $\beta < 1 - \alpha$, $0 \le \alpha \le \frac{1}{2}$ we obtained Telijakovski type estimate. Thus, for this class our results are sharp.

For $f(x) \in \text{Lip}\ 1$ we have the following.

Theorem 2. There exists a function $f(x) \in \text{Lip}\ 1$ and a constant C_9 such that for $0 \le \alpha \le \frac{1}{2}$

$$|S_n^{(\alpha,\alpha)}(f, 0) - f(0)) \ge C_9 \frac{\log n}{n}, \quad n = 6, 8, 10, \ldots$$

2. EXPLICIT FORM OF $S_n^{(\alpha,\alpha)}(f, x)$

Let $\{x_k\}_1^n$ be the zeros of $Q_n(x) \equiv P_n^{(\alpha,\alpha)}(x)$ such that

$$-1 = x_{n+1} < x_n < \dots < x_1 < x_0 = 1. \tag{2.1}$$

From Szasz [18] it follows that

$$S_n^{(\alpha,\alpha)}(f, x) \equiv S_n^{(\alpha)}(f, x) = f(-1)\,\frac{1-x}{2}\,\frac{Q_n^2(x)}{Q_n^2(-1)} + f(1)\,\frac{1+x}{2}\,\frac{Q_n^2(x)}{Q_n^2(1)} + \sum_{k=1}^{n} f(x_k)h_k(x) \tag{2.2}$$

where

$$h_k(x) = \frac{(1-x^2)}{(1-x_k^2)}\,\ell_k^2(x)v_k(x), \quad v_k(x) = 1 - \frac{2\alpha x_k(x-x_k)}{1-x_k^2} \tag{2.3}$$

$$\ell_k(x) = \frac{Q_n(x)}{(x-x_k)Q_n'(x_k)}, \quad k = 1, 2, \dots, n. \tag{2.4}$$

Since $S_n^{(\ )}(f, x)$ is uniquely determined it follows that

$$\frac{1-x}{2}\,\frac{Q_n^2(x)}{Q_n^2(-1)} + \frac{1+x}{2}\,\frac{Q_n^2(x)}{Q_n^2(1)} + \sum_{k=1}^{n} h_k(x) = 1. \tag{2.5}$$

Due to the above identity (2.5) we see that

$$\sum_{k=1}^{n} h_k(x) \le 1, \quad 0 \le h_k(x) \le 1. \tag{2.6}$$

Further, from (2.3) we may write

$$h_k(x) = (1-\alpha)\frac{1-x^2}{(1-x_k^2)}\ell_k^2(x) + \frac{\alpha(1-x^2)^2}{(1-x_k^2)^2}\ell_k^2(x)$$

$$+ \frac{\alpha(1-x^2)Q_n^2(x)}{(1-x_k^2)^2[Q_n'(x_k)]^2}. \tag{2.7}$$

Therefore, owing to (2.6) we conclude that for $0 \le \alpha < 1$,

$$\sum_{k=1}^{n} \frac{1-x^2}{1-x_k^2}\ell_k^2(x) \le \frac{1}{1-\alpha}, \quad \frac{1-x^2}{1-x_k^2}\ell_k^2(x) \le \frac{1}{1-\alpha}. \tag{2.8}$$

3. PRELIMINARIES

Here we state a few well-known properties of $Q_n(x) \equiv P_n^{(\alpha,\alpha)}(x)$. From Szegö [19] (Page 166, formula 7.32.2) it follows that

$$Q_n(1) = \max_{-1\le x\le 1} |P_n^{(\alpha,\alpha)}(x)| = \binom{n+\alpha}{n} \sim n^{\alpha}, \quad \alpha \ge -\frac{1}{2}. \tag{3.1}$$

Also from Szegö [19] (formula 7.32.5, page 165) following asymptotic formula

$$Q_n(x) = P_n^{(\alpha,\alpha)}(x) = \theta^{-\alpha-\frac{1}{2}}\,O(n^{-\frac{1}{2}}), \quad \frac{c}{n} \le \theta \le \frac{\pi}{2} \tag{3.2}$$

$$= O(n^{\alpha}), \quad 0 < \theta \le \frac{c}{n}$$

hold, where c is a fixed positive constant and $x = \cos\theta$. From (3.2) we deduce that

$$(1-x^2)^{\frac{1+2\alpha}{4}} |Q_n(x)| \le c_7 n^{-\frac{1}{2}}, \quad \alpha \ge 0, \ -1 \le x \le 1, \tag{3.3}$$

and

$$(1-x^2)^{\frac{1}{2}} Q_n^2(x) \le c_8 n^{2\alpha-1}, \quad \alpha \ge 0, \ -1 \le x \le 1. \tag{3.4}$$

According to another asymptotic formula ([19], Theorem 8.91, page 236) we have

$$|Q_n'(x_k)| \sim k^{-\alpha-\frac{3}{2}} n^{\alpha+2}, \; k = 1, 2, \ldots, [\tfrac{n}{2}], \tag{3.5}$$

$$|Q_n'(x_k)| \sim (n - k + 1)^{-\alpha-\frac{3}{2}} n^{\alpha+2}, \; k = [\tfrac{n}{2}] + 1, \ldots, n, \tag{3.6}$$

$$1 - x_k^2 \sim (k - \tfrac{1}{2})^2 (n + \tfrac{1}{2})^{-2}, \; k = 1, 2, \ldots, [\tfrac{n}{2}], \tag{3.7}$$

and

$$1 - x_k^2 \sim (n - k + \tfrac{1}{2})^2 (n + \tfrac{1}{2})^{-2}, \; k = [\tfrac{n}{2}] + 1, \ldots, n, \tag{3.8}$$

in the sense that the ratio of these expressions remains between certain positive bounds depending only on α. On using (3.5) - (3.8) we also obtain

$$(1 - x_k^2)^{\frac{3}{2}+\alpha} [Q_n'(x_k)]^2 \sim n, \; k = 1, 2, \ldots, n, \tag{3.9}$$

and

$$\sum_{k=1}^{n} \frac{1}{(1 - x_k^2)[Q_n'(x_k)]^2} \le c_9, \; \alpha \sum_{k=1}^{n} \frac{1}{(1 - x_k^2)^2 [Q_n'(x_k)]^2} \le c_{10}. \tag{3.10}$$

Also from a theorem of P. Erdös [7] it follows that there exists a positive constant c_{11} such that

$$|\ell_k(x)| \le c_{11}, \; k = 1, 2, \ldots, n; \quad -1 \le x \le 1. \tag{3.11}$$

Next, we like to make the following observations:

$$\sin \theta_k \le \sin \theta + \sin \theta_k \le 2 \sin(\frac{\theta + \theta_k}{2});$$
$$\sin \theta \le 2 \sin \frac{\theta + \theta_k}{2} \tag{3.12}$$

and

$$\sin \frac{\theta + \theta_k}{2}) \geq |\sin(\frac{\theta_k - \theta}{2})|. \tag{3.13}$$

Further from (2.8) and (3.11) we have

$$\frac{(1 - x^2)^{1/2} \ell_k^2(x)}{(1 - x_k^2)^{1/2}} \leq c_{11} \frac{(1 - x^2)^{1/2} |\ell_k(x)|}{(1 - x_k^2)^{1/2}} \leq c_{12}. \tag{3.14}$$

Also from Szegö [19] we know that if $x_k = \cos \theta_k$ then for $-\frac{1}{2} \leq \alpha \leq \frac{1}{2}$,

$$\frac{2k - 1}{2n + 1} \pi \leq \theta_k \leq \frac{2k}{2n + 1} \pi, \tag{3.15}$$

and from [10] it is well known that

$$\frac{\Gamma(n + \sigma + \mu + 1)}{\Gamma(n + \sigma + 1)} < c_{13} n^{\mu}, \quad \sigma > -1 \quad \text{and} \quad \mu > -1. \tag{3.16}$$

4. SOME LEMMAS

Here we state and prove a few lemmas which will be used later.

Lemma 4.1. If

$$I_1(x) = \frac{1 + x}{2} \frac{Q_n^2(x)}{Q_n^2(1)} [f(1) - f(x)],$$

$$I_2(x) = \frac{1 - x}{2} \frac{Q_n^2(x)}{Q_n^2(-1)} [f(-1) - f(x)], \tag{4.1}$$

then we have

$$|I_1(x)| \leq c_{14} \omega_f \left(\frac{\sqrt{1 - x^2}}{n}\right), \quad |I_2(x)| \leq c_{15}\, \omega_f \left(\frac{\sqrt{1 - x^2}}{n}\right) \tag{4.2}$$

where C_{14} and c_{15} are absolute constants.

Proof. For $x = \pm 1$ (4.2) follows trivially. So let $-1 < x < 1$. From (4.1), (3.1) and (3.4) we obtain

$$|I_1(x)| \le \frac{1+x}{2}\,\frac{Q_n^2(x)}{Q_n^2(1)}\,\omega_f(1-x) \tag{4.3}$$

$$\le \frac{1+x}{2}\,\frac{Q_n^2(x)}{Q_n^2(1)}\left(1 + n\sqrt{\frac{1-x}{1+x}}\right)\omega_f\left(\frac{\sqrt{1-x^2}}{n}\right)$$

$$\le \left\{\frac{1+x}{2}\,\frac{Q_n^2(x)}{Q_n^2(1)} + \frac{n\sqrt{1-x^2}\,Q_n^2(x)}{2\,Q_n^2(1)}\right\}\omega_f\left(\frac{\sqrt{1-x^2}}{n}\right)$$

$$\le (c_{16} + c_{17})\,\omega_f\left(\frac{\sqrt{1-x^2}}{n}\right)$$

$$\le c_{18}\,\omega_f\left(\frac{\sqrt{1-x^2}}{n}\right).$$

Similarly we have

$$|I_2(x)| \le c_{19}\,\omega_f\left(\frac{\sqrt{1-x^2}}{n}\right).$$

This completes the proof of lemma 4.1.

Lemma 4.2. Let $-1 \le x \le 1$ and let x_j be that zero of $Q_n(x)$ which is nearest to x then for $0 \le \alpha \le \frac{1}{2}$ we have

$$|f(x) - f(x_j)|h_j(x) \le C_{20}\, w_f\left(\frac{\sqrt{1-x^2}}{n}\right) \tag{4.4}$$

where $w_f(\delta)$ is the modulus of continuity of f and C_{20} is an absolute constant.

Proof. For $x = \pm 1$ (4.4) is obviously true. Let $-1 < x < 1$ and note that

$$|x - x_j| \le C_{21} \left[\frac{(1 - x^2)^{1/2}}{n} + \frac{1}{n^2}\right]. \tag{4.5}$$

From (2.3), (3.7), (3.8) we have

$$v_j(x) \le 1 + \frac{|x - x_j|}{1 - x_j^2} \le 1 + 2\,\frac{\left|\sin\frac{\theta + \theta_j}{2}\,\sin\frac{\theta - \theta_j}{2}\right|}{\sin^2\theta_j}$$

$$\le 1 + 2\,\frac{\left|\sin\frac{\theta - \theta_j}{2}\right|}{\sin\theta_j}\left[1 + \frac{\left|\sin\frac{\theta + \theta_j}{2} - \sin\theta_j\right|}{\sin\theta_j}\right] \le C_{22}. \tag{4.6}$$

Therefore on using (4.6), (4.5), (2.6), (3.11) and

$$\frac{\sin\theta}{\sin\theta_j} \le 1 + \frac{|\sin \quad - \sin\theta_j|}{\sin\theta_j} \le 1 + C_{23}$$

we have

$$n|x - x_j| h_j(x) \le C_{21}\left[(1 - x^2)^{1/2} h_j(x) + \frac{1}{n} h_j(x)\right] \tag{4.7}$$

$$\le C_{21}\left[(1 - x^2)^{1/2} + \frac{C_{22}}{n}\,\frac{(1 - x^2)}{(1 - x_j^2)}\, l_j^2(x)\right]$$

$$\le C_{21}\,(1 - x^2)^{1/2}\left[1 + \frac{C_{22}}{n}\,\frac{(1 - x^2)^{1/2}}{(1 - x_j^2)}\right] \le C_{24}(1 - x^2)^{1/2}.$$

Now, on using (4.7) we obtain

$$|f(x) - f(x_j)| h_j(x) \le \left[1 + \frac{n|x - x_j|}{(1 - x^2)^{1/2}}\right] h_j(x) w_f\left(\frac{(1 - x^2)^{1/2}}{n}\right)$$

$$\le [1 + C_{24}]\, w_f\left(\frac{(1 - x^2)^{1/2}}{n}\right).$$

This proves the lemma.

Lemma 4.3. Let $h_k(x)$ be as defined by (2.3). Then for $-1 \quad x \leq 1$, $k = 1, 2, \ldots, n$

$$h_k(x) \leq (1 - 2\alpha) \frac{(1 - x^2)\, l_k^2(x)}{1 - x_k^2} + 4\alpha \frac{\sin^2 \frac{\theta + \theta_k}{2}}{\sin^2 \theta_k} \tag{4.8}$$

$$\frac{(1 - x^2)\, l_k^2(x)}{(1 - x_k^2)}$$

Proof. From (2.3) we note that

$$v_k(x) = 1 - 2\alpha + 2\alpha \frac{(1 - x\, x_k)}{1 - x_k^2}$$

$$\leq 1 - 2\alpha + 2\alpha \frac{(1 - x\, x_k + \sin\theta \sin\theta_k)}{(1 - x_k^2)}$$

$$= 1 - 2\alpha + 2\alpha \frac{(1 - \cos(\theta + \theta_k))}{1 - x_k^2}$$

$$= 1 - 2\alpha + 4\alpha \frac{\sin^2 \frac{\theta + \theta_k}{2}}{\sin^2 \theta_k} .$$

From above and (2.3), (4.8) follows.

Lemma 4.4. Let $0 \leq x \leq 1$ and let x_j be that zero of $Q_n(x)$ which is nearest to x then

$$\sum_{k=1}^{j-1} |f(x_k) - f(x)|\, h_j(x) \leq C_{25} \sum_{1=1}^{j-1} \frac{1}{i^2}\, w_f\left(\frac{i \sin \theta}{n}\right). \tag{4.9}$$

Proof. Since

$$\sin\frac{\theta+\theta_k}{2} \le \sin\theta,\ k = 1, 2 \ldots j-1,$$

it follows that

$$|x_k - x| \le 2\left|\sin\frac{\theta+\theta_k}{2}\sin\frac{\theta-\theta_k}{2}\right| \le \frac{2i\sin\theta}{n},\ k = j-i. \tag{4.10}$$

Further, from (4.8), (2.4) we obtain

$$h_k(x) \le \frac{(1-2\alpha)(1-x^2)^{\frac{1+2\alpha}{2}} Q_n^2(x)}{4(1-x_k^2)^{\frac{3}{2}+\alpha}(Q_n'(x_k))^2}$$

$$\frac{(1-x^2)^{\frac{1-2\alpha}{2}}(1-x_k^2)^{\frac{1+2\alpha}{2}}}{\sin^2\frac{\theta+\theta_k}{2}\sin^2\frac{\theta-\theta_k}{2}}$$

$$+\frac{\alpha(1-x^2)^{\frac{1+2\alpha}{2}} Q_n^2(x)}{(1-x_k^2)^{\frac{3}{2}+\alpha}(Q_n'(x_k))^2}\quad\frac{(1-x^2)^{1/2-\alpha}}{(1-x_k^2)^{1/2-\alpha}\sin^2\frac{\theta-\theta_k}{2}}$$

on using (3.3), (3.9), (3.12) and

$$\left|\sin\frac{\theta-\theta_k}{2}\right| > C_{27}\frac{i}{n}\quad \begin{matrix} k = j-1 \\ k = 1, 2, \ldots, j-1\end{matrix}$$

we obtain

$$h_k(x) \le \frac{C_{28}}{(k-j)^2} + \frac{C_{29}\alpha}{(k-j)^{1+2\alpha}k^{1-2\alpha}}.$$

Now from (4.10) and the above estimate for $h_k(x)$ valid for $k = 1, 2, \ldots, j-1$ we obtain

$$\sum_{k=1}^{j-1} |f(x_k) - f(x)| h_k(x) \tag{4.11}$$

$$\leq C_{31} \sum_{i=1}^{j-1} \left(\frac{1}{i^2} + \frac{\alpha}{i^{1+2\alpha}(j-i)^{1-2\alpha}}\, w_f\left(\frac{i \sin \theta}{n}\right)\right).$$

But

$$\alpha \sum_{i=1}^{j-1} \frac{1}{i^{1+2\alpha}(j-1)^{1-2\alpha}}\, w_f\left(\frac{i \sin \theta}{n}\right) = \sum_{i=1}^{[\frac{j}{2}]} + \sum_{i=[\frac{j}{2}]+1}^{j-1}$$

$$\leq \alpha \sum_{i=1}^{[j/2]} \frac{1}{i^2}\, w_f\left(\frac{i \sin \theta}{n}\right) + \alpha\, w_f\left(\frac{j \sin \theta}{n}\right)$$

$$\sum_{i=[\frac{j}{2}]+1}^{j-1} i^{-(1+2\alpha)}(j-1)^{-(1-2\alpha)}$$

$$\leq \alpha \sum_{i=1}^{[j/2]} \frac{1}{i^2}\, w_f\left(\frac{i \sin \theta}{n}\right) + \frac{4\alpha}{j}\, w_f\left(\frac{j \sin \theta}{2n}\right)$$

$$\leq \alpha \sum_{i=1}^{(j/2)} \frac{1}{i^2}\, w_f\left(\frac{i \sin \theta}{n}\right) + C_{32} \sum_{i=[\frac{j}{2}]+1}^{j-1} \frac{1}{i^2}\, w_f\left(\frac{i \sin \theta}{n}\right)$$

$$\leq C_{33} \sum_{i=1}^{j-1} \frac{1}{i^2}\, w_f\left(\frac{i \sin \theta}{n}\right).$$

(4.10) and the above estimate together yield (4.9). This proves Lemma 4.4.

Lemma 4.5. Let $0 \leq x \leq 1$ and let x_j be that zero of $Q_n(x)$ which is nearest to x then

$$\sum_{k=j+1}^{n} |f(x_k) - f(x)| h_k(x) \le C_{34} (1 - x^2) Q_n^2(x) w_f(1) +$$

$$+ C_{35}\, n(1 - x^2) Q_n^2(x) \sum_{k=j+1}^{[\frac{3n}{4}]} \frac{1}{(k - j)^2 \sin^{1-2\alpha}\theta_k}$$

$$w_f\left(\frac{k - j}{n} \sin \theta_k\right). \tag{4.12}$$

Proof. From (4.8) and

$$\sin \theta_k \le 2\sin \frac{\theta + \theta_k}{2} \le 2 \sin \theta_k, \ k = j + 1, \ldots, [\tfrac{n}{2}] \tag{4.13}$$

it follows that

$$h_k(x) \le (1 + 2\alpha) \frac{(1 - x^2)\, l_k^2(x)}{1 - x_k^2}, \quad k = j + 1, \ldots, [\tfrac{n}{2}]$$

$$= \frac{(1 + 2\alpha)(1 - x^2) Q_n^2(x) (1 - x_k^2)^{1/2 + \alpha}}{4 (1 - x_k^2)^{\frac{3}{2} + \alpha} (Q_n'(x_k))^2 \sin^2 \frac{\theta + \theta_k}{2} \sin^2 \frac{\theta - \theta_k}{2}}.$$

On using (3.9), (4.13) and

$$\left|\sin \frac{\theta_k - \theta}{2}\right| > C_{36} \frac{k - j}{n}, \ k = j + 1, \ldots, n \tag{4.14}$$

it follows that

$$h_k(x) \le \frac{C_{37}\, n(1 - x^2) Q_n^2(x)}{(k - j)^2 \sin^{1-2\alpha} \theta_k}, \ k = j + 1, \ldots, [\tfrac{n}{2}]. \tag{4.15}$$

Also, on using (4.13) once more

$$|f(x) - f(x_k)| \quad w_f(|x - x_k|) \quad w_f\left(2\left|\sin \frac{\theta + \theta_k}{2} \sin \frac{\theta_k - \theta}{2}\right)\right.$$

$$\le C_{38}\, w_f\left(\frac{k-j}{n}\sin\theta_k\right),\ k = j+1, \ldots, \left[\frac{3n}{4}\right]. \quad (4.16)$$

Therefore, from (4.15) and (4.16) we obtain

$$\sum_{k=j+1}^{(n/2)} |f(x) - f(x_k)|\, h_k(x) \quad (4.17)$$

$$\le C_{39}\, n(1-x^2)\, Q_n^2(x) \sum_{k=j+1}^{[\frac{n}{2}]} \frac{1}{(k-j)^2} \sin^{1-2\alpha}\theta_k$$

$$w_f\left(\frac{k-j}{n}\sin\theta_k\right).$$

Next, we note that

$$C_{40} < \sin\theta_k \le C_{41},\ k = \left[\frac{n}{2}\right] + 1 \ldots \left[\frac{3n}{4}\right] \quad (4.18)$$

where C_{40}, C_{41} are absolute positive constant independent of k and n. Therefore, on using (4.8), (3.8), (4.16) and

$$C_{40} \le \sin\theta_k \le 2\sin\frac{\theta+\theta_k}{2} \quad (4.19)$$

we obtain

$$h_k(x) \le C_{41}(1-x^2)l_k^2(x) \le \frac{C_{41}(1-x^2)\, Q_n^2(x)}{4(Q_n'(x_k))^2 \sin^2\frac{\theta+\theta_k}{2}\sin^2\frac{\theta-\theta_k}{2}}$$

$$\le \frac{C_{42}\,(1-x^2)\, Q_n^2(x)}{(Q_n'(x_k))^2 \sin^2\frac{\theta-\theta_k}{2}} \le \frac{C_{43}n(1-x^2)\, Q_n^2(x)}{(k-j)^2} \quad (4.20)$$

$$\le C_{44}\frac{n(1-x^2)\, Q_n^2(x)}{(k-j)^2\sin^{1-2\alpha}\theta_k}$$

on using (4.16) and (4.20) we obtain

$$\sum_{k=[\frac{n}{2}]+1}^{[3n/4]} |f(x) - f(x_k)| h_k(x) \tag{4.21}$$

$$\leq \sum_{k=[\frac{n}{2}]+1}^{[3n/4]} \frac{C_{44} n(1 - x^2)\, Q_n^2(x)}{(k - j)^2 \sin^{1-2\alpha}\theta_k} w_f\left(\frac{k - j}{n} \sin\theta_k\right).$$

In the case $k = [\frac{3n}{4}] + 1 \ldots n$, $0 \leq x \leq 1$ we have

$$|x - x_k| > \frac{1}{4}$$

and on using (4.8) we obtain

$$h_k(x) \leq \frac{16(1 - 2\alpha)\, (1 - x^2)\, Q_n^2(x)}{(1 - x_k^2)\, (Q_n'(x_k))^2} +$$

$$+ \frac{64\alpha\, (1 - x^2)\, Q_n^2(x)}{(1 - x_k^2)^2\, (Q_n'(x_k))^2}$$

on using (3.10) and

$$|f(x) - f(x_k)| \leq w_f(|x - x_k|) \leq 2w_f(1)$$

we obtain

$$\sum_{k=[\frac{3n}{4}]+1}^{n} |f(x) - f(x_k)| h_k(x) \leq C_{45}\, w_f(1)\, (1 - x^2) Q_n^2(x) \tag{4.23}$$

on combining (4.17), (4.21), (4.23) we obtain (4.12). This proves the lemma.

5. PROOF OF THEOREMS

Proof of Theorem 1 is an easy consequence of lemma 4.1, 4.2, 4.4, and (4.5). For

$$f(x) - S_n^{(\alpha,\alpha)}(f, x) = (f(x) - f(1)) \frac{1 + x}{2} \frac{Q_n^2(x)}{Q_n^2(1)}$$

$$+ (f(x) - f(-1) \frac{(1 - x)}{2} \frac{Q_n^2(x)}{Q_n^2(-1)} + \sum_{k=1}^{n} (f(x) - f(x_k))h_k(x)$$

hence an application of the above lemmas yield the required result.

Proof of Corollary 1. Following the hypothesis of Theorem 1, (3.3) we have

$$C_4(1 - x^2)\, Q_n^2(x)\, w_f(1) \le \frac{C_{46}}{n} w_f(1)$$

$$\le C_{47} \sum_{i=[\frac{3n}{4}]+1}^{n} \frac{1}{i^2} w_f\left(\frac{i^2}{n^2}\right).$$

Also on using $(0 \le \alpha \le \frac{1}{2})$

$$\begin{cases} \sin\theta < \sin\theta_k,\ k = j + 1, \ldots, [\frac{n}{2}];\ \sin\theta_k > C, \\ k = [\frac{n}{2}] + 1, \ldots, [\frac{3n}{4}] \\ \sin\theta_k = \sin\theta_k - \sin\theta + \sin\theta \\ \le \sin\theta + 2\left|\sin\frac{\theta - \theta_k}{2}\right| \le \sin\theta + 2\frac{|k - j|}{n} \end{cases} \tag{5.1}$$

we obtain

$$C_3\, n(1 - x^2)^{\frac{1+2\alpha}{2}}\, Q_n^2(x) \sum_{k=j+1}^{[\frac{3n}{4}]} \frac{1}{(k - j)^2} \left(\frac{\sin\theta}{\sin\theta_k}\right)^{1-2\alpha}$$

$$w_f\left(\frac{k - j}{n}\right) \sin\theta_k)$$

$$\leq C_{48} \sum_{k=j+1}^{[\frac{3n}{4}]} \frac{1}{(k-j)^2} [w_f (\frac{k-j}{n} \sin\theta) + w_f (\frac{k-j}{n})^2)].$$

From (1.7) and the above estimates the Corollary 1 follows.

Proof of Corollary 2. Following the hypothesis of Theorem 1, (3.3) we have

$$C_4 (1 - x^2) Q_n^2(x) w_f(1)$$

$$\frac{C_{49}}{n} (1 - x^2)^{\frac{1-2\alpha}{2}} w_f(1)$$

$$< C_{50} \sum_{i=[\frac{3n}{4}]+1}^{n} \frac{1}{i^2} (\sin\theta)^{1-2\alpha} w_f(\frac{i}{n})$$

$$< C_{51} \sum_{i=[\frac{3n}{4}]+1}^{n} \frac{1}{i^2} w_f (\frac{i}{n} (\sin\theta)^{1-2\alpha}).$$

Also, on using (3.3), (5.1) and well known properties of modulus of continuity we have

$$C_3 n(1 - x^2)^{\frac{1+2\alpha}{2}} Q_n^2(x) \sum_{k=j+1}^{[\frac{3n}{4}]} \frac{1}{(k-j)^2} (\frac{\sin\theta}{\sin\theta_k})^{1-2\alpha}$$

$$w_f (\frac{k-j}{n} \sin\theta_k)$$

$$\leq C_{52} \sum_{k=j+1}^{[\frac{3n}{4}]} \frac{1}{(k - j^2)} w_f (\frac{k-j}{n}) (\sin\theta)^{1-2\alpha} (\sin\theta_k)^{2\alpha}).$$

From (1.7) and the above estimates Corollary 2 follows as well. Proof of Corollary 3 follows directly from (1.7) and def. of $f \in \text{Lip}\,\beta$, $\beta < 1 - \alpha$. The details are omitted.

Proof of Theorem 2. Let $f(x) = |x|$, $x = \cos\theta = 0$,

$= \frac{\pi}{2}$ and $n = 6, 8, 10, \ldots$. Then from (2.2) and (2.5) it follows that

$$S_n^{(\alpha,\alpha)}(f,\ 0) - f(0) = \frac{Q_n^2(0)}{Q_n^2(1)} +$$

$$+ \sum_{k=1}^{n} \frac{1}{1 - x_k^2} \left(1 + \frac{2\alpha x_k^2}{1 - x_k^2}\right) \frac{Q_n^2(0)}{|x_k|\,[Q_n'(x_k)]^2} \tag{5.2}$$

$$\geq \sum_{k=1}^{\frac{n}{2}} \frac{Q_n^2(0)}{(1 - x_k^2)x_k[Q_n'(x_k)]^2} .$$

From Szegö [19], p. 81 and p. 169 we have

$$|Q_n(0)| = |P_n^{(\alpha,\alpha)}(0)| = \frac{\Gamma(2\alpha + 1)\Gamma(n + \alpha + 1)}{\Gamma(\alpha + 1)\Gamma(n + 2\alpha + 1)} \binom{\frac{n}{2} + \alpha - \frac{1}{2}}{\frac{n}{2}} ,$$

$$-\frac{1}{2} < \alpha \leq \frac{1}{2} . \tag{5.3}$$

Thus using inequality (3.16) we have from (5.3),

$$|Q_n(0)| > C_{46} n^{-\alpha} \binom{\frac{n}{2} + \alpha - \frac{1}{2}}{\frac{n}{2}} \tag{5.4}$$

$$> C_{47}\, n^{-\frac{1}{2}},\ 0 \leq \alpha \leq \frac{1}{2}.$$

Now on using (5.4), (3.15) and (3.5) we obtain from (5.2) for $0 \leq \alpha \leq \frac{1}{2}$,

$$S_n^{(\alpha,\alpha)}(f,\ 0) - f(0) \geq c_{48} \sum_{k=1}^{\frac{n}{2}} \frac{n^{-1}\ k^{3+2\alpha}}{\cos\theta_k\ \theta_k^2\ n^{4+2\alpha}}$$

$$\geq C_{49} \sum_{k=1}^{\frac{n}{2}} \frac{k^{1+2\alpha}}{n^{3+2\alpha} \cos \theta_k}$$

$$\geq C_{49}\, n^{-3-2\alpha} \sum_{k=[\frac{n}{4}]+1}^{\frac{n}{2}} \frac{k^{1+2\alpha}}{\cos\theta_k}$$

$$\geq C_{50}\, n^{-2} \sum_{k=[\frac{n}{4}]+1}^{\frac{n}{2}} \frac{1}{\cos \theta_k}$$

$$= c_{50}\, n^{-2} \sum_{k=[\frac{n}{4}]+1}^{\frac{n}{2}} \frac{1}{2 \sin(\frac{\theta + \theta_k}{2}) \sin(\frac{\theta - \theta_k}{2})}$$

$$\geq C_{51}\, n^{-2} \sum_{k=[\frac{n}{4}]+1}^{\frac{n}{2}} \frac{1}{(\frac{\theta - \theta_k}{2})}$$

$$\geq C_{52}\, n^{-1} \sum_{k=[\frac{n}{4}]+1}^{\frac{n}{2}} \frac{1}{2n - 4k + 1}$$

$$\geq C_{53} \frac{\log n}{n}, \quad n \geq 6,$$

from which the theorem follows.

REFERENCES

1. Berman, D. L., "A study of the process of the Hermite-Fejér interpolation" (in Russian) Doklady Akad. Nauk USSR, 187, 1969, pp. 241-244 (Soviet Math. Dokl. 10 1969, pp. 813-816).

2. Berman, D. L., "On an everywhere divergent process of the Hermite-Fejér interpolation" (in Russian), Izv. Vyssh. Vceb. Zav., Matematika, 92, 1970, pp. 3-8.

3. Berman, D. L., "A study of interpolation processes constructed on extended systems of nodes", Izv. Vyssh. Vceb. Zav., Matematika, 105, 1971, pp. 22-31.

4. Bojanic, R., "A note on the precision of interpolation by Hermite-Fejér polynomials", Proceedings of the 1969 conference on constructive theory of functions, Akademiai Kiadó, Budapest, 1972, pp. 69-76.

5. Bojanic, R., Prasad, J., and Saxena, R. B., "An upper bound for the rate of convergence of Hermite-Fejér process on the extended Chebyshev nodes of second kind", J. Approx. Theory 26(3), 1979, pp. 195-203.

6. Egerváry, E., and Turán, P., "Notes on interpolation V", (On the stability of interpolation) Acta Math. Acad. Sci. Hung. 9, 1958, pp. 259-267.

7. Erdös, P., "On the maximum of the fundamental functions of the ultraspherical polynomials", Annals of Math. 45(2), 1944, pp. 335-339.

8. Fejér, L., "Über interpolation", Nachrichten d.k. Gesellschaff zu Gottingen, 1916, pp. 66-91.

9. Mills, T. M., and Varma, A. K., "On a theorem of E. Egerváry and P. Turán on the stability of interpolation", J. Approx. Theory 2(3), 1974, pp. 275-282.

10. Natanson, I. P., "Constructive function theory", (English transl.), (FUPC., New York, 1964).

11. Prasad, J., and Saxena, R. B., "Degree of convergence of quasi-Hermite-Fejér interpolation", Publications De L'Institute Math. Nouvelle Serie 19(33), 1975, pp. 123-130.

12. Prasad, J., and Varma, A. K., "A study of some interpolatory processes based on the roots of Legendre polynomials", J. Approx. Theory, 31(3), 1981, pp. 244-252.

13. Prasad, J., "On the rate of convergence of interpolation polynomials of Hermite-Fejér type", Bull. Austral. Math. Soc. 19, 1978, pp. 29-37.

14. Saxena, R. B., "A note on the rate of convergence of Hermite-Fejér interpolation polynomials", Can. Math. Bull. 17(2), 1974, pp. 299-301.

15. Saxena, R. B., and Mathur, K. K., "The rapidity of convergence of quasi-Hermite-Fejér interpolation polynomials", Acta Math. Acad. Sci. Hung. 28, 1976, pp. 343-347.

16. Szabados, J., "On the convergence of Hermite-Fejér interpolation based on the roots of the Legendre polynomials", Acta Math. Acad. Sci. Hung. 34, 1973, pp. 367-370.

17. Szabados, J., "On Hermite-Fejér interpolation for the Jacobi Abscissas", Acta Math. Acad. Sci. Hung. 23 (3-4), 1972, pp. 449-462.

18. Szasz, P., "On quasi-Hermite-Fejér interpolation", Acta Math. Acad. Sci. Hung. 10, 1959, pp. 413-439.

19. Szegö, G., "Orthogonal polynomials", Amer. Math. Soc. Colloq. Pub. 23, 1959.

20. Varma, A. K., and Prasad, J., "A contribution to the problem of L. Fejér on Hermite-Fejér interpolation", J. Approx. Theory 28, 1980, pp. 185-196.

21. Vértesi, P. O. H., "On the convergence of Hermite-Fejér interpolation", Acta Math. Acad. Sci. Hung. 22, 1971, pp. 151-158.

USING INCLUSION THEOREMS TO ESTABLISH THE SUMMABILITY OF ORTHOGONAL SERIES

B. E. Rhoades

Indiana University

Let $\{\phi_n(x)\}$ be an orthogonal system of real functions, with respect to a distribution $d\mu(x)$, and defined over $[a, b]$. We shall be interested in matrix transformations of

$$s_n(x) = \sum_{k=0}^{n} c_k \phi_k(x) \tag{1}$$

for

$$\{c_n\} \in \ell^2. \tag{2}$$

A number of papers have been written concerning the Cesaro summability of orthogonal series (see, e.g. [2], [4], [5], [8], [11], [17], [18], [20], and [21]) and some of these results appear in [1].

The purpose of this paper is to show that inclusion theorems can be used to obtain new proofs for the Norlund summability of orthogonal series. It has been shown in [1] that Abel and Cesaro summability are equivalent a.e. for orthogonal series satisfying (2). The first theorem of this paper shows that the same is true for the generalized Abel and Cesaro methods. This result is then used to show that a wide class of Norlund and Cesaro matrices are equivalent a.e. for orthogonal series. The other theorems are generalizations of known results concerning the Norlund summability of orthogonal series.

S. P. Singh et al. (eds.), Approximation Theory and Spline Functions, 441–453.

An infinite matrix is called regular if it maps a convergent sequence into a convergent sequence with the same limit. A matrix A is included in a matrix B, written $A \subset B$ if every sequence to which A assigns a limit is assigned the same limit by B. A Norlund matrix A is a lower triangular matrix with entries $a_{nk} = p_{n-k}/P_n$, where $\{p_n\}$ is an arbitrary real or complex sequence satisfying $P_n = \sum_{k=0}^{n} p_k \neq 0$ for each n. Necessary and sufficient conditions for the regularity of A are: (i) $p_n = o(P_n)$ and (ii) $P_n^* = O(|P_n|)$, where $P_n^* = \sum_{k=0}^{n} |p_k|$. The choice $p_n = \binom{n+\alpha-1}{\alpha-1}$ generates the Cesaro matrix of order α, written (C, α). The terms of the $(C, 1)$ transform of a sequence $\{s_k\}$ are $\sigma_n = (n+1)^{-1} \sum_{k=0}^{n} s_k$.

The Abel method, (A), is a series to function method defined as follows. A series $\sum u_k$ is Abel summable to s if

$$a(x) = \sum_{k=0}^{\infty} u_k x^k$$

converges for each $0 \leq x < 1$ and $\lim_{x \to 1-} a(x) = s$.

A series $\sum u_k$ is generalized Abel summable, (GA) to s, if $a(x)$ converges for $|x| < \rho$ for some $\rho < 1$, $a(x)$ is analytic for $0 \leq x < 1$ and $\lim_{x \to 1-} a(x) = s$.

From [3, Theorem 18], $(N, p) \subset (GA)$ for every regular Norlund matrix.

Lemma 1. If $\sum_{n=0}^{\infty} u_n$ is (GA)-summable, and $\sum_{n=0}^{\infty} n u_n^2 < \infty$, then $\sum_{n=0}^{\infty} u_n$ is (A)-summable.

Proof. $\sum_{n=0}^{\infty} u_n$ (GA)-summable implies $a(x)$ has a positive radius of convergence, is analytic for $0 \le x < 1$, and $\lim_{x\to 1-} a(x)$ exists.

$$|a(x)| \le |u_0| + \sum_{k=1}^{\infty} \sqrt{k}\, |u_k| \left|\frac{x^k}{\sqrt{k}}\right|$$

$$\le |u_0| + \left(\sum_{k=1}^{\infty} k u_k^2\right)^{1/2} \left(\sum_{k=1}^{\infty} x^{2k}/k\right)^{1/2} < \infty$$

since $\sum_{k=1}^{\infty} x^{2k}/k$ has radius of convergence one. Therefore $a(x)$ converges for $0 \le x < 1$. Since $a(x)$ is absolutely convergent for $|x| < \rho$ for some $\rho < 1$, $\sum_{k=0}^{\infty} u_k x^k$ is the analytic extension of $a(x)$ along $[0, 1)$. Therefore the series is Abel summable.

Lemma 2. If $\sum_{n=0}^{\infty} u_n$ is (GA)-summable, then $\sum(\sigma_n - \sigma_{n-1})$ is (GA)-summable.

Proof. Define

$$b(x) = \sum_{n=0}^{\infty} (\sigma_n - \sigma_{n-1})x^n = (1 - x) \sum_{n=0}^{\infty} \sigma_n x^n,$$

and

$$a(x) = \sum_{n=0}^{\infty} u_n x^n.$$

Since $a(x)$ is (GA)-summable, it has a positive radius of convergence and is analytic for $0 \le x < 1$.

$$a(x) = \sum_{n=0}^{\infty} (s_n - s_{n-1})x^n, \quad s_{-1} = 0$$

$$= (1 - x) \sum_{n=0}^{\infty} s_n x^n = (1 - x) \sum_{n=0}^{\infty} [(n + 1)\sigma_n - n\sigma_{n-1}]x^n$$

$$= (1 - x)^2 \sum_{n=0}^{\infty} (n + 1)\sigma_n x^n,$$

so that

$$\sum_{n=0}^{\infty} (n + 1)\sigma_n x^n = \frac{a(x)}{(1 - x)^2} .$$

For $|x| < \rho$,

$$\sum_{n=0}^{\infty} \sigma_n x^{n+1} = \int_0^x \frac{a(t)}{(1 - t)^2}\, dt.$$

Hence,

$$b(x) = \begin{cases} \dfrac{1 - x}{x} \displaystyle\int_0^x \frac{a(t)}{(1 - t)^2}\, dt, & |x| < \rho,\ x \neq 0, \\ a(0) & , \quad x = 0. \end{cases}$$

Since $a(t)$ is analytic for $0 \le x < 1$, the above representation defines $b(x)$ for $0 \le x < 1$, and b is analytic for $0 \le x < 1$.

It remains to show that $\lim_{x \to 1-} b(x)$ exists. If the integral exists as $x \to 1-$, then $b(1-) = 0$. If the integral fails to exists, then, by L'Hospital's rule, $b(1-) = a(1-)$.

Theorem 1. (GA) and (C, α) summability are equivalent a.e. for each series (1) satisfying (2).

Proof. Since $(C, \alpha) \subset (A) \subset (GA)$, the first part is trivial. To prove the converse, suppose (1) is (GA)-summable a.e. on a set E. As in [1, p. 111], $\sum_{k=1}^{\infty} k[\sigma_k(x) - \sigma_{k-1}(x)]^2$ converges a.e. From Lemma 2, the (GA)-summability of (1) a.e. implies the (GA)-summability of $\sum(\sigma_k(x) - \sigma_{k-1}(x))$ a.e. From Lemma 1, $\sum(\sigma_k(x) - \sigma_{k-1}(x))$ is (A)-summable a.e. Thus, from

[1, Theorem 2.2.6], (1) is (C, 1)-summable a.e. on E. The remainder of the proof is the same as that in [1].

Define $N^\alpha = \{(N, p) \mid (N, p)$ is regular and $(N, p) \supset (C, \alpha)$ for some $\alpha > 0\}$.

Corollary 1. Let $(N, p) \in N^\alpha$. Then (N, p) and (C, α) are equivalent a.e. for such series (1) satisfying (2).

Proof. Suppose (1) is summable (C, α) a.e. for some $\alpha > 0$. By hypothesis $(C, \alpha) \subset (N, p)$, so (1) is summable (N, p) a.e.

Conversely, if (1) is summable (N, p) a.e. then since $(N, p) \subset (GA)$, so (1) is summable (C, α) a.e. for each $\alpha > 0$ by Theorem 1.

Let $\{n_k\}$ be an arbitrary sequence of integers satisfying

$$1 < q \leq \frac{n_{k+1}}{n_k} \leq r, \quad k = 0, 1, 2, \ldots, \tag{3}$$

where q and r are positive constants.

Theorem 2. Let $(N, p) \in N^\alpha$. Then a series (1), satisfying (2), is (N, p)-summable a.e. on a set E if and only if, for each $\{n_k\}$ satisfying (3), $\{s_{n_k}(x)\}$ converges a.e. on E.

This result is a direct consequence of Corollary 1 and Theorem 2.7.3 of [1].

Let $M = \{\{p_n\} \mid \{p_n\}$ is positive, monotone and $P_n \to \infty\}$. Define $M^\alpha = \{\{p_n\} \in M \mid \lim np_n/P_n = \alpha \geq 0\}$ and $BVM^\alpha = \{\{p_n\} \in M^\alpha \mid \{S_n\} \in BV\}$, where $S_n = P_n^{-1} \sum_{k=0}^{n} P_k/(k + 1)$.

Meder [12] showed that, if a series (1) satifies (2), $\{p_n\} \in BVM^\alpha$, $\alpha > \frac{1}{2}$, and $\{n_k\}$ satisfies (3), then (1) is (N, p)-summable a.e. if and only if $\{s_{n_k}(x)\}$ converges a.e. Meder [13] also proved this result for the class

$\overline{M}^{\alpha} = \{\{p_n\} \in M \mid \lim n(p_{n-1} - p_n)/p_n = 1 - \alpha\}$, $\alpha > \frac{1}{2}$. These results were then generalized by Kopec [7] to the class $\overline{M}^{\alpha}$, where $M^{\alpha} = \{\{p_n\} \mid p_n > 0$ and $\alpha \le \underline{\lim}\, np_n/P_n \le \overline{\lim}\, np_n/P_n = \beta < \infty\}$ and $\overline{M}^{\alpha} = \{\{p_n\} \in M^{\alpha} \mid n(p_{n-1} - p_n)/p_n = O(1)\}$.

Corollary 2. [7, Theorem 3] Let $\{p_n\} \in \overline{M}^{\alpha}$ for some $\alpha > \frac{1}{2}$, $\{n_k\}$ satisfy (3). Then (1), satisfying (2), is (N, p)-summable a.e. on a set E if and only if $\{s_{n_k}(x)\}$ converges a.e. on E.

From Theorem 2 it is sufficient to show that $(N, p) \supset (C, \frac{1}{2} + \theta)$ for some $\theta > 0$.

In a recent paper the author established the following results.

Lemma 3. [16] Let $\{p_n\}$ be a real or complex sequence satisfying $P_n \neq 0$ for each n. Then

$$\frac{n^{\alpha}}{|P_n|} \sum_{k=0}^{n-1} \frac{|P_k|}{(k+1)^{1+\alpha}} = O(1), \quad 0 \le \alpha < 1, \tag{4}$$

and

$$T_n^* = \frac{1}{|P_n|} \sum_{k=0}^{n} (k+1)|\Delta p_k| = O(1), \tag{5}$$

imply $(N, p) \supset (C, \alpha + \beta)$ for each β satisfying $0 < \beta < 1/\mu$, where

$$\mu = \overline{\lim}\, \frac{(n+1)^{\alpha}}{|P_n|} \sum_{k=0}^{n-1} \frac{|P_k|}{(k+1)^{1+\alpha}}.$$

To establish the inclusion it is sufficient to show that $\{p_n\}$ satisfies conditions (4) and (5) with $\alpha = \frac{1}{2}$.

Using a theorem of Stolz [6, p. 77] with $x_n = \sum_{k=0}^{n} P_k(k+1)^{-3/2}$,

$y_n = n^{-1/2} P_n$, and, from [14, Lemma 1], the fact that $y_n \to \infty$,

$$\mu = \overline{\lim} \frac{x_n}{y_n} \leq \overline{\lim} \frac{x_n - x_{n-1}}{y_n - y_{n-1}} = \overline{\lim} \frac{p_n(n+1)^{-3/2}}{P_n/\sqrt{n} - P_{n-1}/\sqrt{n-1}}$$

$$= \overline{\lim} \left[(n+1)^{3/2} \left[\frac{1}{\sqrt{n}} - \frac{1}{\sqrt{n-1}} \left(1 - \frac{p_n}{P_n}\right)\right]\right]^{-1}$$

$$= \overline{\lim} \left[(n+1)^{3/2} \left[\frac{1}{\sqrt{n}} - \frac{1}{\sqrt{n-1}}\right] + \left(\frac{n+1}{n-1}\right)^{1/2} \left(\frac{(n+1)p_n}{P_n}\right)\right]^{-1}$$

$$\leq 1/(\beta + \alpha),$$

where $\alpha = \underline{\lim}\, np_n/P_n$ and $\beta = \lim(n+1)^{3/2}[n^{-1/2} - (n-1)^{-1/2}] = -\frac{1}{2}$. Thus μ is finite and (4) is satisfied.

To establish (5) again use Stolz's theorem with $x_n = \sum_{k=0}^{n} (k+1)|\Delta p_k|$ and $y_n = P_n$. Then

$$\overline{\lim}\, T_n^* = \overline{\lim}\, x_n/y_n \leq \overline{\lim} \frac{(n+1)|\Delta p_n|}{p_n} = O(1).$$

<u>Theorem 3</u>. Let $(N, p) \supset (C, 1)$. If (1), with $\{c_n\}$ defined by

$$c_n = \int_a^b f(x)\phi_n(x)\,dx, \quad f \in L^2[a, b], \tag{6}$$

is (N, p)-summable to f a.e., then $\{s_{n^k}(x)\}$ is (N, p)-summable to f a.e. for each positive integer k.

Proof. Suppose (1) is (N, p)-summable a.e. By Corollary 1, (1) is summable (C, α) a.e. for each $\alpha > 0$. From a result [11] with $\alpha = 1$, $\{s_{n^k}(x)\}$ is (C, 1)-summable a.e for each positive integer k. Since $(N, p) \supset (C, 1)$, the result follows.

Corollary 3. [19] Let $\{p_n\}$ be a nonnegative nondecreasing sequence such that $p_n \to \infty$ and $np_n/P_n = 0(1)$. If (1), with $\{c_n\}$ defined by (6) is (N, p)-summable to f a.e., then $\{s_{n^k}(x)\}$ is (N, p)-summable to f a.e. for each integer $k \geq 2$.

Since the conditions on $\{p_n\}$ imply $(N, p) \supset (C, 1)$, the result follows from Theorem 3.

Corollary 4. Let (N, p) be any regular Norlund method. If (1), satisfying (2), is summable (N, p) a.e., then it is (A)-summable a.e.

The proof is immediate from Theorem 1, since $(N, p) \subset (GA)$ and $(C, \alpha) \subset (A)$.

Theorem 1 of [22] is a special case of Corollary 4 in which the $\{p_n\}$ are nonnegative.

Corollary 5. Let $(N, p) \in N^{\alpha}$. Then (N, p) is a.e. equivalent to the Abel and Cesaro methods for each series (1) satisfying (2).

This equivalence is a by-product of the proof of Theorem 1 and Corollary 2.

Corollary 6. [22, Theorem 2] If $\{p_n\}$ is positive, monotone, and satifies (4) with $\alpha = 0$, then (N, p) is equivalent a.e. to the Abel and Cesaro methods for series (1) satisfying (2).

Proof. From Corollary 5 it is sufficient to show that $(N, p) \in N^{\alpha}$. If $\{p_n\}$ is monotone decreasing then (5) is satisfied, and $(N, p) \supset (C, \alpha)$ for some $\alpha > 0$ by Lemma 3. If $\{p_n\}$ is monotone increasing, then $(C, 1) \subset (N, p)$.

Theorem 4. Let $(N, p) \in N^{\alpha}$. If (1), satisfying (2), is (N, p)-summable to s(x) a.e. then

$$\sum_{k=0}^{n} [s_{n_k}(x) - s(x)]^2 = o(n), \tag{7}$$

for $\{n_k\}$ an arbitrary convex sequence.

Proof. From Corollary 4, if (1) is (N, p)-summable a.e. to $s(x)$, then it is (C, α) summable to $s(x)$ for each $\alpha > 0$. From Theorem 2.6.4 of [1], (7) is satisfied.

Corollary 7. [12, Theorem 10] If (1), satisfying (2), is (N, p)-summable to $s(x)$ a.e., and if $\{p_n\} \in M^\alpha$, $\alpha > 0$, then (7) is satisfied for $\{n_k\}$ an arbitrary convex sequence.

Proof. From Theorem 4 it is sufficient to show that $(N, p) \in N^\alpha$. Since $\{p_n\} \in M^\alpha$, this follows as in the proof of Corollary 6.

We now generalize Theorem 2.8.2 of [1].

Theorem 5. If $\{v(n)\}$ is a positive monotone increasing sequence satisfying $v(n) = o(\log \log n)$, then there exists an orthonormal series $\sum_{n=1}^{\infty} c_n \Phi_n(x)$ which is nowhere (GA)-summable, although its coefficients satisfy

$$\sum_{n=1}^{\infty} c_n^2 v^2(n) < \infty. \tag{8}$$

Proof. As in the proof of [1, Theorem 2.8.2], it is shown that (8) is satisfied and yet $\{s_{2^n}(x)\}$ diverges a.e. From Theorem 1, the series can be (GA)-summable at most in a set of measure zero. By appropriate alteration of the points of the set, one can make the series nowhere (GA)-summable.

Theorem 3 of [12] is now an immediate corollary. Simply choose $\{v(n)\}$ and $\Phi_n(x)$ so that Theorem 5 is satisfied. Then, since $(N, p) \subset (GA)$, the series is not (N, p)-summable. For, if it were, then one would obtain a contradiction to Theorem 5.

Weighted mean matrices are lower triangular, matrices with entries p_k/P_n, where $p_0 > 0$, $p_n \geq 0$ for $n > 0$,

$P_n = \sum_{k=0}^{n} p_k$. The condition for regularity is $P_n \to \infty$. The

The choice $p_n = (n+1)^{-1}$ gives the logarithmic method, denoted either by (ℓ) or $(R, 1)$.

Theorem 6. Let B be any matrix which is stronger than (C, α) for some $\alpha > 0$. If

$$\sum_{n=3}^{\infty} c_n^2 (\log \log n)^2 < \infty \tag{9}$$

then (1) is summable B a.e.

Proof. Condition (9) implies, by Theorem 2.8.1 of [1], that (1) is summable (C, α) a.e. for each $\alpha > 0$. Since $B \supset (C, \alpha)$ for some $\alpha > 0$, (1) is summable B.

The special case of Theorem 6 when $B = (\ell)$ is weaker than Theorem 3 of [11] since, in that Theorem condition (9) is replaced by the weaker condition $\sum_{n=s}^{\infty} c_n^2 (\log \log \log n)^2 < \infty$.

A sequence $\{s_n\}$ is said to be absolutely summable by a sequence-to-sequence matrix method, written $\{s_n\}$ is $|A|$-summable, if the sequence of matrix transforms is of bounded variation.

The following result appears in [16].

Lemma 4. Let (N, p), (N, q) be two regular Norlund methods satisfying $(N, q) \subset (N, p)$. Then $|N, q| \subset |N, p|$.

Define

$$A_m = \sum_{v=2^m+1}^{2^{m+1}} a_v^2.$$

Theorem 7. Let $(N, p) \in N^{\alpha}$ for some $\alpha > \frac{1}{2}$. If

$$\sum_{n=0}^{\infty} A_m < \infty, \tag{10}$$

then (1) is summable $|N, p|$ a.e. on $[0, 1]$.

Proof. From [8], (10) implies (1) is summable $|C, \alpha|$ a.e. on $[0, 1]$ for each $\alpha > \frac{1}{2}$. By hypothesis $(C, \alpha) \subset (N, p)$ for some $\alpha > \frac{1}{2}$. The absolute inclusion follows from Lemma 4, since (C, α) is a Norlund method.

Corollary 8. [15, Theorem 2] Let $\{p_n\} \in \overline{M}^{1/2}$ with $0 < \overline{\lim}\, n|p_n - p_{n-1}|/p_n = \overline{\beta} < \frac{1}{2}$. If (10) is satisfied then (1) is summable $|N, p|$ a.e. on $[0, 1]$.

Proof. From Theorem 7 and Lemma 4, it is sufficient to show that $(N, p) \supset (C, \frac{1}{2} + \theta)$ for some $\theta > 0$, which follows from the proof of Corollary 2.

Remarks.

1. In Corollary 3 the hypothesis $np_n/P_n = O(1)$ is not needed.
2. In Corollary 3 the result is also true for $k = 1$.
3. A different proof of Corollary 2 appears in [15].
4. In Corollary 8 the condition $\overline{\beta} < \frac{1}{2}$ is not needed. It is sufficient that $\{n|p_{n-1} - p_n|/p_n\}$ be bounded.
5. The (N, p) method with $p_n = (n + 1)^{-1}$ is called the harmonic method, is denoted by $(H, 1)$, and satisfies the condition $(H, 1) \subset (C, \alpha)$ for each $\alpha > 0$. Consequently, if (1), satisfying (2), is $(H, 1)$ summable a.e., then $\{s_{n_k}(x)\}$ converges a.e. for each lacunary sequence $\{n_k\}$ satisfying (3). The proof of the converse is an open question.

REFERENCES

1. Alexits, G, "Convergence problems of orthogonal series", Pergamon Press, Oxford, 1961.

2. __________, "Ein summationssatz fur orthogonalreihen", Acta Math. Acad. Sci. Hung 7, 1956, pp. 5-9, [MR 18, 124].

3. Hardy, G. H., "Divergent series", Oxford, 1949.

4. Jurcenko, A. M., "Summation of orthogonal series by Cesaro methods", Vestnik Moscow Univ. Ser. I Math. Mech. 25, 1970, pp. 3-9, [MR 42, #8160].

5. __________, "Summability of quasi-orthogonal sequences by Cesaro methods", Anal. Math. 1, 1975, pp. 231-247, [MR 52, #14822].

6. Knopp, K., "Theory and application of infinite series", Blackie & Son Ltd., London, 1947.

7. Kopec, J., "On some classes of Norlund means", Bull. Acad. Polon. des Sci. 16, 1968, pp. 93-98, [MR 37, #1843].

8. Leindler, L., "Uber die absolute Summierbarkeit der orthogonalreihen", Acta. Sci. Math. 22, 1961, pp. 243-268, [MR 24, #A2782].

9. __________, "Über approximation mit orthogonalreihenmitteln unter strukturellen Begingungen", Acta Math. Acad. Sci. Hung. 15, 1964, pp. 57-62, [MR 28, #4297].

10. Matsumura, Y., "Note on the summability of orthogonal series", J. Osaka Inst. Sci. Tech. 3, 1951, pp. 21-24, [MR 15, 119].

11. Meder, J., "On the summability almost everywhere of orthnormal series by the method of first logarithmic means", Rozprawy Mat. 17, 1959, pp. 3-33, [MR 21, 7387].

12. __________, "On the Norlund summability of orthogonal series", Ann. Polon. Math. 12, 1963, pp. 231-256, [MR 26, #2771].

13. __________, "Further results concerning Norlund summability of orthogonal series", Ann. Polon. Math. 16, 1965, pp. 237-265, [MR 30, #5121].